Recent Research in Psychology

Gerhard H. Fischer
Donald Laming
Editors

Contributions to Mathematical Psychology, Psychometrics, and Methodology

With 47 Illustrations

Springer-Verlag
New York Berlin Heidelberg London Paris
Tokyo Hong Kong Barcelona Budapest

Gerhard H. Fischer, Ph.D.
Professor
Department of Psychology
University of Vienna
Liebiggasse 5
A-1010 Vienna, Austria

Donald Laming, Ph.D.
Department of Experimental Psychology
Cambridge University
Cambridge, England CB2 3EB

Library of Congress Cataloging-in-Publication Data
Contributions to mathematical psychology, psychometrics, and methodology / Gerhard H. Fischer, Donald Laming, editors.
p. cm. — (Recent research in psychology)
Papers presented at the European Mathematical Psychology Group, 22nd EMPG Meeting in Vienna, Austria, Sept. 9–12, 1991.
Includes bibliographical references and index.

ISBN-13: 978-0-387-94169-1 e-ISBN-13: 978-1-4612-4308-3
DOI: 10.1007/978-1-4612-4308-3

1. Psychology—Mathematical models—Congresses. 2. Psychometrics—Congresses. 3. Psychology—Research—Methodology—Congresses. I. Fischer, Gerhard H. II. Laming, D. R. J. (Donald Richard John) III. European Mathematical Psychology Group Meeting. (22nd : 1991 : Vienna, Austria) IV. Series.
BF39.C598 1993
150'.1'51—dc20 93-31777

Printed on acid-free paper.

Production managed by Hal Henglein; manufacturing supervised by Genieve Shaw.
Camera-ready copy prepared by the editors.

9 8 7 6 5 4 3 2 1

Preface

The European Mathematical Psychology Group (EMPG) – no society, just an informal group of scientists from various fields interested in the application of mathematical reasoning to psychology – came into being in 1971, and two meetings were held that year. Since then, a tradition has developed of a yearly gathering. After a while, the papers presented at these meetings were collected and published in volumes of proceedings. The first such volume, *Trends in mathematical psychology* (Amsterdam: North-Holland / Elsevier) was edited by E. Degreef & J. van Buggenhaut in 1984. It was followed by E. E. Roskam & R. Suck (Eds.), *Progress in mathematical psychology* (Amsterdam: North-Holland / Elsevier, 1987). The third and fourth volumes were published by Springer-Verlag New York within its series "Recent Research in Psychology" (E. E. Roskam (Ed.), *Mathematical psychology in progress*, 1989, and J.-P. Doignon & J.-Cl. Falmagne (Eds.), *Mathematical psychology. Current developments,* 1991). The present volume, summarizing research presented at the 22nd. EMPG Meeting in Vienna, Austria, September 9th to 12th, 1991, continues this tradition.

To the best knowledge of the editors, the EMPG Meeting in Vienna was the largest ever held: A total of (exactly) 100 participants registered, including a handful of students from the University of Vienna and some last-minute registrations of guests from Eastern European countries who took the opportunity of establishing contacts with Western scientists. The relatively large number of participants was partly due to travel support for invited speakers and to further financial support for guests from Eastern Europe, granted by the Federal Ministry of Science and Research, by the University of Vienna, and by the Municipality of Vienna. On behalf of all the participants supported in this way, the editors would like to express deep gratitude to these institutions for their help in making the meeting a success.

Of a total of 55 papers read at the meeting, 44 were later submitted for the proceedings volume; of these, 28 (or app. 64%) survived a meticulous process of review and revision, comparable to the procedures followed by international scientific journals. This selection of work is now presented in this volume. As its title and, even more, its table of contents reveals, it contains not only contributions to mathematical psychology in the narrow sense but also work in psychometrics and general methodology. The common element in all of the contributions, however, is the attempt to deal with scientific problems in psychology with rigorous mathematical reasoning. The contents are representative of the interests and scientific work within the EMPG.

The editors also decided to set a high standard with respect to the appearance of the book: All chapters are set in a uniform layout using LaTeX as the text editing software. Since, however, about half of the authors had no access to

LaTeX (or TeX), their manuscript files had to be transformed into LaTeX – a task that, unfortunately, can be done only by hand. This and all the other typesetting was carried out expertly and untiringly by Ms. Elisabeth Seliger. The editors wish to extend their gratitude to her for her loyal help throughout the whole editorial process. Finally, thanks also go to the Department of Psychology of the University of Vienna for practical assistance. Both the organizing of the meeting and the editing of the proceedings volume required unnumbered hours of secretarial support, especially by Ms. Karin Waldherr. Neither the conference nor this volume of proceedings would have been possible without all that devoted assistance.

April 30, 1993

Gerhard H. Fischer
Vienna

Donald Laming
Cambridge

Contents

Part II: Choice and Reaction Time

Part III: Social Systems

Part IV: Measurement and Psychometrics

Part V: Methodology

Part I

Perception, Learning, and Cognition

1

Metrics in Color Spaces: Im Kleinen und im Großen

Tarow Indow[1]

ABSTRACT Colors can be represented as points in various three-dimensional spaces. It has been a tradition in color science to regard just-noticeable differences in a space denoted (x, y, Y) as line-elements and the Munsell solid defined in another space as a structure in which supra-threshold color differences are represented. These ideas and also the relationship between these two "metrics" are discussed with special attention to two different modes of appearance, aperture and surface colors.

1.1 Color spaces

Color can appear in various different modes. In this article we are concerned only with two basic modes. One is the appearance of light presented in a small opening in a dark surround (*aperture color mode*). No matter where the source of light is, we see the aperture as being filled with color floating in the air. The other is the appearance as color of something we are perceiving around us (*surface color mode*). This is a result of the light reflected by the surface. We can touch surface colors, but we feel as if our fingers will go through aperture colors. Brightness changes from dim to bright for aperture colors and blackish to whitish for surface colors. When we observe a brown surface through a tube having black inside, we see the opening as being filled with a dark reddish yellow aperture color. In order to be brown, it is necessary to have a blackish touch in it. When the surround of an opening is white and illuminated, the appearance of light coming through it is not of aperture color mode. It looks as if a patch of surface color has been placed on the white surface. In this article, this appearance will be called *simulated surface color*. Compared with a real colored patch, this appearance does not have texture.

The number of colors we can discriminate from each other is in the order of 7 million, and each color j can be represented as a point F_j in a three-dimensional space (*color space*). The best known spatial representation of surface colors is the Munsell color system, which is physically displayed by standard color chips arranged according to the basic attributes of color: hue, brightness, and saturation. This system was defined and standardized by visual assessment and serves as a framework for specifying surface colors. By comparing a sample j with

[1]School of Social Sciences, University of California, Irvine, CA 92717, U.S.A., Department of Cognitive Sciences, University of California, Irvine

the standard chips under standard illumination, we can determine the attribute values of the sample and its position F_j in this space. There are color spaces of a different nature in which F_j is defined from the spectral composition of the light (colorimetry). The attributes of a color, no matter in what mode it appears, are determined by the distribution of radiant energy $P(\lambda)$ over wavelength λ. The CIE (Commission Internationale de l'Eclairage) colorimetric systems were developed on the basis of color matching experiments. The most widely used one is the (x, y, Y) space in which brightness is represented by Y, and hue and saturation on the plane called ***chromaticity diagram***, (x, y). This plane is perpendicular to the Y axis. We now have instruments to obtain $P(\lambda)$ and $F(x, y, Y)$ for any color stimulus.

There are an infinite number of possible patterns $P(\lambda)$ that are perceived as the same color. These are said to be *metameric* with each other. Color space is defined so that all metameric patterns for a color j are represented by the same point F_j. If colors attributed to shine of a surface, e.g., gold, are excluded, three axes are necessary and sufficient for this specification of distinguishable colors, possibly 7 million in number. This is due to the phenomenological fact that we see three attributes in color and to the empirical fact that any $P(\lambda)$ can be matched by an additive mixture of three suitably chosen monochromatic lights. The latter is due to the physiological fact that there are only three kinds of cones in our retina. Spatial representation of color presents interesting problems to mathematical psychologists. Though a space is used in both cases, (x, y, Y) or the Munsell color solid, it is not clear what mathematical properties and metric we can assign to the space.

1.2 Problems

A. H. Munsell[1] was an artist and art teacher in Boston. His original system, *Book of Colors*, was published in 1929. The Optical Society of America (OSA) established a subcommittee in 1937 to examine the color spacing in this book and, after extensive visual checking by 41 observers, a final report was published in 1943. The Munsell system we use today is based on this report. Standard color chips are arranged according to cylindrical coordinates (Figure 1.1). The vertical axis represents lightness (the change from black to white) denoted as Value (V). Polar angle and distance respectively represent Hue (H) and saturation denoted as Chroma (C).

Standard chips were selected so that two adjacent chips in each coordinate have a constant perceptual difference. No attempt has been made to equate these constant differences on respective attributes. Hence, the system only has local uniformity between two neighboring chips along H, V, C taken separately. The local uniformity is necessary to make it easier for users to interpolate in as-

[1] All the original sources are referred to and listed in Indow (1988).

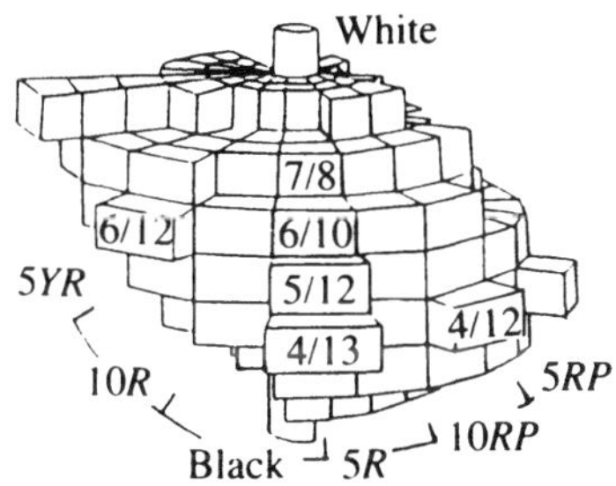

FIGURE 1.1. Munsell solid

signing values of $H\ V/C$ to a given sample. Standard chips in each attribute are coded by integers. For example, ten regions of H are denoted by five principal hue names (R, Y, G, B, and P) and their combinations. Each of these 10 regions is sub-divided into 10 intervals of H, numbered 0 to 10, where 5 is given to the most representative color of that hue. If two colors j and k differ in one attribute only, e.g., in C and the difference is not too large, we may take $|C_j - C_k|$ to be the size of that difference we perceive between these colors, d_{jk}. In general, however, colors differ in more than one attribute, and neither Munsell nor the OSA sub-committee discussed how multi-attribute color differences should be represented in the Munsell color space. Standard chips are displayed in a physical, Euclidean space. This is not because it is guaranteed that color difference is representable by Euclidean metric. This is a problem that was not examined by the sub-committee. The global structure was not adequately studied either. By definition, lightness V and saturation C are represented by axes orthogonal to each other. It is not clear, however, whether this definition is appropriate for representing multi-attribute color differences.

Color matching for the CIE spaces[2] is carried out in the following way. A light of spectral composition $P(\lambda)$, either coming directly from a source or reflected by a surface, is presented on one side of a bipartite target and three primary color stimuli on the other side. The primary stimuli must be linearly independent in the sense that none of them can be matched by a mixture of the remaining two. When the intensities of the primary stimuli have been appropriately adjusted by the observer, the two sides in the target appear exactly the same (complete match). The property of $P(\lambda)$ related to color perception is then specified by a point F having these three intensities as coordinates. For some colors, it is necessary to desaturate by mixing a primary stimulus with $P(\lambda)$, and this mixture is then matched by a mixture of the remaining two. The intensity of the desaturating primary is defined as a coordinate of negative value. Although primaries can be any color stimuli, in order to reduce the necessity of desaturation it is

[2] All the original sources referred to in this and the next sections are given in Wyszecki & Stiles (1982).

better to use three monochromatic lights appropriately separated on the spectrum. The CIE standardized a system for a 2 degree target in 1931 and a system for a 10 degree target in 1964. Both are based on two matching experiments.

In the present article, only the 1931 system will be examined, because most of the results relevant to our discussion are given by this system. The two color-matching experiments for this system used three primaries of different wavelengths. However, when both sets of data were transformed to a common set of primaries (700, 546.1, 435.8 nm), the agreement was impressive. This fact implies that perception of color at this level is based on a linear system and coordinate axes can be freely transformed in a color space. This is one of the laws formulated by H. G. Grassmann, a mathematician, in 1853. The unit representing the intensity of each primary stimulus as coordinate can also be freely chosen. The CIE adopted the (x, y, Y) space by choosing a set of coordinate axes and units that are convenient for the following colorimetric calculations:

$$X \propto \int P(\lambda)\bar{x}(\lambda)d\lambda, \quad Y \propto \int P(\lambda)\bar{y}(\lambda)d\lambda, \quad Z \propto \int P(\lambda)\bar{z}(\lambda)d\lambda, \tag{1.1}$$
$$x = X/(X+Y+Z), \quad y = Y/(X+Y+Z), \quad z = 1 - x - y,$$

where $\propto$ means that the two sides are proportional, and $\bar{x}(\lambda)$ etc. are color matching functions. Values of X etc. are defined so that they take positive values only. To specify the color perception produced by a surface under an illumination having the energy distribution $E(\lambda)$, $P(\lambda)$ is replaced by $R(\lambda)E(\lambda)$ in (1.1), where $R(\lambda)$ is the spectral reflectance of the surface. The axis Y called *luminance* is so chosen that the brightness of colors is determined solely by this variable. The *chromaticity diagram* (x, y) represents stimuli of the same luminance, but varying in hue (direction) and saturation (distance) around the achromatic point of that brightness. The color matching functions are tabulated, but not expressed in equations. It was a difficult problem in the precomputer period to carry out the numerical integrations in (1.1). The three variables, X, Y, Z, were selected so as to minimize the computational labor. Hence, these three variables do not directly represent processes in our visual system.

The representation $F(x, y, Y)$ is a unique specification of the color F produced by a light with energy distribution $P(\lambda)$. Munsell standard chips, when observed under a CIE standard light such as C or D65, are specified as points in this CIE space. In the (x, y, Y) space, relationships between colors are not represented. We can, however, say that it has a topology because, if color j is obtained by mixing color stimuli i and k, then F_j is always located between F_i and F_k. In other words, color F_j is closer to F_i than is F_k. A large number of attempts have been made by color scientists to transform the (x, y, Y) space into a color space in which we can define a distance d_{jk} that represents the perceptual difference between colors j and k. This effort is called the search for a *uniform color scale* (UCS) (Wyszecki & Stiles, 1982, Chapter 6). Most attempts use as the measure of perceptual difference either the just-noticeable difference Δ or the Munsell system. Let $\Delta(F; \pi)$ be the distance from F, colors within which are judged indiscriminable from F with probability π, then its size on the (x, y, Y) changes with F and with direction from F. Hence, a number of attempts have been made to find a space in which all $\Delta(F;\pi)$ turn out to have the same size. When Munsell standard chips having the same value of V are represented as F in the chromaticity diagram (x, y), their structure is distorted compared to the Munsell notation. So, various transformations have been tried to restore the desired structure. In the following sections, two problems will be discussed; whether Δ's or the Munsell system can be used as a measure of distance, and the relationship between the two.

1.3 Discrimination ellipsoids in the (x, y, Y) space

It is generally supposed that the $\Delta(F_0;\pi)$, taken in various directions from F_0, form an ellipsoid in the (x, y, Y) space and, when Y is held constant, an ellipse in the chromaticity plane (x, y), and that the value of π affects only the size. In most studies using aperture colors, $\Delta(F_0;\pi)$, in each direction from F_0, is taken to be the standard deviation of repeated matches (Wyszecki & Stiles, 1982, 5.4). The subject matches the comparison color F to F_0 in the way described earlier. If change of F is restricted to a particular direction, a unidimensional distribution of matches is generated, and defining $\Delta(F_0;\pi)$ by its standard deviation is equivalent to put $\pi = .68$. When F is adjusted in the three directions in each match, matches $F(x, y, Y)$ follow a three-dimensional (3-D) normal distribution around F_0. Its vertical slice in any direction gives a unidimensional normal distribution, and its horizontal slice at a fixed height of density gives an ellipsoid $\Delta(F_0;\pi)$. According to the height at which the distribution is sliced, the size of ellipsoids and the value of π are changed.

The procedure stated above cannot be employed for determining the region of surface colors F that are indistinguishable from a standard surface color F_0, because a surface color cannot be continuously adjusted while being observed. Indow and Morrison (1991) developed a procedure to obtain $\Delta(F_0;\pi)$ by means of the method of constant stimuli. Comparison colors F_j, $j = 1, 2, \ldots, n$, differed from the standard color $F_0(x_0, y_0, Y_0)$ in various directions and by various distances. Let P_j be the probability that F_j is discriminated from F_0, and d_j be the distance between F_0 and F_j. If the distribution of F is 3-D normal, the psychometric function $P(d)$ for F at various distances from F_0 in a direction Λ is a sigmoid (integrated normal), the parameter values of which change according to Λ. If the argument d is replaced by $\bar{d}_j = d_j/\Delta(F_0;\pi)$, distances from F_0 measured by the just-noticeable difference in the direction of F_j, then difference due to Λ is eliminated and we have one $P(d)$ for all F_j. Such an ellipsoid $\Delta(F_0;\pi)$ that gives the best fitted sigmoid to the data P_j including all directions can be obtained under the constraint that its two principal axes are in the chromaticity diagram (x, y) at the level of Y_0 and that the third axis is perpendicular to this plane, i.e., in the direction of Y. This method was shown to give satisfactory results to sets of data for painted color samples.

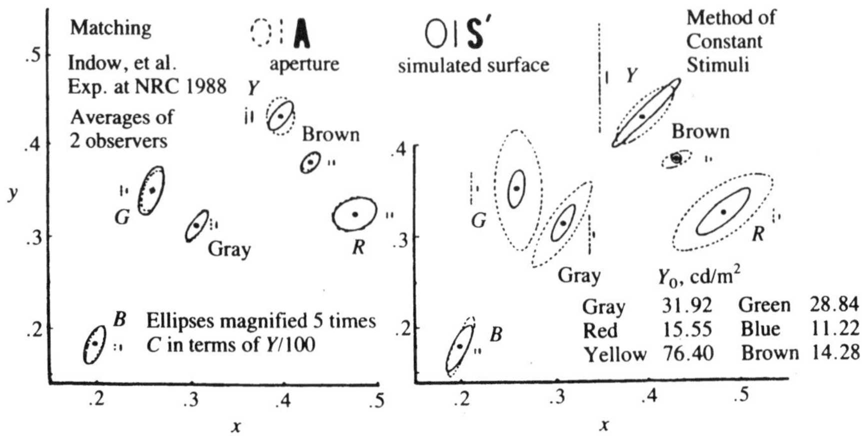

FIGURE 1.2. Discrimination ellipsoids

An experiment was performed with six standard colors, F_0. Stimuli were displayed on a color monitor, once on the dark surround (aperture color mode) and once on the achromatic surround brighter than both F_0 and F_j (simulated surface color mode). Five of the F_0 were colors recommended by CIE to be used in studies of color differences, and the other one was *brown*, because of its drastic change of appearance depending on whether the surround is bright or dark. The luminance of each F_0 was set at the level in which the color appeared most saturated. With each F_0 of respective modes of appearance, four ellipsoids were obtained by two subjects and by two methods, the method of uni-dimensional matchings in respective directions ($\pi = 0.68$) and the method of constant stimuli ($\pi = 0.5$) (Indow et al., 1992). On the basis of these two studies:

(1) The region of F indiscriminable from F_0 is ellipsoidal in the (x, y, Y) space for F_0 of three modes, aperture (A), surface (S), and simulated surface (S′). The average ellipsoids for the two subjects in the second study are shown in Figure 1.2. Ellipses on the chromaticity diagram are magnified 5 times, and the bars represent the lengths of the third axes. These ellipses are similar in relative size and orientation to those in other studies with colors of A- and S-modes. Absolute sizes depend upon experimental condition and criterion (π). If the plane (x, y) is transformed to another plane by affine and projective transformations, though shapes and orientations are changed, ellipses remain ellipses. Various CIE spaces are related to each other by these transformations.

(2) The matching procedure gave the same ellipsoids for colors of A-mode and S′-mode.

(3) The method of constant stimuli gave larger ellipsoids and smaller false discrimination probabilities for A-mode colors than for S′-mode colors. The false discrimination means that a "different" judgment is given to the identical pair (F_0, F_0). So far as these 12 sets of data are concerned, these two differences are essentially due to the parallel shift of the sigmoid functions $P(\bar{d})$ between the two modes.

These findings suggest that, when some internal process ξ occurs corresponding to a color stimulus, its distribution is normal in a 3-D color space that can be obtained from the (x, y, Y) space by affine and projective transformations. Its parameter values depend upon the color stimulus. When a pair (F_0, F) is presented, there occurs a unidimensional normal variable ζ, being the difference between ξ and ξ_0 corresponding to F and F_0. Suppose there is a threshold ε for detection of differences. If and only if $|\zeta| > |\varepsilon|$, the difference is detected in the method of constant stimuli. If distributions of ξ_j for the respective F_j differ only in location, having the same variance in terms of $\bar{d}$, we can expect $P(\bar{d})$ to be a sigmoid. Its slope and location on $\bar{d}$ will be determined by the variance and ε respectively. For the matching experiment, the same model predicts that the distribution of matches should be independent of ε. Finding (3) suggests that the distribution of internal process ξ underlying perception of a color F will be the same irrespective of its mode of appearance, A or S′, but the criterion $|\varepsilon|$ will be different. It will be larger for the A-mode than for the S′-mode, which may be related to the fact that A-mode colors are less "tangible". In spite of that the subject perceives the "brown" stimulus as a very different color depending on its mode of appearance, the ellipsoid turned out almost the same. This suggests that discrimination of colors is rooted at a deeper level than determination of color appearance. A challenging problem for psychophysicists is to find a space in which the distributions of ξ are the same for all colors. Because of its ellipsoidal form, $\Delta(F;\pi)$ has often been taken equivalent to a Riemannian line element ds. There are charts giving coefficients in the first fundamental tensor from which we can estimate for any aperture color F, in the chromaticity diagram at $Y = 47cd/m^2$, the ellipse $\Delta(F)$ that embraces about 40% of 2-D matches with F. This is equivalent to define the ellipse by the distance of one standard deviation of the 1-D normal distribution obtained by slicing the 2-D normal distribution vertically in every direction. If the (x, y) plane is transformed to a Riemannian surface on these bases, the equivalent Euclidean curved surface is extremely complicated (Wyszecki and Stiles, 1982, Figure 5 (8.4.4)). More will be said in the last section about the introduction of a metric in small, e.g., the metric ds.

1.4 The mathematical nature of Munsell space

The Munsell standard chips have been selected and arranged as shown in Figure 1.1 so that colors differ from one to the next by an equal perceptual step in each attribute. The basic procedure involves matching of uni-attribute color differences of medium size between two neighboring chips in each axis taken separately. When we see the display in Figure 1.1, however, we are tempted to define from this color solid perceptual differences of more general nature across different attributes. Then, the problem of geometry of the Munsell solid comes to the fore. Since 1952, multidimensional scaling (MDS) has been applied a number of times to the Munsell system (Table 1, Indow 1988). Some studies are with methodological interest of MDS and some with the purpose of making explicit the geometrical nature of the Munsell solid.

Let a symmetric matrix $\mathbf{D} = (d_{jk})$ be given in which d_{jk} are scaled values of perceptual differences between j and k, $j, k = 1, 2, \ldots, n$. The origin of the scale is defined so that $d_{jj} = 0$. If data are given in all cells of $\mathbf{D}$, $n \times n$, it is called *complete,* and otherwise *incomplete.* Taking $\mathbf{D}$ as the input, MDS constructs a configuration of points $\{F_j\}$ in an m-dimensional space R^m in such a way that inter-point distances between F_j and F_k, $\hat{d}_{jk}$, reproduce d_{jk} as closely as possible. The data $\mathbf{D}$ can be incomplete. The dimensionality m, coordinates of F_j, and the functional relationship between d and $\hat{d}$ are given as the output. The nature of space R has to be specified in advance. In most studies, because the Munsell solid is displayed in a Euclidean space (E), $\{F_j\}$ is embedded in E^3, and Euclidean inter-point distances $\hat{d}$ and data d are expected to be proportional, $d \propto \hat{d}$. The expectation is fulfilled in our experiments if d is scaled in the following way. The Munsell gray series is a well established scale of equal-appearing intervals for achromatic surface colors on a gray background of medium lightness level. The most extensive series of Munsell gray chips covers from 0.5 V to 9.5 V, where V (Value) means the Munsell unit for lightness. Its unidimensionality is a matter of perceptual fact, and the equality of intervals, at least in its middle part, has been thoroughly checked by the CIE sub-committee and in other studies. Its relation with the luminous reflectance Y (0 to 1.0) is approximated by a function (the simplest form),

$$V = 2.5(100Y/Y_0)^{1/3} - 1.6,$$

where Y_0, the luminance of the reference white surface, is often set equal to 1.0. When a pair of colors (j, k) are presented, the subject selected, using an apparatus of slide rule type, such a pair of grays V_a and V_b on the gray background that appeared to have the same size of difference. Hence, the color difference is matched by a gray difference $|V_a - V_b|$. With each pair (j, k), matched gray differences were obtained in various parts of the gray series and the mean is defined as d_{jk}. In Figures 1.3, parts of $\{F_j\}$ in Figures 6 and 8 in Indow (1988) and the relationship between d and $\hat{d}$ are given. The configuration was constructed by a MDS program (DMRPD) in a 3-D Euclidean space and also in a 3-D elliptic space with the optimized curvature. The part shown in Figure 1.3 is the region from $5R$ to $5Y$ of $\{F_j\}$ between $4V$ to $8V$. The axis representing V always appears, not only in this study but also in others, to be orthogonal to the plane

(H, C). Hence, $\{F_j\}$ is vertically projected to this plane. Munsell chips of the same value of C are connected by curves denoted by $2, 4, \ldots, 12$, and those of the same H notation are connected by line segments. Some local irregularities of color spacing will be noticed. These deviations from the Munsell structure will be easily amended by changing notation of Munsell chips. The global structure of $\{F_j\}$ exhibits two systematic deviations from the Munsell notation. One is that the most representative blue $5B$ in the Munsell notation is actually greenish. The other is the following trend in $\{F_j\}$ constructed in E^3 (A). In all hue directions, the first C step from gray is larger, and C steps at the periphery, from $10C$ to $12C$ in this example, are smaller than other chroma steps in the middle (more clearly in Indow, 1988, Figure 5). This is the distortion we have when concentric circles on an elliptic surface are projected to E^2 with the achromatic points in the center. For this reason, $\{F_j\}$ was embedded in an elliptic space and the deviation in this $\{F_j\}$ is less apparent (B). Scaled perceptual color differences d and obtained inter-point distances $\hat{d}$ are proportional with the scatter as shown in Figure 1.3, where RMS means root-mean-square. Although an additional parameter (curvature) was estimated in B, the reproducibility of d by $\hat{d}$ was not improved. However, the values of RMS, 0.20 and 0.24, are sufficiently small because these are only about 3 times the just-noticeable lightness difference under most favorable observing conditions. Hence, we can conclude that multi-attribute color differences of surface colors can be represented by inter-point distances of $\{F_j\}$ in R^3, either Euclidean or elliptic, provided the degree of scatter as shown in Figure 1.3 is disregarded and the following constraint is taken into account. When color differences are not too large, we can clearly feel how large or small they are. However, when two colors are too different, such as red and green, they are simply "different" and we cannot evaluate the size of their difference. In perceiving color differences, we are not far from the stage of primitive tribes who count numbers "one, two, and many". For this reason, pairs of colors, (j, k), presented to the subjects, were limited within the range that $d_{jk} < 3.5\ V$ and remaining cells of $\mathbf{D}$ were left empty (incomplete). Hence, we cannot claim that the Munsell solid is embedded in E^3 as a whole, but we can say that surface colors are represented in a three-dimensional manifold with a locally Euclidian (or elliptic) metric. In the Munsell notation, no relationship was given between units of different attributes. We now have a unified unit for all the attributes, and it has become apparent that one Munsell unit of V is approximately equivalent to two Munsell units of C. The units of H and C are interlocked in the (H, C) plane. The 2:1 relationship between V and C Munsell units is in agreement with the impression that color practitioners have had for a long time.

The hue of each aperture color can be decomposed into components. Orange is reddish and yellowish. These components are regarded as two opponent processes, red(R)–green(G), and yellow(Y)–blue(B). None of these four hues can be described in terms of any other color names than its own. Curves were ob-

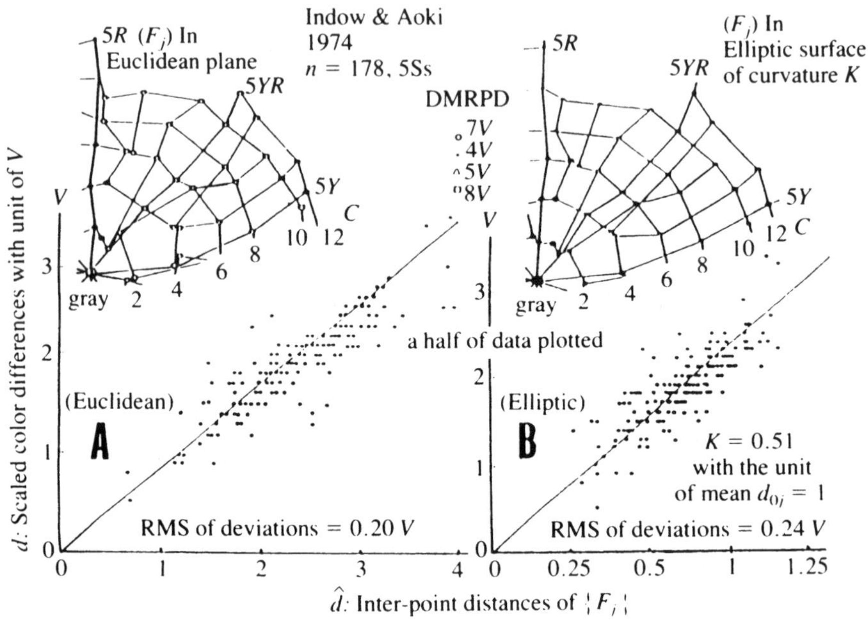

FIGURE 1.3. Region from R to Y of $\{F_j\}$ and reproducibility of MDS

tained, first by Jameson and Hurvich in 1955, that show how these components in aperture color due to monochromatic lights change as functions of the wavelength λ. The curves called *chromatic response functions*, $H_\alpha(\lambda)$, $\alpha = 1$ to 4, were obtained by the method of cancellation (Wyszecki and Stiles, 1982, 5.13). This method cannot be applied to surface colors. However, if judgments about components α, degrees of redness or yellowness, etc., in individual Munsell chips j are given as data, we can define in the (H, C) plane a set of such radial vectors f_α from the achromatic point that represent directions of pure red, pure yellow, etc. Munsell used five principal hues; purple (P) in addition to the above mentioned four. The subjects can assess components α in each Munsell chip j in terms of either four or five principal hue names. Each radial vector f_α was defined such that (contravariant) coordinates of points F_j on f_α reproduce the assessments of the principal hue α in these colors j. Using $\{F_j\}$ and vectors f_α, both defined in the same space, we can obtain H_α-curves for surface colors that have the same meaning as $H_\alpha(\lambda)$-curves for aperture colors. In H_α-curves for surface colors, the argument is not λ but the Munsell H. What color is called the most representative blue, for example, may be different according to

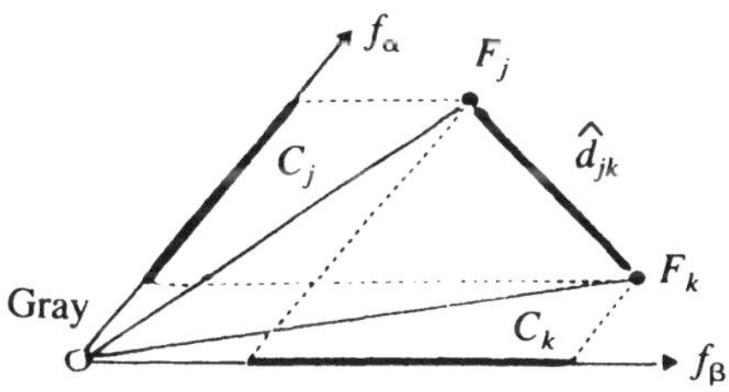

FIGURE 1.4. Color difference and hue vectors

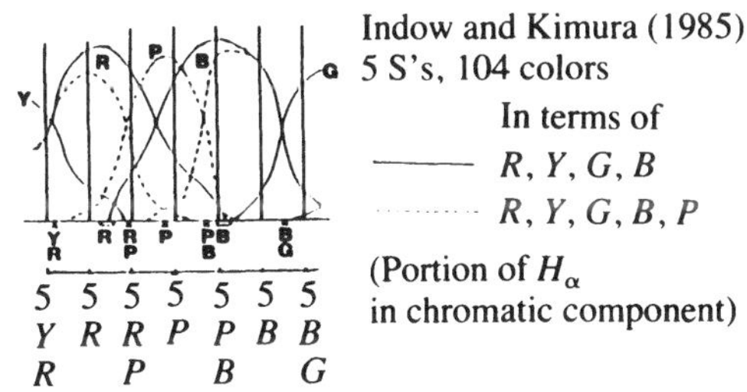

FIGURE 1.5. H_α in most saturated colors from 5*BG* to 5*YR*

the individual. Hence, vectors f_α and f_β were determined individually. On the other hand, $\{F_j\}$ was assumed to be the same for all the subjects. When the five hue names were used, individual differences for red, green and blue vectors are small whereas those for yellow and purple are large. This finding observed in two studies (Indow, 1980, 1987) is interesting in the view that we have only three different cones in the retina. Yellow, though perceptually unique, is obtained by simultaneous excitations of red and green cones. Perceptually, purple has reddish and bluish tints. Hence, cone activities for these two hues must be more heterogeneous than for the other three hues. When four color names were used as in the chromatic response functions, the H_P-curve was evenly divided into H_R- and H_B-curves as shown in Figure 1.5 (Indow, 1980, 1987). Individual differences for f_R and f_B became larger than when f_P was included. Purple is not indispensable but plays some role in purifying concepts of red and blue. The assertion that Munsell 5B is not real blue but greenish is due to the position of F_j for 5B chips (j) and to the H_B-curve (Figure 1.5). Whether the Munsell color space is regarded as Euclidean or elliptic is related to the combination rule, how d_{jk} is generated from the differences of coordinates of F_j and F_k on the two principal hue vectors f_α flanking these two points, and also how the saturation of a color j increases as a function of the radial distance from the gray point to F_j (Figure 1.4).

1.5 The relationship between two metrics

As stated at the end of Section 1.2, color scientists have tried to transform the (x, y, Y) to a space having a uniform color scale. Some used the Munsell system as a criterion on the assumption that the current Munsell solid is a metric representation for color differences. According to this criterion, if its global structure is represented without distortion in a space, the space is thought to have a color metric in the large. Some used discrimination ellipses for aperture colors as a criterion. If discrimination ellipses for all colors turn out as circles of the same size in a space, the space is thought to have a metric in the small that is valid at any location and for any direction. It was discussed in Section 1.3 that discrimination ellipsoids around a color F of three different modes (A, S, and S′) seem to have approximately the same form though the size depends upon the mode and the method by which just-noticeable differences $\Delta(F;\pi)$ are defined. Herein the form means orientation and relative sizes between the three principal axes.

It is well-known for mathematical psychologists that deriving a scale for supra-threshold difference from just-noticeable differences is a complicated problem even on a unidimensional continuum. Let s be a unidimensional change of stimulus from an arbitrary origin and $u\{s\}$ be a scale for supra-threshold difference. Let $s'(s;\pi)$ be a stimulus that is separated from s by the size of one jnd $\Delta(s;\pi)$. Then, the difference on u between s and $s'(s;\pi) = s + \Delta(s;\pi)$ is defined by

$$d(\pi) = u\{s'(s;\pi)\} - u\{s\}, \quad 0 < \pi < 1,$$

where $d(\pi)$ is a strictly monotone function of π and is independent of s. It was shown by Luce (1963) that such an interval scale $u\{s\}$, a strictly monotone function of s that is independent of π, can be uniquely defined (the revised Fechner problem). The condition that $\Delta(s;\pi)$ has to satisfy for the existence of $u\{s\}$ was discussed by Falmagne (1971) and Krantz (1971). To define line element ds in a color space is more than what was discussed above. Suppose that $\Delta(s;\pi)$ are given for points F' between F_0 and F, where s is the distance from F_0 to F', and $\Delta(s;\pi)$ satisfy the condition referred to above, then $u\{s\}$ will be an interval scale by which perceptual color differences from F_0 in the direction of F are uniquely defined. In order to interpolate between two adjacent Munsell standard chips, perhaps 7 to 12 jnd's have to be concatenated. In contrast to the case of a unidimensional continuum, the distance s under discussion has to be defined in theory along the geodesic, the shortest path from F_0 to F. Perhaps it is misleading to call $\Delta(F,\pi)$ in a CIE space a line element ds. It should be interpreted as the region around F within which colors are not discriminated from F with the criterion π, and no other connotation should be attached to it. If we look for a space in which all $\Delta(F,\pi)$ are represented by line-segments of the same length, it will be more productive to think of a color space that is related to physiological or cognitive models for color perception (e.g., Nagy et al, 1979). It is a different problem, however, whether or not supra-threshold color differences are also represented by line-segments of longer length in this space. It was discussed in Section 1.4 that the Munsell solid is represented in a 3-D manifold with a locally Euclidean or elliptic metric. However, this conclusion is based upon a presupposition that data d_{jk} behave as a metric. I have encountered experimental results against this interpretation. Once $\{F_j\}$ is obtained in a space R^3, we can take a series of collinear points $(F_1, F_2, \ldots F_k, \ldots)$ in which all d's

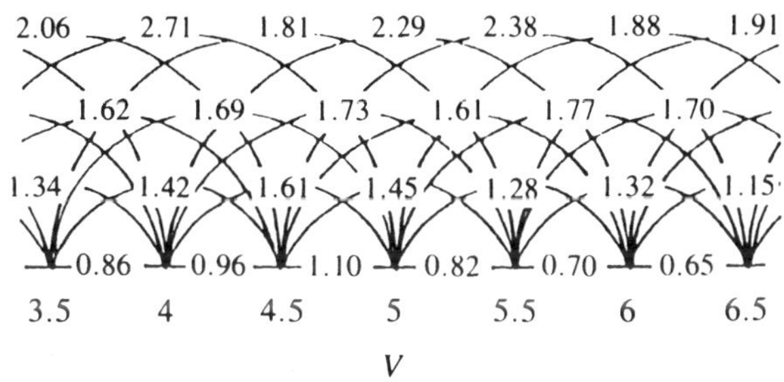

FIGURE 1.6. Subadditivity in scaling of grays

between points are available as data. If d behaves as a metric, then d_{1k} has to be equal to the sum of $d_{12} + d_{23} + \ldots + d_{k-1,k}$. Denote the sum of this type by Σ_{1k}. It was discovered that this is not the case and, d_{1k} is always a power function of Σ_{1k} having the exponent not equal to unity but in the order of 0.5. The same subadditivity was found even if F_j's are grays rather than colors. The perceptual difference d_{jk} between two grays (j, k) was scaled in the same way. Namely, for (j, k), the subject selected, in a region of the Munsell gray series in which neither j nor k is, two such grays that appear to be different with the same size as (j, k). A part of an extensive matching experiment between two gray pairs is shown in Figure 1.6. The fact of subadditivity, $d_{1k} < \Sigma_{1k}$, is easy to see.

The situation is different in judgments on perceptual distances between two points (j, k) presented in the visual space. The additivity clearly holds for perceptually collinear points in various directions. In this case, d_{1k} and Σ_{1k} coincide very well and, if a power function is fitted, the exponent is unity (Indow, 1991, Figure 13). Each scaled value d_{jk} in this experiment was obtained through systematic judgments on ratios between perceptual distances to other points from j and from k. In our early experiments on Munsell colors, the color difference of (j, k) was scaled in this way. The subject assigned positive values representing ratios of two perceived color differences, (i, j) over (i, k), where a number of colors can be taken as i $(\neq j, k)$. Judgments about such ratios were shown to be consistent. When $\mathbf{D} = (d_{jk})$, in which d_{jk} were thus scaled, was used as the input to MDS, the relation between d and $\hat{d}$ was slightly curvilinear, convex downward (Indow 1990, Figure 3). In my own experience, the proportionality between d and $\hat{d}$ as shown in Figure 1.3 is obtained when color differences are scaled by matching with gray intervals and the range of color differences to be judged is constrained as stated before. Then, the scatter of points around the proportionality line in Figure 1.3 and also in results of other studies (Indow, 1988) may be ascribed, at least partially, to the subadditivity in d. For collinear F_1, F_2, F_3 in $\{F_j\}$, $\hat{d}_{13} = \hat{d}_{12} + \hat{d}_{23}$ on the abscissa, but $d_{13} < d_{12} + d_{23}$ on the ordinate. A similar relationship will hold for non-collinear F's also. The problem of subadditivity will be discussed elsewhere. Herein, it will suffice to point out that there is a problem in regarding the scaled color difference d as a metric in the large, no matter whether they are uni-attribute or multi-attribute differences and even if all color differences are perceptually "palpable". However, it is still true that the Munsell solid, if appropriately amended, is a coherent and compact display of the global structure inherent in surface colors. We can predict from this system the following perceptual facts: how the total area is divided into subregions of respective principal hues, how a color difference of medium size in one region corresponds to color differences in other regions, how color differences are related to differences in their components, etc. Most models of human color vision are based on data obtained with aperture colors. We need a model for surface colors because colors we see around us, except the blue sky, are of surface color mode. Any model for surface color perception (e.g., Seim & Valberg, 1986) must take into account the information inherent in the Munsell solid or in some other systems of surface colors. Hence, these systems and their relations to colorimetric color spaces must be carefully examined, theoretically and experimentally, from the viewpoint of mathematical psychology. What was questioned in this article is whether color difference d_{jk} of medium size can be decomposed to the level of jnd's between F_j and F_k, or d_{ij} and d_{jk} can be concatenated to give a larger color difference d_{ik}. Except Krantz (1971), Suppes et al., 1989) and Drösler (the presentation in this meeting and 1990), not many mathematical psychologists have taken color as their targets of study. It is the purpose of my presentation to make clear that color science is a fertile field because data are rich and fruitful consequences are to be expected.

1.6 References

Drösler, J. (1990). *Quantitative psychology.* Toronto: Hogrefe & Huber.

Falmagne, J.C. (1971). The generalized Fechner Problem and discrimination. *Journal of Mathematical Psychology, 8,* 22-43.

Indow, T. (1980). Global color metrics and color appearance systems. *Color Research and Application, 5,* 5-12.

Indow, T. (1987). Psychologically unique hues in aperture and surface colors. *Farbe, 34,* 253-260.

Indow, T. (1988). Multidimensional studies of Munsell color solid. *Psychological Review, 95,* 456-470.

Indow, T. (1991). A critical review of Luneburg's model with regard to global structure of visual space. *Psychological Review, 98,* 430-453.

Indow, T., & Morrison, M.L. (1991). Construction of discrimination ellipsoids for surface colors by the method of constant stimuli. *Color Research and Application, 16,* 42-56.

Indow, T., Robertson, A.R., Grunau, M. von, & Fielder, G.H. (1992). Discrimination ellipsoids of aperture and simulated surface colors by matching and paired comparison. *Color Research and Application, 17,* 6-23.

Krantz, D.H. (1971). Integration of just-noticeable differences. *Journal of Mathematical Psychology, 8,* 591-599.

Luce, R.D. (1963). Psychophysical scaling. In R.D. Luce, R.R. Bush, & E. Galanter (Eds.) *Handbook of mathematical psychology* (Vol. 1, pp. 245-307). New York: Wiley.

Nagy, A.L., Eskew, R.T. Jr., & Boynton, R.M. (1987). Analysis of color-matching ellipses in a cone-excitation space. *Journal of the Optical Society of America, A4,* 756-768.

Seim, T., & Valberg, A. (1986). Towards a uniform color space: A better formula to describe the Munsell and OSA color scales. *Color Research and Application, 11,* 11-24.

Suppes, P., Krantz, D.M., Luce, R.D., & Tversky, A. (Eds.) (1989). Color and force measurement. In *Foundations of measurement* (Vol. 2, pp. 226-298). New York: Academic Press.

Wyszecki, G., & Stiles, W.S. (1982). *Color science: Concepts and methods, quantitative data and formulae.* New York: Wiley.

2

Color Similarity Represented as a Metric of Color Space

Jan Drösler[1]

ABSTRACT The traditional vector space representation of colors is based on the empirical conditions of Grassmann's (1853) laws: If the data of color matching exhibit a *Grassmann structure* , the corresponding representation theorem is interpretable in the color laboratory. The theory encompasses the phenomena of metameric stimuli, color matching, and *visually invariant* change of primary colors. It does not apply to the phenomena of color adaptation and of visual similarity of pairs of colors. The present study develops a representation for visual similarity of colors by specializing the Grassmann structure. If, in such a structure, (i) pairs of stimuli form an additive conjoint structure with respect to detectability, if (ii) the structure contains a smallest element, if (iii) (temporal) adaptation to color forms a group with respect to composition, and if (iv) the spectral locus is an invariant with respect to color adaptation, then there exists a mapping from pairs of colors into the real numbers which has the properties of a metric and is invariant with respect to projective hyperbolic transformations. The map is interpreted as *dissimilarity of colors*. The proof of the map's existence leads to analytic expressions for color matching functions. These are used to identify the analytic expression for the surface of the color solid. This color cone, in turn, is employed as a geometrical invariant with respect to the automorphisms of color space. They are interpreted as the effects of color adaptation. The major result states that color space possesses a projective hyperbolic metric. Since all assumptions are formulated in empirical terms, they can be tested experimentally. Available data do not contradict these assertions.

The psychology of color vision is distinguished by a long tradition of richly structured theory based on readily reproducible data. Still, there are some defects: Several areas of color vision stand isolated from each other. The theory of color matching, for instance, is largely independent of the theory of context effects. There are also areas in which theorizing proceeds in a pragmatic, *ad hoc* fashion, as is the case with the scaling of color difference. Theories of color similarity, for this reason, are neither connected to the theory of context effects nor to that of color matching.

The present study aims at developing a unified theory of color vision which is reducible to a few assumptions. These assumptions will be formulated in empirical terms. Its main effort is a theoretical conceptualization of larger areas as compared to color matching alone.

---

[1]Universität Regensburg, Institut für Psychologie, Universitätsstr. 31, 8400 Regensburg, Germany; e-mail: jan@rpss1.psychologie.uni-regensburg.de

2.1 Grassmann Structures

The traditional approach to color vision is based on Grassmann's (1853) laws stating that the properties of visual color matching under superposition of stimuli are compatible with the theoretical properties of a positive cancellation semigroup. Moreover, the properties of visual intensity control in color matching are compatible with the theoretical properties of scalar vector multiplication. Using the notation of Krantz (1975) the results are as follows.

Definition 1 *A* Grassmann structure *is a quadruple* $\langle A, \oplus, *, \sim \rangle$ *such that* A *is a set,* $\oplus$ *is a function on* $A \times A$*,* $*$ *is a function on* $\mathbb{R}^+ \times A$*, and* $\sim$ *is a binary relation on* A*, satisfying Axioms 1 - 5.*

1. $\langle A, \oplus \rangle$ is a commutative cancellation semigroup.
2. $*$ is a scalar multiplication on $\langle A, \oplus \rangle$.
3. Laws of equivalence: $\sim$ *is an equivalence relation on* A*; that is, for all* $a, b, c \in A$*:*

   (3i) $a \sim a$;

   (3ii) if $a \sim b$*, then* $b \sim a$;

   (3iii) if $a \sim b$ and $b \sim c$*, then* $a \sim c$.
4. Laws of additivity: *For all* $a, b, c \in A$*,* $a \sim b$ *if and only if* $a \oplus c \sim b \oplus c$.
5. Laws of scalar multiplication: *For all* $a, b \in A$*,* $r \in \mathbb{R}^+$*: if* $a \sim b$*, then* $r * a \sim r * b$.
6. Laws of trichromacy:

   (6i) For any $a_0, a_1, a_2, a_3 \in A$*, there exist positive numbers* r_i, u_i*,* $i = 0, 1, 2, 3$*, such that* $r_i \neq u_i$ *for at least one* i*, and such that*

   $$\sum_{i=0}^{3} r_i * a_i \sim \sum_{i=0}^{3} u_i * a_i.$$

   (We use the summation notation for sums involving $\oplus$*.)*

   (6ii) There exist $a_1, a_2, a_3 \in A$ *such that, for any positive* $r_i, u_i, i = 1, 2, 3$*, if*

   $$\sum_{i=1}^{3} r_i * a_i \sim \sum_{i=1}^{3} u_i * a_i,$$

   then $r_i = u_i$ *for* $i = 1, 2, 3$.

A set $a_1, \ldots, a_3$ satisfying the condition of (6ii) is called a basis, *or a* set of primaries.

Theorem 1 *Let $\langle A, \oplus, *, \sim \rangle$ be a Grassmann structure. Then there exists a vector space V over $\mathbf{R}$, a convex cone $C \in V$, and a function ϕ from A onto C, such that, for all $a, b \in A$, $r \in \mathbf{R}^+$ and $v \in V$:*

(i) $\phi(a \oplus b) = \phi(a) + \phi(b)$;
(ii) $\phi(r * a) = r\phi(a)$;
(iii) $a \sim b$ *if and only if* $\phi(a) = \phi(b)$;
(iv) *there exist* $c, d \in A$ *such that* $v = \phi(c) - \phi(d)$.

Proof: see Krantz (1975).

The map ϕ is a homomorphism of the Grassmann structure $\langle A, \oplus, *, \sim \rangle$ onto $\langle C, +, \cdot, = \rangle$, where C is a convex cone in V; property (iv) guarantees that V is a minimal vector space, since any element in V is generated as a difference of elements $\phi(c), \phi(d)$ in C.

Theorem 2 *Let $\langle A, \oplus, *, \sim \rangle$ be a Grassmann structure, with two homomorphisms ϕ, ϕ' onto convex cones C, C' in vector spaces V, V', satisfying* (i) - (iv) *of Theorem 1. Then there exists a nonsingular linear transformation T of V onto V' such that, for all $a \in A, T(\phi(a)) = \phi'(a)$.*

The map ϕ is unique up to nonsingular linear transformations.

Proof: see Krantz (1975).

Grassmann's achievements are clear: For stimuli which are represented by equal vectors, metameric matches are predicted.

The automorphisms of color scaling with composition form the general linear group. Some of its members are interpretable as representing a visually invariant *change of primary colors* in a color matching experiment. Some other members are interpreted by v. Kries (1905) as representing the temporal context effect of *color adaptation.*

2.2 Problem of the Present Study

There are shortcomings in the vector space representation. It neither represents the visual similarity of colors nor the independence which the gamut of blue (respectively yellow) colors empirically exhibit from that of the red (respectively green) colors, for instance, in hue cancellation experiments.

What is missing is the identification of the proper subgroup of visual automorphisms which leaves invariant the visual dissimilarity of any pair of colors. The present study will try to define this group of automorphisms by means of a study of color adaptation in an empirically well-founded way.

Well-foundedness refers to the question whether the mathematics applied can be, in essence, derived from those qualitative propositions which constitute the set of empirical observations. Observance of this restriction, although introduced by Grassmann (1853), is by no means standard in color science. Some contemporary discussants, however prominent, fall back far behind Grassmann's well-foundedness. Among these are the trial and error data fittings of "uniform color spaces" as well as proponents of "neural networks" for color vision.

The present line of theorizing is encouraged by the fact that more powerful tools have been used in color theory for a long time: Color coordinates are stated as product integrals of the stimulus spectrum $U(\nu)$ with three linearly independent color matching *functions* $\bar{r}(\nu), \bar{g}(\nu), \bar{b}(\nu)$,

$$\begin{aligned} R &= \int_{\nu_1}^{\nu_2} \bar{r}(\nu)\, U(\nu)\, d\nu, \\ G &= \int_{\nu_1}^{\nu_2} \bar{g}(\nu)\, U(\nu)\, d\nu, \\ B &= \int_{\nu_1}^{\nu_2} \bar{b}(\nu)\, U(\nu)\, d\nu. \end{aligned} \tag{2.1}$$

The metameric relation can hold for two stimuli $U_1(\nu)$ and $U_2(\nu)$: If their visual coordinates $[R_1, G_1, B_1]$ and $[R_2, G_2, B_2]$ are equal, then they are visually indistinguishable.

For technical reasons, the present notation refers to the light stimulus' frequency ν instead of the traditionally used wavelength of light λ. The two are connected by means of the velocity of light. Moreover, ν is proportional to the scale of "lines per centimeter" which is sometimes used in describing the color stimulus.

The notation $F(\nu)$ refers to the spectrum of the physical stimulus, not to the spectrum of its temporal modulation, that is, to the time course of stimulus presentation.

If the functional notation of equation (2.1) is to be well defined, a Hilbert space representation of color must be assumed, at least tacitly. As this particular functional notation is undoubtedly empirically interpretable, the Hilbert space assumptions are likely to prevail. The present study will lay out the assumptions in the open and will try to exploit some of their hitherto unnoticed ramifications.

The effect $h(t)$ of any linear system upon the input $u(t)$ can be represented by the convolution in the time domain,

$$\int_0^t h_i(\tau) u(t-\tau) d\tau = y_i(t), \quad i = 1, 2, 3,$$

where $h_i(t)$ are the system weighting functions in Grassmann's three dimensions. In the present context a study of the frequency domain is more enlightening. After Fourier-Transformation the convolution integral above is represented as a product in the frequency domain,

$$H_i(\nu) U(\nu) = Y_i(\nu), \quad i = 1, 2, 3. \tag{2.2}$$

Here the $H_i(\nu)$ are the components of the systems spectral characteristic, and ν denotes frequency proportional to lines per centimeter.

Color can be regarded as the output $Y_i(\nu)$ of a linear system, the input of which is the stimulus $U(\nu)$. It would be possible to determine the system's characteristic $H(\nu)$ if analytic expressions for the color matching functions $F_i(\nu)$ were known.

Experiments for determining the so called *color matching functions* $F_i(\nu)$ proceed by using three fixed lights (which may be monochromatic) and find their metameric adjustment to a fourth monochromatic light successively for all frequencies. In the sequel the notation $F(\nu)$ (without subscript) will refer to the (vectorial) sum of all three $F_i(\nu)$.

Representing these fixed monochromatic stimuli by Dirac impulses in the frequency domain, i. e.,

$$\delta(\nu - \nu^*) = \begin{cases} \infty & \text{for} \quad \nu = \nu^*, \\ 0 & \text{for} \quad \nu \neq \nu^*, \end{cases}$$

$$\int_{-\infty}^{\infty} \delta(\nu - \nu^*) d\nu = 1,$$

the system input can be written

$$U_1(\nu) = F_r(\nu)\,\delta(\nu - \nu_r) + F_g(\nu)\,\delta(\nu - \nu_g) + F_b(\nu)\,\delta(\nu - \nu_b),$$

substituting the frequencies ν_r, ν_g, ν_b of the narrow-band primaries for ν^*. The terms $F_i(\nu)$, $i = r, g, b$, are weighting factors which permit a three dimensional variation of the comparison stimulus $U_1(\nu)$. They constitute, for each ν, a particular physical level of the respective primary. The variable monochromatic stimulus of frequency ν^* in a color matching experiment is

$$U_2(\nu^*) = \delta(\nu - \nu^*).$$

Substituting in equation (2.2) the last equation yields, for all $\nu \in (\nu_a, \nu_b)$, the three color matching functions,

$$\int_{\nu_a}^{\nu_b} F(\nu^*)\delta(\nu - \nu^*)d\nu =$$

$$\int_{\nu_a}^{\nu_b} F(\nu^*)\delta(\nu - \nu_r)d_\nu + \int_{\nu_a}^{\nu_b} F(\nu^*)\delta(\nu - \nu_g)d_\nu + \int_{\nu_a}^{\nu_b} F(\nu^*)\delta(\nu - \nu_b)d_\nu. \quad (2.3)$$

Equality holds in (2.3) because empirically any color elicited by a monochromatic test stimulus $\delta(\nu - \nu^*)$ can be matched by a weighted superposition of three fixed stimuli $\delta(\nu - \nu_r), \delta(\nu - \nu_g), \delta(\nu - \nu_b)$. The weighting factors F_i emerge as functions of ν^* as the monochromatic test stimulus is varied over the visible range of frequencies. Simplifying (2.3) and writing ν for ν^* yields

$$F(\nu) = F_r(\nu) + F_g(\nu) + F_b(\nu). \tag{2.4}$$

In this way it can be shown that any triplet of color matching functions $F_i(\nu)$ obtained with three monochromatic primaries determines the spectral characteristic $H(\nu)$ of color vision in the sense a linear system, and that the right-hand side of equation (2.4) can be regarded as a base in a suitable three-dimensional space of functions. Both relationships provide worthwhile reasons for ascertaining analytic expressions for the color matching functions $F(\nu)$.

2.3 Extensions of Theory: Color Space a Hilbert Space

The present approach will employ a special Hilbert space representation which is a richer theoretical structure. It is distinguished from the vector space representation in that the points of its basic set are *functions*, for which a norm as well as a scalar product are defined.

Definition 2 *A function* $f : \mathbb{R} \to \mathbb{R}$ tends to zero rapidly at infinity *if, for each* $m \in \mathbb{N}^+$, *the function*

$$\begin{aligned} g &: \mathbb{R} \to \mathbb{R} \\ x &\mapsto |x|^m f(x) \end{aligned}$$

is bounded for $|x|$ *sufficiently large.*

Definition 3 *Let* $\mathcal{S}$ *be a non-empty set of functions*

$$s : \mathbb{R} \to \mathbb{C},$$

which

(i) are infinitely differentiable, and
(ii) tend rapidly to zero at infinity.

Then $\mathcal{S}$ *is a Schwartz* space.

For a Schwartz space one can define a *scalar product*

$$\begin{aligned} \langle\ \rangle : \mathcal{S} \times \mathcal{S} &\to \mathbb{C} \\ (r, s) &\mapsto \int_{-\infty}^{\infty} r\bar{s}\, dx. \end{aligned}$$

In particular, a *norm* exists for all $r \in \mathcal{S}$,

$$\begin{aligned} |\ | : \mathcal{S} &\to \mathbb{R} \\ r &\mapsto \langle r\bar{r}\rangle^{1/2}. \end{aligned}$$

The norm of a stimulus is interpreted here as its electromagnetic radiant power. A Schwartz space is a particular Hilbert space (cf. Lang, 1969).

Obviously, an extension of the representational range increases the set of possible theoretical statements. These theoretical tools include a wealth of Hilbert space operators besides the norm and scalar product already mentioned. The question is whether additional isomorphic empirical foundations can be operationalized.

2.4 Further Empirical Restrictions

(i) A Conjoint Structure

Obviously, a richer theoretical framework can be based only on a richer empirical data structure. To support such an enhancement, detectability data are introduced. Color mixing data will only support a vector space representation. It will be shown, however, that detectability data can serve as the empirical foundation of a Hilbert space representation. Comparisons of detectability supply the qualitative basis for the numerical structure of a space of functions.

There is no doubt about the physical radiant stimuli being representable as elements of a Hilbert space of infinite dimensionality. Let A be a set of radiant stimuli tending to zero rapidly at infinity.

For each $f(t) \in A$ the physical energy is kept constant at the value

$$\int_{-\infty}^{\infty} |f(t)|^2 dt = 1.$$

Define

$$\mu_\nu = \int_{-\infty}^{\infty} \omega |F(\omega)|^2 d\omega.$$

Let the positive sqare root of

$$D_t^2 = \int_{-\infty}^{\infty} \tau^2 |f(\tau)|^2 \, d\tau \tag{2.5}$$

be the *duration* of the temporal energy distribution $f(t)$, and the positive square root of

$$D_\nu^2 = \int_{-\infty}^{\infty} \omega^2 |F(\omega)|^2 \, d\omega - \mu_\nu \tag{2.6}$$

the *spectral width* of its Fourier transform $F(\nu)$. Thus each $f(t) \in A$ is identified with a pair (D_t, D_ν).

Let $\succsim$ be the judgment of the experimental subject "is at least as detectable as". It is well known (cf. Krantz et al., 1971, p. 257 f) that, if the relational system $\langle A, \succsim \rangle$ fulfills empirically the attributes of weak ordering, independence, the Thomsen condition, restricted solvability, and the Archimedean axiom, then all $(D_t, D_\nu) \in A$ can be scaled numerically on the basis of an empirical ordering alone (by a procedure called *conjoint measurement*) on log-interval scales ϕ_1, ϕ_2 for each component.

The present development assumes that the detectability of the stimuli forms conjoint a structure with respect to their attributes of duration and spectral width. Then, the scale values obey the representation condition for all (D_{t_1}, D_{ν_1}), $(D_{t_2}, D_{\nu_2}) \in A$,

$$(D_{t_1}, D_{\nu_1}) \succsim (D_{t_2}, D_{\nu_2}) \quad \text{iff} \quad \phi_1(D_{t_1})\, \phi_2(D_{\nu_1}) \geq \phi_1(D_{t_2})\, \phi_2(D_{\nu_2}). \tag{2.7}$$

This condition is critical for the present operationalization. It is of practical significance that, if the empirical assumptions for the conjoint structure are fulfilled, the numerical values D_t and D_ν of equations (2.3) and (2.6) can be used as psychological scale values ϕ_1 and ϕ_2 for detectability. This is so because the psychological structure has been shown to be isomorphic to the respective numerical structure. In the same manner Grassmann suggested using the numerical values of equations (2.3) as scale values for the visual components of the color vector.

(ii) Minimal Uncertainty

If it can be shown empirically that the weak order $\langle A, \succsim \rangle$ possesses a smallest element, then for this response a visual $F_1(\nu)$ can be derived from a slight generalization of the *minimal uncertainty theorem* (cf. Papoulis, 1962, p. 63 f), using equations (2.5) and (2.6). The result is a Gaussian spectrum

$$F_1(\nu) = \frac{1}{\sqrt{2\pi}}\, e^{-\dfrac{(\nu-\mu)^2}{2}}$$

with free parameters μ and a scale factor for ν. This undetermined scale factor for frequency (and its reciprocal for time) is of great importance. It enables the frequency of the light in the Tera-Hertz (10^{12} Hz) range to be used for scaling purposes even if the actual frequency of biological oscillations in the visual system remains unknown. The epistemological problem of identifying the ultimate smallest element is solved by constructing stimuli which, according to the ramifications of the theory to be developed, should lead to optimal detectability.

Thus one (functional) coordinate of visual space has been reached from purely qualitative assumptions. From Grassmann's algebraic approach the three dimensionality of color space will be retained. An orthogonal basis of *three* dimensions is sought. The functional coordinates $F_2(\nu), F_3(\nu)$ for the space can be derived from $F_1(\nu)$ by finding two solutions of the integral equation

$$\frac{1}{\sqrt{2\pi}} \int_{-\infty}^{\infty} e^{-\dfrac{(\nu-\mu)^2}{2}}\, F_i(\nu)\, d\nu = 0, \quad i = 2, 3. \tag{2.8}$$

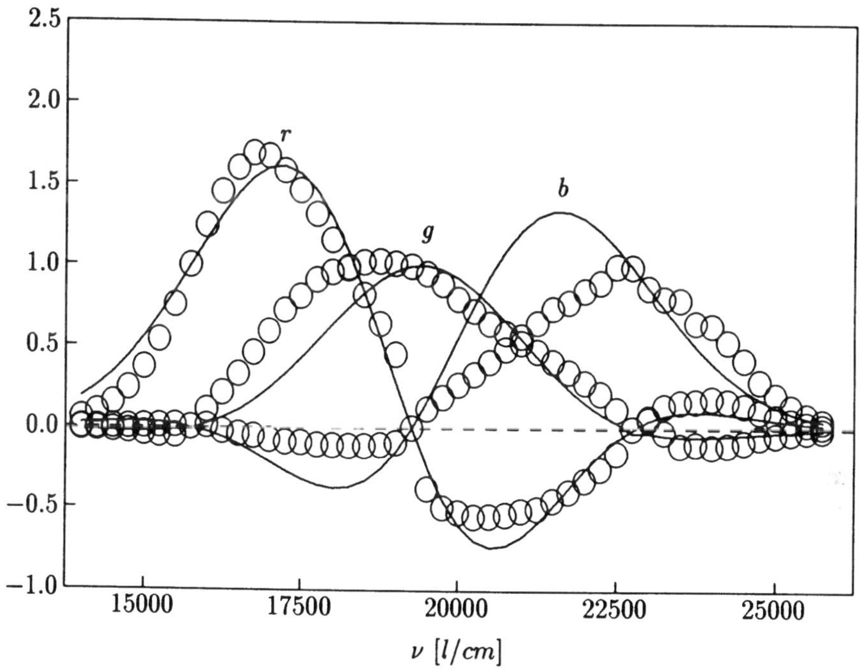

FIGURE 2.1. A comparison of linear combinations of the color matching functions $F_1(\nu), F_2(\nu), F_3(\nu)$ (lines) with Stiles' 2° measurements of his subject # 1 (circles)

Two such solutions are

$$\begin{aligned} F_2(\nu) &= -(\nu - \mu)\, F_1(\nu) \text{ and} \\ F_3(\nu) &= [(\nu - \mu)^2 - 1]\, F_1(\nu). \end{aligned} \tag{2.9}$$

The functions $F_1(\nu), F_2(\nu), F_3(\nu)$ form an orthogonal basis of color space and are interpreted as color matching functions.

2.5 A First Empirical Test

Before further derivations arc given, some reference to suitable experiments is in order. Detectability experiments have been conducted by Regan & Tyler (1971). Their results are in accord with equation (2.7). Now an empirical test of the functions $F_i(\nu)$, $i = 1, 2, 3$, is in place. If they form a valid basis for color space, experimental data from color matching experiments should be describable as linear combinations of $F_1(\nu), F_2(\nu), F_3(\nu)$.

At the frequencies $\nu_1 = 15500$ l/cm, $\nu_2 = 19000$ l/cm, $\nu_3 = 22500$ l/cm of narrow-band primary stimuli which Stiles used in his 2° color matching experiments linear combinations of the theoretical color matching functions require either zero or unit contributions from the primaries.

Parameters were estimated by solving the corresponding matrix equation

$$\boldsymbol{A}\boldsymbol{F} = \boldsymbol{I},$$

that is,

$$\begin{bmatrix} a_{11} & a_{12} & a_{13} \\ a_{21} & a_{22} & a_{23} \\ a_{31} & a_{32} & a_{33} \end{bmatrix} \begin{bmatrix} |F_1(\nu_1)|^2 & |F_1(\nu_2)|^2 & |F_1(\nu_3)|^2 \\ |F_2(\nu_1)|^2 & |F_2(\nu_2)|^2 & |F_2(\nu_3)|^2 \\ |F_3(\nu_1)|^2 & |F_3(\nu_2)|^2 & |F_3(\nu_3)|^2 \end{bmatrix} = \begin{bmatrix} 1 & 0 & 0 \\ 0 & 1 & 0 \\ 0 & 0 & 1 \end{bmatrix},$$

for $\boldsymbol{A}$.

Parameter estimates were optimized graphically by three dimensional plotting. For each of Stiles' subjects the chi-square sample statistics were plotted in three dimensions against μ and the scale factor for ν. The pair of parameter values at the minimum were chosen. All values were near $\hat{\mu} = 19500$ and a scale factor for ν of 1/1875, that is, dividing all frequencies by 1875 and inserting $\mu = 19500/1875$ made equations (2.8) and (2.10) applicable.

Figure 2.1 shows a fit of one of Stiles' subjects from his original 2° experiment (Trezona, 1987).

2.6 Still More Empirical Restrictions

(iii) The Temporal Context Effect of Color Adaptation

V. Kries (1905) tried to interpret some members of the group of automorphisms of color space as representing adaptation to color. His approach represents chromatic adaptation as *motion* in color space. This amounts to postulating a group structure for adaptation effects and their composition. There is empirical evidence for this assumption in the data by Burnham et al. (1957). Figure 2.2 demonstrates how a set of binocular color matches changes position in color space with adaptation of one eye to a specific color.

On the other hand this group cannot be the linear group of automorphisms. All previous studies, including the classical one by v. Kries, overlook a decisive matter: Any linear transformation of the set of colors predicts that adaptation will render some stimuli invisible and will make visible some stimuli which were invisible before.

Such predictions are contradicted by all available experimental evidence. Thus, linear theories of adaptation are contrary to fact. If a group of automorphisms is to be the representation of adaptation, linear theories, moreover, are ruled out by contradiction to Grassmann's postulates because they could lead to negative physical stimulus energies.

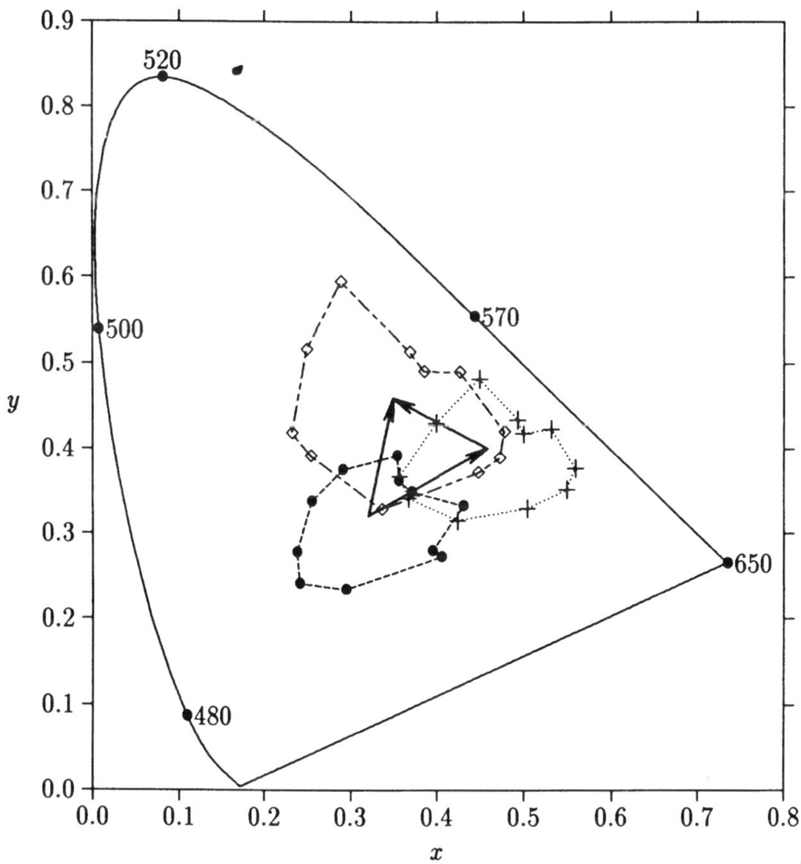

FIGURE 2.2. Matches for a set of colors (filled circles) after adaptation to illuminant R (open squares) and after adaptation to illuminant S (crosses). Data from Burnham et al., 1957. The arrows indicate the representation of adaptation as motion in color space.

(iv) The Surface of the Color Solid as an Invariant

Yilmaz (1962) postulated that adaptation always will transform one color to another color, never to a non-color. For the bounding surface of the color solid this has a particular consequence: any motion representing an adaptation to color will transform the boundary surface of the color solid into itself. The present study will regard the boundary of the color solid as an invariant with respect to the motions which represent adaptation. The resulting set of motions serves to identify that specific subgroup of the linear group which represents the geometry of color space.

The bounding surface of the color cone is best appraised if projective coor-

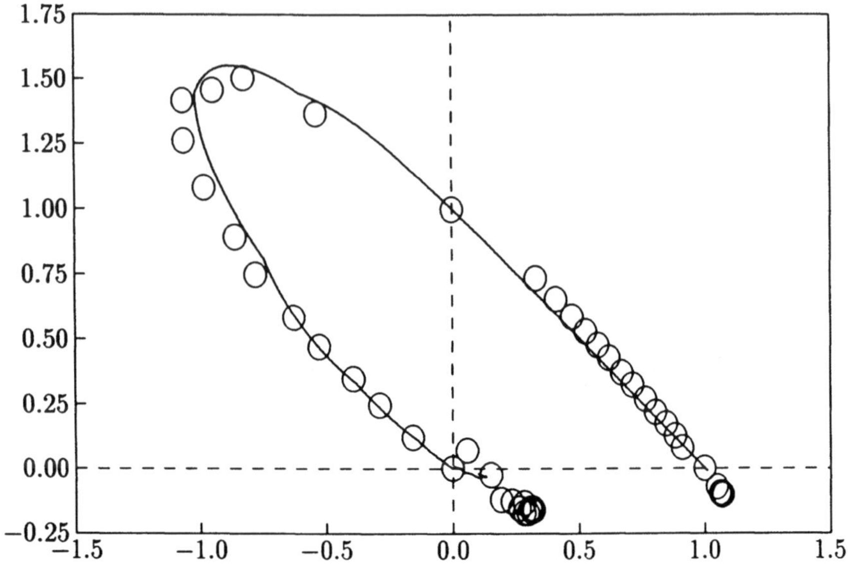

FIGURE 2.3. The 2° measurements of Stiles' subject # 1 in chromaticity coordinates (circles) compared with the theoretical spectrum locus developed here (line)

dinates are used. In color science they are called chromaticity coordinates. The solutions $x(\nu), y(\nu), z(\nu)$ are

$$x(\nu) = \frac{|F_1(\nu)|}{|F_1(\nu)| + |F_2(\nu)| + |F_3(\nu)|}$$

etc. Figure 2.3 shows a comparison of Stiles' measurements of his subject # 1 in chromaticity coordinates and the theory developed here.

The projective coordinates permit a rational representation because the transcendental terms cancel out after substitution from equation (2.10):

$$\begin{aligned} y(\nu) &= \frac{1}{(\nu-\mu)^2 - (\nu-\mu)}, \\ x(\nu) &= \frac{(\nu-\mu)}{(\nu-\mu) - (\nu-\mu)^2}, \\ z(\nu) &= \frac{(\nu-\mu)^2 - 1}{(\nu-\mu)^2 - (\nu-\mu)}. \end{aligned}$$

The analytic expression for the spectral locus is found by eliminating the frequency variable ν from the first two of the equations above. The result is

$$x^2 + x\,y - y = 0. \tag{2.10}$$

In projective coordinates the equation for the spectrum locus is a quadratic. The purple line is represented by the straight line joining the two points on the quadric, which represent the limits of the visible spectrum near $\nu = 13750$ and $\nu = 25750$ 1/cm. The quadric as an invariant induces well-known geometrical properties on the projective space at hand.

In homogeneous coordinates equation (2.10) has the coefficient scheme

$$\begin{bmatrix} 1 & \frac{1}{2} & 0 \\ \frac{1}{2} & 0 & -\frac{1}{2} \\ 0 & -\frac{1}{2} & 0 \end{bmatrix}. \tag{2.11}$$

Since in the Grassmann representation there is an equivalent projective transformation belonging to any admissible linear transformation of coordinates

$$\begin{aligned} x^* &= \frac{a_1x + b_1y + c_1}{a_3x + b_3y + c_3}, \\ y^* &= \frac{a_2x + b_2y + c_2}{a_3x + b_3y + c_3}, \end{aligned} \tag{2.12}$$

it is possible to transform equation (2.10) to principal axes by a projective transformation which produces coefficient scheme

$$\begin{bmatrix} \frac{1}{A^2} & 0 & 0 \\ 0 & \frac{1}{B^2} & 0 \\ 0 & 0 & -1 \end{bmatrix}. \tag{2.13}$$

Transforming to principal axes does not determine all eight of the free parameters of equation (2.12) (one of the original nine parameters can be fixed without loss of generality as the scale factor of the other parameters in (2.12)).

It should be noted that principal axes play a purely technical *rôle* in the present development. They are introduced as practical frames of reference for simplifying the description of color adaptation effects. Unlike v. Kries's (1905) classical study, they are not expected here to admit empirical interpretation.

The unknown parameters a_i, b_i, c_i, $i = 1, 2, 3$, can be found more easily if the inverse projective transformation is calculated and then inverted. Let A and B be the lengths of the principal axes. Then,

$$\frac{x^{*2}}{A^2} + \frac{y^{*2}}{B^2} = 1,$$

with coefficient scheme (2.13). Applying equations (2.12) results in the following coefficient scheme,

$$\begin{bmatrix} a_1^2 + a_2^2 - a_3^2 & \frac{1}{2}(a_1b_1 + a_2b_2 - a_3b_3) & \frac{1}{2}(a_1b_1 + a_2b_2 - a_3b_3) \\ \frac{1}{2}(a_1b_1 + a_2b_2 - a_3b_3) & b_1^2 + b_2^2 - b_3^2 & \frac{1}{2}(b_1c_1 + b_2c_2 - b_3c_3) \\ \frac{1}{2}(a_1b_1 + a_2b_2 - a_3b_3) & \frac{1}{2}(b_1c_1 + b_2c_2 - b_3c_3) & c_1^2 + c_2^2 - c_3^2 \end{bmatrix}. \tag{2.14}$$

A comparison of the coefficient schemes (2.13) and (2.14) supplies six equations for the eight unknown parameters, from which four can be determined. In order to obtain the remaining four parameters, two point transformations need to be fixed. In a principal axes transformation it is feasible to transform the original conic's foci into the foci of the resulting ellipse.

The procedure of finding the parameters has been described in detail because it is essentially the way to determine the parameters of those members of the subset of projective transformations which will describe visual adaptation to color.

Any projective automorphism which leaves the spectrum locus invariant has a set of four free parameters. Thus, a complete description of the effect of color adaptation requires two steps: After fixing the spectrum locus by chosing of a particular admissible frame of reference, two pairs of points are chosen to represent the particular effect of the adaptation stimulus. The coordinates introduced for the purpose of parameter estimation could be those of the following points in the chromaticity diagram:

1. the achromatic color before color adaptation,
2. the color to which the originally achromatic color is visually equivalent after color adaptation,
3. the color which becomes achromatic after color adaptation,
4. the achromatic color after color adaptation.

Under these conditions the equations (2.12) are determined up to multiple solutions which result from multiple square roots. The uniqueness can be achieved by excluding those solutions which do not preserve the sense of orientation.

2.7 The Metric of Color Space

The invariant ellipse in projective coordinates determines the metric of the geometry prevailing in color space. The metric which in any projective plane leaves the spectrum locus invariant under motion is the projective hyperbolic metric. Motion can be interpreted as the effect of color adaptation. The invariance proposition means that under adaptation colors remain colors and non-colors remain non-colors. The metric distance measure is the logarithmic cross ratio of four colors in projective coordinates r, g, b, interpretable as *dissimilarity of colors.*

Let $\approx \subseteq A^2 \times A^2$ be a binary relation which represents the *visual equivalence in dissimilarity of two pairs of colors* (y_i, y_j) and (y_l, y_m). There is a well-known theorem in projective geometry which says that

$$(y_i, y_j) \approx (y_l, y_m) \quad \text{iff} \quad \frac{y_i - s_j}{y_j - y_i} \div \frac{s_i - s_j}{y_j - s_i} = \frac{y_l - s_m}{y_m - y_l} \div \frac{s_l - s_m}{y_m - s_l}. \tag{2.15}$$

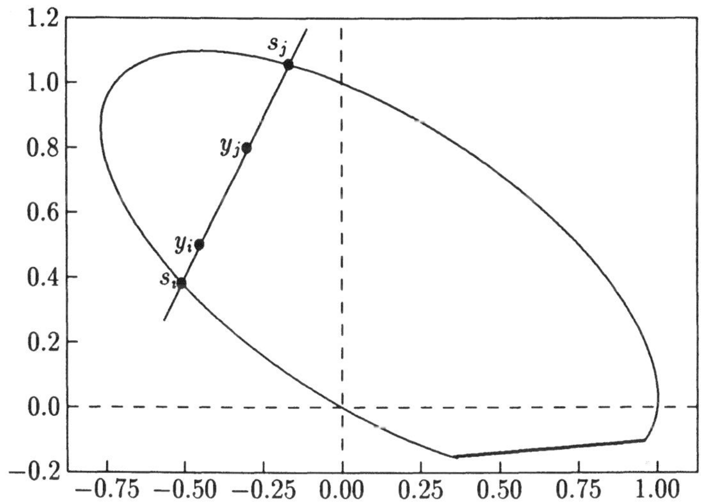

FIGURE 2.4. The theoretical chromaticity diagram showing two colors y_i, y_j and the intersections of their connecting line with the fundamental quadric, the spectrum locus. All four points enter into the calculation of the logarithmic cross ratio which represents visual dissimilarity.

Here s_i and s_j, respectively s_l and s_m, are the intersections of the connecting line of y_i and y_j, respectively y_l and y_m, with a fundamental quadric in the projective plane (Figure 2.4). Any determination of the distance between colors only depends on the coordinates of the colors concerned. The two colors at which the extended connecting line of the two colors intersects the boundary of the color cone are also included in the calculation. The choice of plane, which in color science is called a *chromaticity diagram,* is arbitrary because they all are congruent in projective geometry, that is, they transform into each other by projective automorphisms.

The *cross ratio* in (2.15) is the numerical projective non-Euclidean hyperbolic invariant. Its negative logarithm possesses the properties of a metric. It carries an arbitrary scale factor. The argument in formula (2.15) corresponds to the well known "center of gravity rule".

Since the metric is invariant with respect to projective transformations, its empirical validation can be conducted within any projective plane of the color space, e.g., the chromaticity diagram. Since in color science, for practical reasons, the discussion is mostly confined to the chromaticity diagram, the automorphisms will be studied here. Wright's (1941) dashes provide experimental data for a validation (Figure 2.5). Their length is predicted with high precision by (2.15) (Pearson's correlation coefficient $\hat{\rho} = 0.66$ globally, for most subsets $\hat{\rho} > 0.95$), if the theoretical cross ratios from equation (2.15) are correlated

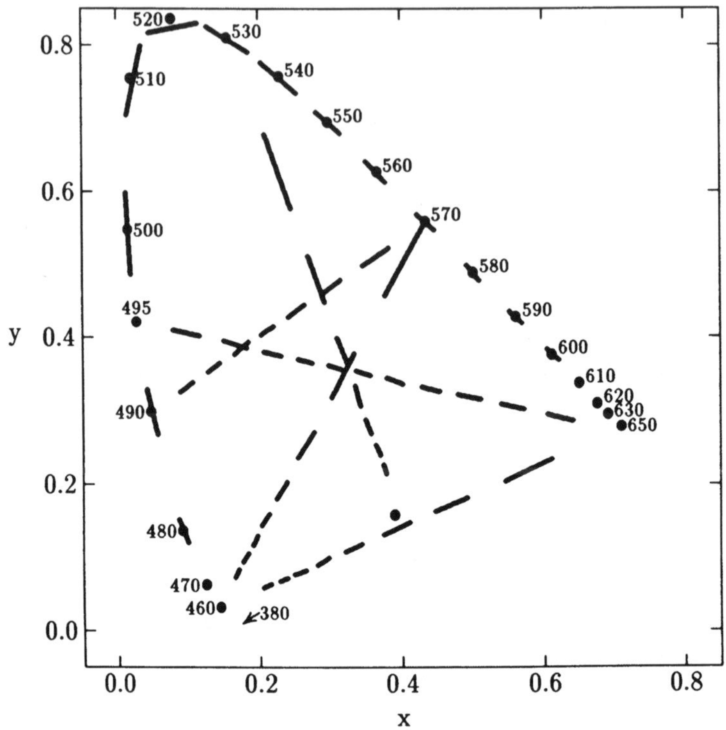

FIGURE 2.5. Wright's (1941) dashes showing experimental data for pairs of colors with unit dissimilarity as projected onto the CIE chromaticity diagram

with empirical cross ratios, computed from the data that lead to Figure 2.5. The projective character of the metric introduced here allows one to treat all the available data in the plane of the chromaticity diagram, disregarding their individual luminances.

This development amounts to a rather informal proof of the following representation theorem.

Theorem 3 *Let $\langle A, \oplus, *, \sim\rangle$ be a Grassmann structure, $\approx$ an equivalence relation on $A \times A$, $\succsim$ a weak ordering on $A \times A$; then, for all $(a,b),(c,d) \in A \times A$, there exists a map*

$$\varphi : A \times A \to \mathbb{R}$$

with

$$\begin{aligned} \varphi(a,b) &\geq 0, \\ \varphi(a,b) &= \varphi(b,a), \\ \varphi(a,b) &\leq \varphi(a,b) + \varphi(b,c), \end{aligned}$$

such that

$$(a,b) \approx (c,d) \ \text{iff} \ \varphi(a,b) = \varphi(c,d),$$

if the following axioms hold:

(i) *the relational system $\langle A, \succsim \rangle$ is an additive conjoint structure;*

(ii) *the additive structure contains a smallest element;*

(iii) *the additive conjoint mappings of stimulus pairs into the reals for different energy levels is consistent with the Grassmann structure's energy parameter r of the stimulus singletons;*

(iv) *the extremal of the convex hull of the Grassmann scaling results is an invariant under its admissible transformations.*

The map φ is invariant with respect to projective metric hyperbolic transformations.

2.8 A Signal Processing Analysis of Color Matching

The conclusions drawn from the present approach bear on the overall characteristic of the processing of colors by the visual system. It has been suggested that the system performs an imperfect Fourier analysis of the stimulus, because it allegedly uses an unsuitable system of reference functions and stops its evaluation of the stimulus after only three Fourier components have been considered.

The present approach leads to a different appraisal. Color vision can be represented by the effect of a single bandpass filter on the stimulus and on two of its frequency derivatives. Formally, this can be seen by introducing the color matching functions proposed here into the mechanism of determining the color coordinates of a stimulus.

The first color coordinate G is calculated from visual detectability data, albeit based empirically on integration over the visible interval of frequencies $[\nu_0, \nu_1]$,

$$G = \int_{\nu_0}^{\nu_1} F_1(\nu)U(\nu)d\nu.$$

For the second color coordinate B one gets, using equations (2.1),

$$B = \int_{\nu_0}^{\nu_1} F_1'(\nu)U(\nu)d\nu = F_1(\nu)U(\nu)\Big|_{\nu_0}^{\nu_1} - \int_{\nu_0}^{\nu_1} F_1(\nu)U'(\nu)d\nu.$$

Since $F_1(\nu)$ functions assume very small numerical values at the boundaries ν_1 and ν_0 of the visually effective domain of frequencies, this expression reduces up to an arbitrarily close approximation to

$$B = -\int_{\nu_0}^{\nu_1} F_1(\nu)U'(\nu)d\nu.$$

From (2.1), the third color coordinate is

$$R = \int_{\nu_0}^{\nu_1} F_1(\nu)'' U(\nu) d\nu = F_1'(\nu)U(\nu)\Big|_{\nu_0}^{\nu_1} - \int_{\nu_0}^{\nu_1} F_1(\nu)' U'(\nu) d\nu$$
$$= (F_1'(\nu)U(\nu) - F_1(\nu)U'(\nu))\Big|_{\nu_0}^{\nu_1} - \int_{\nu_0}^{\nu_1} F_1(\nu) U''(\nu) d\nu.$$

Again, because $F_1(\nu)$ assumes very small numerical values at the boundaries, this is, up to an arbitrarily close approximation,

$$R = \int_{\nu_0}^{\nu_1} F_1(\nu) U''(\nu) d\nu.$$

These simple transformations demonstrate that the color visual system operates in a manner that can be represented by an evaluation of the stimulus together with its first and second derivatives with respect to frequency, in each instance using the identical spectral weighting function of $F_1(\nu)$. Differentiation with respect to light frequency is most likely an impossible task for a biological system. Frequency differentiation as an operator in the Hilbert space of stimuli, however, can be equivalently represented as multiplication by the independent variable in the time domain. Moreover, time (and accordingly frequency) enter the present discussion with values determined only up to an unknown scale factor.

2.9 Conclusion

The result of the present study demonstrates that, by considering empirical restrictions in color vision, a simple theory which employs only two numerical parameters can be developed. Even if its empirical correspondence with the experimental observations is not as close as that of some traditional approaches which use dozens of parameters, it is distinguished by transparency in structure and comprehensiveness: color vision is characterized as the effect of a three component operator in the Hilbert space of stimuli. Its first component carries out a filtering of the radiant stimulus. The other two components bandfilter the radiant stimulus' first and second derivative with time. These operators impose a projective hyperbolic metric upon color space which leads to an empirically well founded appraisal of visual color difference.

2.10 References

Burnham, R. W., Evans, R. M., & Newell, S. M. (1957). Prediction of color appearance with different adaptation illuminations. *Journal of the Optical Society of America, 47*, 35-42.

Grassmann, H.G. (1853). Zur Theorie der Farbenmischung. *Poggendorffs Annalen der Physik und Chemie, 89*, 69-84.

Krantz, D. H. (1975). Color measurement and color theory: I. Representation theorem for Grassmann structures. *Journal of Mathematical Psychology, 12,*, 283-303.

Krantz, D. H., Luce, R. D., Suppes, P., & Tversky, A. (1971). *Foundations of measurment* (Vol. 1). New York: Academic Press.

Von Kries, J. (1905). Die Gesichtsempfindungen. In W. Nagel (Ed.), *Handbuch der Physiologie des Menschen* (Vol. 3, pp. 109-279). Braunschweig: Vieweg.

Lang, S. (1969). *Real analysis.* Reading, Mass.: Addison-Wesley.

Papoulis, A. (1962). *The Fourier integral and its applications.* New York: McGraw-Hill.

Regan, D., & Tyler, C. W. (1971). Temporal summation and its limit for wavelength changes: An analog of Bloch's law for color vision. *Journal of the Optical Society of America, 61,* No. 10, 1414 - 1421.

Trezona, P. W. (1987). Individual observer data for the 1955 Stiles-Burch 2° pilot investigation. *Journal of the Optical Society of America, 4,* No. 4, 769-782.

Wright, W. D. (1941). The sensitivity of the eye to small colour differences. *Proceedings of the Physical Society of London, 53,* 93-99.

Yilmaz, H. (1962). On color perception. *Bulletin of Mathematical Biophysics, 24,* 5-29.

3

The Gravity Lens Illusion and its Mathematical Model

Seiichiro Naito[1] and James B. Cole[2]

ABSTRACT A small dot near a large circle is perceived as being displaced towards the circle. We call this the Gravity Lens Illusion because it can be described using a geometric model in which the large circle induces changes in the geometrical properties of a surrounding "perception manifold" involving an analogy with general relativity theory in which a massive object changes the geometry of physical space. We regard perceived objects and their orientations as contours and vectors on the perception manifold. In order to compare two objects, they must be moved to a common location. This transport can be carried out either by the mathematical process of Lie dragging or by parallel transport. The latter process assumes that the perception manifold is Riemannian. We follow the Schwarzschild model of general relativity to model the geometry of the perceptual manifold. We find that a dot is easily displaced, while a line segment resists being shifted in a perpendicular direction.

3.1 Introduction

We perceive the shapes and orientations of static figures by locating the positions of "important" points on their contours and determining the directions of their tangents. When we compare different objects located in different places, or different elements of the same object at different locations, we need to perform some kind of comparison process. In this paper we seek to elucidate the mathematical nature of the perceptive comparison process by studying the strengths and characteristics of illusions. Coren (1978) and Robinson (1972) gave systematic lists of illusory figures and the associated theories. These illusions include over- or under-estimates of angle, length, or size. In many illusions, we find reliable evidence indicating that a small dot located close to a figure is perceived as displaced toward the figure; Yokose (1952, 1953), Obonai (1954), Uchiyama (1961), Taylor (1962), and Coren (1986). The magnitude of displacements, however, are small and difficult to see. These phenomena can be explained by hypothesizing that the space around a figure apparently shrinks. This has been proposed in some papers such as Watson (1977, 1978). Unfortunately, as pointed out by Coren (1978), the nature of the illusion is complicated and a variety of illusion

---

[1]NTT Basic Research Laboratories, 3-9-11 Midori-cho Musashino-shi Tokyo 180, JAPAN; naito@apollo3.ntt.jp

[2]Naval Research Laboratory, Acoustics Division, Code 5126 Washington, D.C. 20375-5000 USA; cole@umdhep.umd.edu

The authors thank Professor Patric Cavanagh, Professor Ramachandran, Professor Ken Nakayama, Professor Azriel Rosenfeld, and Dr. Takao Sato for their valuable advice.

types contradict the predictions of the theories. A single theory can not cover all of them. Our theory follows the analogy of general relativity in which the gravity of a massive object induces distortions in the surrounding space. In this paper, a newly found illusory figure which produces a strong dot displacement is introduced. Then, a theory which explains the illusion is proposed. The theory uses metric distortion and does not use the actual or physical displacement of dots. The ideas that geodesics are not straight and that the distance shrinks in a particular orientation play an important role. This was tried by Watson (1977, 1978), though in a simple way, as a weak gravity model. The idea was extended to a tensor field by Lehman (1981). His rigorous calculations prove that an appropriate metric tensor field explains some types of illusions which are made of line segments. Both authors' theories, however, are linear. We try a nonlinear theory attempting to explain the finding in Section 3.5. Watson hypothesized an intuitive potential function. Lehman determined the basic field so as to fit the measurement results of the "T" and "L" shape illusions. A fundamentally interesting problem, however, is the underlying mechanism by which the field is created. For a model of the mechanism, we propose a theory following the full relativistic theory using the Schwarzschild space-time model.

3.2 Gravity Lens illusion

The new illusory figure is shown in Figure 3.1.

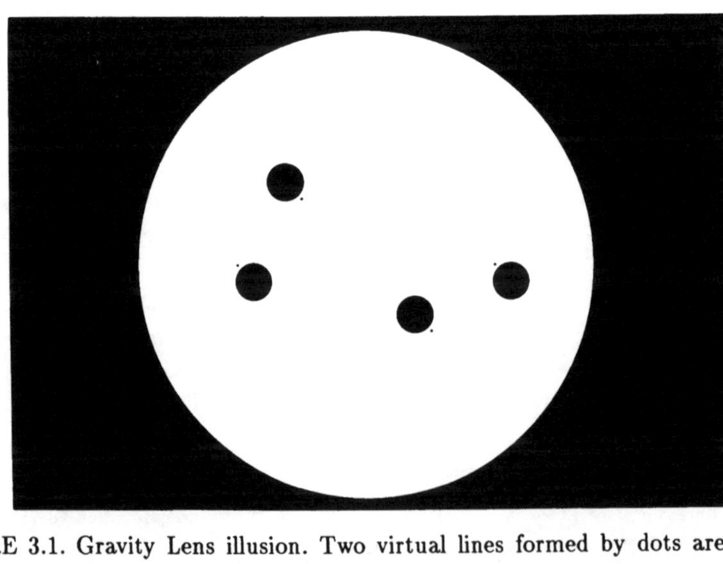

FIGURE 3.1. Gravity Lens illusion. Two virtual lines formed by dots are actually parallel.

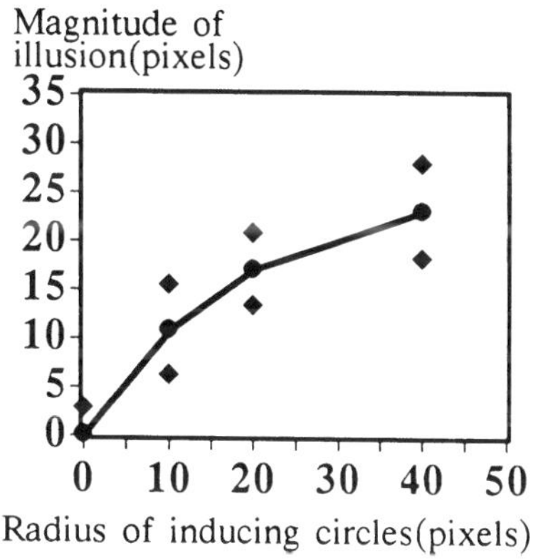

FIGURE 3.2. Size of inducing figure

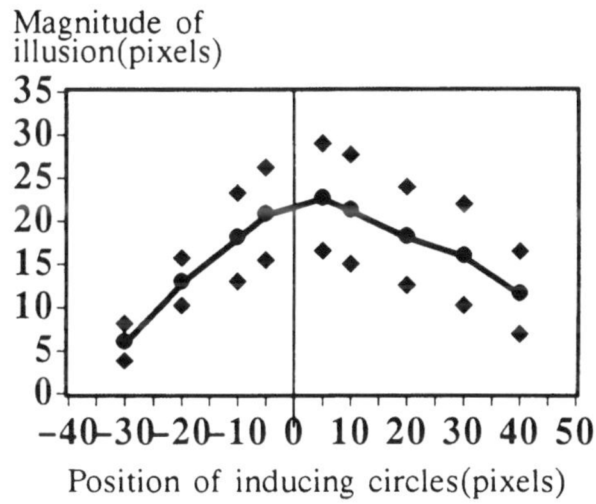

FIGURE 3.3. Distance to inducing figure

There are 4 small dots and 4 large filled circles. The small dots are test dots. The filled circles are inducing figures. Although the virtual line connecting the two test dots on the left and the one connecting the two test dots on the right are actually parallel, they appear to be inclined to each other. The white circle helps to eliminate absolute clues about the orientation. By analogy with gravity, where a large mass distorts the space around it, a figure distorts perceptual space. As discussed in a later section, the illusion is well explained by general relativity, so we call this the "Gravity Lens" illusion. The white background circle in Figure 3.1 subtends 11.9° visual angle when observed from a distance of 120cm; its radius is 500 pixels on the screen. The inducing circle has a 40 pixel radius. The radius of the test dots is 2 pixels. The distance between the center of a test dot and the center of an inducing circle is 45 pixels, and the distance from the center of a test dot to the border of an inducing circle is therefore 5 pixels. The distance between the left 2 test dots is 200 pixels. So is the distance between the right ones. The distance between the uppermost test dot and the lowermost test dot is 400 pixels.

The magnitude of the illusion, that is, the magnitude of perceptual displacement of the test dots, is measured in the following way. Subjects are asked to change the physical position of test-pairs, so that the 2 virtual line segments formed by the test dots look parallel. The positions of the test dots are adjustable in the $-45°$ direction, while the orientation of the virtual line segments is approximately 45°. The inducing figure moves together with the corresponding test dot. Suppose that a subject moves the uppermost test dot 1 pixel down and 1 pixel to the right; then the leftmost test dot automatically moves in the opposite direction, 1 pixel up and 1 pixel to the left. The rightmost and lowermost pair of test dots moves simultaneously in a symmetric way, so as to decrease the

TABLE 3.1. Properties for Gravity Lens Illusion

Properties which affect the illusion	Null properties which do not affect the illusion
Size of inducing figure (Figure 3.2)	Filled or unfilled inducing figure
Distance between test dots and inducing figure (Figure 3.3)	Luminous contrast of inducing figure to background
Frontal plane angle of whole figure (Anisotropy)	Color contrast of inducing figure to background
Depth between inducing figure and test dot	Texture embedded in inducing figure
Equiluminance condition	Real or subjective inducing contour

angle between the virtual line segments. This change in dot position makes the virtual line segments look nearly parallel. If, at this moment, the subject sees the lines as parallel, the magnitude of the illusion will be read as $\sqrt{2}$ pixels.

The first important results of this measurement are shown in Figure 3.2. The size of inducing figures is varied, using 40, 20, or 10 pixels for the radius, while the distance between the center of a test dot and a border of an inducing figure is kept constant at 5 pixels. When there is no inducing figure, we have no illusion. The data is derived from 27 trials by 5 subjects. The flanking diamonds show the overall standard deviation, while the circles show the average magnitude of the illusion. In the case of a 40 pixel radius, the magnitude of the illusion is significant compared to the case with no inducing figure ($F(1,4) = 90.7$, $p < 0.001$). The result shows that a larger inducing circle creates a larger illusion, at least in this certain range.

Figure 3.3 shows the data from a second experiment in which the test dots are displaced consistently toward the center of the corresponding inducing circle. The most important result is that the magnitude of the illusion reaches its maximum just outside the border of the inducing circles and gradually decreases either inside or outside the inducing circle. While the distance between the center of a test dot and the border of an inducing circle is changed, the radius of the inducing circle is kept constant at 40 pixels. In Figure 3.3, on the horizontal axis, zero distance means that the test dots are on the borders of the inducing figures. A value of -40 for the position of the inducing circle means that the test dots are at the centers of the inducing circles. Plus 5 is the particular example shown in Figure 3.1. When the distance is $+40$, the inducing figures are far away from the test dots. This immediately reminds one of the gravity potential.

One of the surprising findings was that, even when the inducing circle is an unfilled contour circle, the displacement is maintained in magnitude and direction. Moreover, the illusion is independent of color, luminance contrast, or embedded texture, and even some kinds of subjective contour create the same illusory effects. The properties which affect the illusion and the null properties which do not affect it are listed in Table 3.1 The list of null properties suggests that the illusion may occur after figure-ground segregation. Generally illusions disappear when the figure is presented in equiluminous red and green color, and the magno-cellular system stops working (Livingston, 1978). It is interesting to note, however, that under equiluminous red-green (figure-ground) conditions, the magnitude of the illusion was unaffected in Figure 3.1. Then, we used a variant of Figure 3.1, where a test dot is flanked by 2 different inducing figures; filled and unfilled on opposite sides of the virtual line segment. In this variant, the effects of the two inducing figures cancel out, and no illusion is seen. In the equiluminous condition, however, the illusion was still observed. Another major finding was that a dot was easily displaced, while a line segment could not be rotated by inducing figures. This is an important experimental result and an important ingredient of the theory discussed further in a later section. More detailed psychophysical measurements of the illusion and discussions about the psychological model are reported in Naito (1991).

3.3 A mathematical framework for visual perception

For the sake of simplicity, let us consider only monocular vision. Binocular depth vision, accommodation, or focussing in depth can be constructed later upon the basis of the monocular model.

Perceptual manifold
If one's eye fixates a point in the visual field, one sees a scene. In the fovea, one gets a detailed image with fine resolution, while in peripheral areas the image is blurred. A small region of the scene on the fovea can be mapped one-to-one into a region of 2-D Euclidean space by orthogonal or perspective projection from three to two dimensions. We assume it to be possible to set up a local Euclidean coordinate system. If the fixation point is shifted to a neighboring point, overlap of the visual field makes us to transform from one local coordinate system to another. The visual field forms a manifold S^2, the surface of a sphere.

Fiber bundle
The set of tangent spaces $\{TS^2\}$, roughly equivalent to images taken by a camera, at all possible points on S^2 forms a simple tangent bundle. This bundle is not necessarily trivial; $TS^2 \neq R^2 \times S^2$. The structural group, possibly an element of $GL(2, R) - \{0\}$, may represent the global topology of the bundle. A very important question arises when the bundle is the image in area 17 or at some level higher in the visual cortex at each particular moment of eye fixation. The bundle may constitute a higher level intermediate visual perception. The cross-section of the fibers passing through the origin of the coordinate system corresponds to tracking a target by eye movement.

Non-uniform metric

When we view an illusory figure, we perceive distorted information: distances, angles, or sizes are over- or under-estimated. This simply means that the metric on the perceptual manifold is not uniform. It changes depending on the surrounding figures. From a mathematical point of view, objects such as vectors, orientations, or small distances at different locations on a manifold can only be compared after they have been transported to a common location. We hypothesize that some of the visual illusions are strongly related to the geometrical properties of the transportation process.

Riemann tensor

We use two methods to compare objects at different positions: Lie dragging and parallel transport. The basic mathematical idea is fully given in textbooks such as Schutz (1988). We follow his book for notation and definitions. Lie dragging is a procedure by which curves with parameters (λ, μ) and their vectors $\left(\frac{\partial}{\partial\lambda}, \frac{\partial}{\partial\mu}\right)$, can be transported along a parameterized congruence of curves. Lie dragging induces the idea of a Lie bracket, $\left[\frac{\partial}{\partial\lambda}, \frac{\partial}{\partial\mu}\right] = \frac{\partial}{\partial\lambda}\frac{\partial}{\partial\mu} - \frac{\partial}{\partial\mu}\frac{\partial}{\partial\lambda}$. It can be shown that this is the same as the Lie derivative: $\pounds_{\frac{\partial}{\partial\lambda}}\frac{\partial}{\partial\mu}$. The vectors need not be coordinate bases, and instead of vanishing, the Lie derivative measures the difference of the two bases from the real coordinate basis vectors, $\pounds_{\frac{\partial}{\partial\xi}}\frac{\partial}{\partial\zeta} = \left[\frac{\partial}{\partial\xi}, \frac{\partial}{\partial\zeta}\right] = 0$. To carry out parallel transport, $\nabla_{\frac{\partial}{\partial\xi}}\frac{\partial}{\partial\zeta} = 0$, one must know the affine connections Γ^k_{ij}, $\nabla_{\bar{e}_i}\bar{e}_j = \Gamma^k_{ji}\bar{e}_k$), which prescribe how basis vectors change as they move through the manifold. The affine connections determine the geodesic line $X^i(s)$ that obeys $\frac{d^2X^p}{ds^2} + \Gamma^p_{ij}\frac{dX^i}{ds}\frac{dX^j}{ds} = 0$, where s is the line parameter. Moreover, affine connections are related to the metric tensor g, by $\Gamma^p_{ij} = \frac{1}{2}g^{pk}\left\{g_{ki,j} + g_{jk,i} - g_{ij,k}\right\}$. Now, using the Lie derivative and parallel transport, we can define the Riemann tensor: $R\left(\frac{\partial}{\partial\lambda}, \frac{\partial}{\partial\mu}\right) = \left[\nabla_{\frac{\partial}{\partial\lambda}}, \nabla_{\frac{\partial}{\partial\mu}}\right] - \nabla_{\left[\frac{\partial}{\partial\lambda}, \frac{\partial}{\partial\mu}\right]}$. It can be equivalently represented by components $R^l_{kij} = \left\{R(\bar{e}_i, \bar{e}_j)\bar{e}_k\right\}\tilde{\omega}^l$, where the $\tilde{\omega}^l$ are duals to the $\bar{e}_l$.

Relativistic characteristics of perceptual manifold

Let us examine the speed of transport on the perceptual manifold. This is the transport of the fixation point. It is limited by the speed of eye movement. Eye movement speed is finite even in a rapid saccadic eye movement. We cannot see a whole manifold or a whole scene at once. This leads to the basic hypothesis that our perceptual manifold should, at least, follow special relativity. In addition, as demonstrated in the previous section, the metric on the manifold is changed by neighboring inducing figures. We therefore hypothesize that the perceptual manifold should follow general relativity theory. We examine Einstein's field equation, $R_{ij} - \frac{1}{2}g_{ij}R = \kappa T_{ij}$, where R_{ij} and R are the Ricci tensor and Ricci scalar obtained by contraction of the Riemann tensor, and T_{ij} is the stress energy tensor. Given T_{ij}, which depends on the physical configuration, we can derive a metric g_{ij} which explains the distorition of the illusory figure.

3.4 A mathematical model of the illusion

Watson (1977, 1978) tried a weak gravity field model. He hypothesized a metric g, $g_{tt} = -1$, $g_{xx} = 1 + h$, $g_{yy} = g_{zz} = 1$. He examined the configuration of a line on the y-axis. The linear potential h is given intuitively, $h = \frac{\partial^2 f}{\partial x^2}$, and f is an exponential-type potential function. Then he obtained a geodesic line which is an extension of a line crossing the y-axis at an acute angle. He found that the extended geodesic was bent in such a way that the Poggendorff illusion was reasonably predicted. However, this did not entirely follow Einstein's equation. The linear potential model cannot predict complicated illusory phenomena.

Schwarzschild space-time

Our inducing figure is a simple circle. Fortunately, a very similar physical configuration has been studied where a single spherically symmetric star creates a metric distortion around it: Schwarzschild time-space. The advantage is that all functions are described by only one parameter, namely, by the distance from the center of the star, r. This relativistic gravitation idea is discussed in detail in Misner (1973). Although the Schwarzschild model is 3-dimensional, the illusion is 2-dimensional. We prefer to use a 3-D theory, since our preliminary experiment, where the inducing spheres and test dots were shown at different depths in a binocular stereo display, suggested that the illusion would also exist in 3 dimensions. Let us simplify the problem by only thinking about one inducing figure, so that we can apply the Schwarzschild model to our illusion. Suppose that there is an inducing sphere of radius D with uniform mass density. The stress energy tensor T^i_j is given by

$$T^i_j = \begin{bmatrix} -\rho & & & \\ & p & & \\ & & p & \\ & & & p \end{bmatrix} \begin{matrix} t : \text{time} \\ r : \text{radius} \\ \phi \\ \theta \end{matrix} , \tag{3.1}$$

where ρ is analogous to physical mass density and is supposed to be constant, and p is ananlogous to physical pressure which is a function of r only. Any ij represents t, r, ϕ, or θ.

Let D be the radius of the sphere. The solution of Einstein's field equation outside the inducing figure ($D \leq r < \infty$) is

$$g_{ij} = \begin{bmatrix} -\left(1-\frac{\alpha}{r}\right) & & & \\ & \frac{1}{1-\frac{\alpha}{r}} & & \\ & & r^2 & \\ & & & r^2\sin^2\theta \end{bmatrix}, \tag{3.2}$$

where α is a constant named the Schwarzschild radius. Inside the figure ($0 \leq r \leq D$) the solution is

$$g_{ij} = \begin{bmatrix} -\frac{\left(3\sqrt{1-\frac{D^2}{a^2}}-\sqrt{1-\frac{r^2}{a^2}}\right)^2}{4} & & & \\ & \frac{1}{1-\frac{r^2}{a^2}} & & \\ & & r^2 & \\ & & & r^2\sin^2\theta \end{bmatrix}, \tag{3.3}$$

where a is constant: $a^2 = \frac{D^3}{\alpha}$.

Explanation of the illusion: Gravity Lens effect

Now our illusion can be explained by interpreting the Schwarzschild solutions. A well-known phenomenon, the gravity lens effect, explains the illusion. In Figure 3.4, the geodesic line passing near the border of the inducing circle should be bent at an angle $\delta\Psi = \frac{2\alpha}{D}$, if the geodesic lines were photon trajectories. Let us interpret the geodesics as perceptually straight lines. Due to the non-Euclidean geometry of the perceptual manifold, straight geodesics (Figure 3.4) eventually intersect when they are extended and therefore appear to be inclined to each other. This explains the illusion.

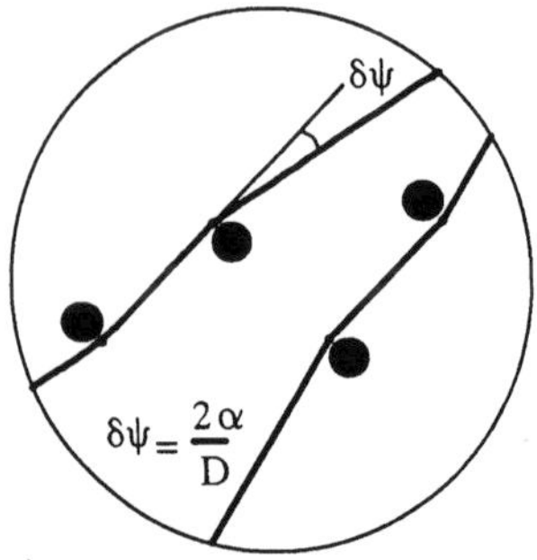

FIGURE 3.4. Explanation of illusion by gravity lens effect

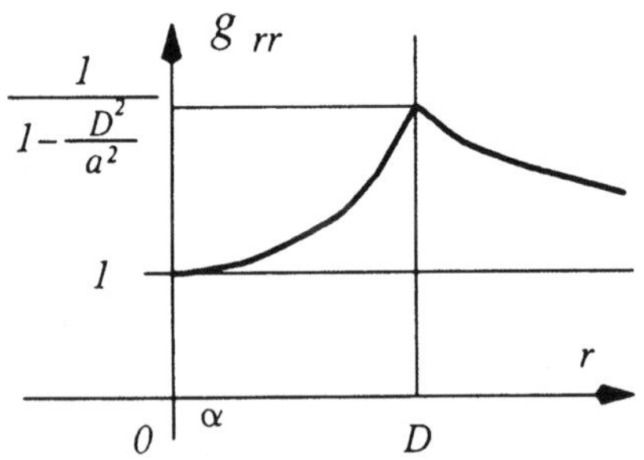

FIGURE 3.5. Explanation of illusion by metric distortion

Explanation of the illusion: Metric distortion

More generally, the illusory effect can be explained by the metric. In Figure 3.5, the outline of the metric g_{rr} which is a function of the radius r, represents shrinkage in the radial direction. The constant D is the radius of the sphere. As shown in Figure 3.5, g_{rr} is greater than 1. This simply means that perceptual distance on the figure is longer than an internal parameter distance. This is interpreted as shrinking of space. Figure 3.5 shows that the shrinking is a maximum at the border of the figure, and gradually tends to zero both inside and outside. At the center of the figure, there is no shrinkage. If the eye travels over the inducing figure, it proceeds more rapidly than might be expected. This is consistent with the experimental results that the test dots, positioned near the inducing circles, tend to be displaced toward the figures.

3.5 Discussion

Comparison of psychophysical data and theory

The properties in Table 3.1 suggest that the psychophysical entity corresponding to mass density ρ and/or pressure p in the theory is the perceived figure portion of the figure-ground segregation, not the black or white color value of the physical figure pixel. Since the α is proportional to the total mass, a is a constant. In general, α is small; if $\alpha \ll D$, then $D \ll a$. The test dot position, $P(r)$, corresponds to a function obtained by a line integral of g_{rr} from infinity to the test dot along the parameter r. The perceptual displacement of the test dot corresponds to the derivative of the function $P(r)-r$; it is proportional to $g_{rr}-1$. The magnitude of illusion on the border of the inducing figure is proportional to $D^2/(a^2-D^2)$. A large inducing figure, a large D, creates a large illusion. This corresponds to Figure 3.2. For more detailed graphical comparison, we add an attenuation that compensates for the difference between 2D area and 3D volume, when the radius D changes. The $g_{rr}-1$ in Figure 3.5 corresponds to Figure 3.3. To compare them numerically, a factor or ratio constant β is introduced, since a metric value and the magnitude of the illusion can not be compared directly. We can then match the parameters quantitatively. Observing Figure 3.3, let the magnitude of illusion be 25 pixels on the border of the inducing figure, when the radius D is 40 pixels. Then $\tan(\delta\psi) = 0.25$, since the distance between test dots before displacement is 200 pixels. These parameters give $\alpha = 5$ pixels, $a = 113$ pixels, and $\beta = 0.0057$. These are reasonable values, since $\alpha < D < a$. Let us examine the case $r = 80$, which corresponds to the rightmost point in Figure 3.3. Using the parameters obtained above, the magnitude of the illusion is estimated as 11.7 pixels. This is exactly the same as the measurement value in Figure 3.3.

Metric field acting on a point and acting on a line

There is a simple way to eliminate the illusion completely. As shown in Figure 3.6, the illusory effect is completely lost if the two test dots are connected with a real line, regardless of how weak the line is. The phenomenon may be simply summarized as follows: a metric field induces a displacement on a point, but does not act on a line segment in a direction perpendicular to its length. Moreover, it is particularly interesting that if extra dots are added to the imaginary line segments formed by the two points of test dots, they then look like subjective line segments. We found that a subjective line formed with 6 or more equally spaced dots behaves like a real line segment, which means that they are not displaced. A line segment may be more special than a dot (Hubel, 1977). Recently, von der Heydt (1984) reported neurons that respond to subjective contours in area 18 in the visual cortex of monkeys, and, interestingly, he did not find such neurons in area 17. Like the Müller-Lyer illusion, a line segment may shrink or expand, but it does not rotate. This is intriguing and may reveal a deeper mechanism of geometrical illusion.

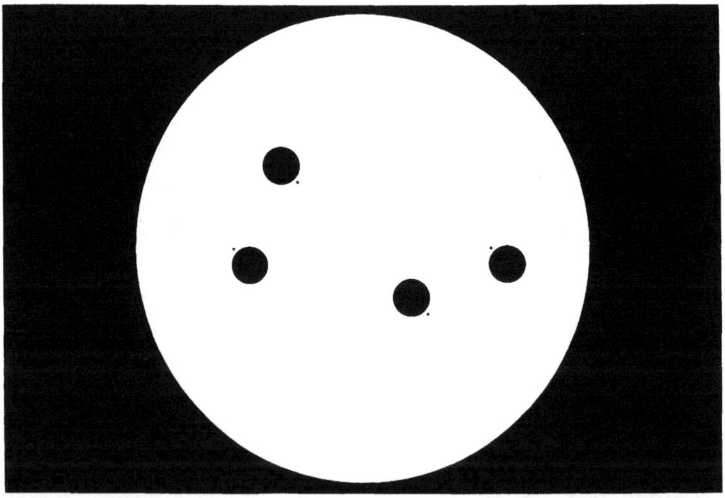

FIGURE 3.6. Gravity Lens illusion. With line segments, the illusion is lost.

Hypothetical theory for line segment

The newly discovered phenomenon in Figure 3.6 raises a basic problem for conventional theories which use the Newtonian potential or the weak gravity model. Since they are linear theories and the behavior of the dots are linearly additive, then line segments should behave similarly, because a line is formed by a series of dots. These theories can not predict the phenomenon demonstrated above. Morinaga and Ikeda (1965) summarized an interesting finding "Paradox of displacement". Using a common inducing figure, displacements of a dot and of a line were in opposite directions. They tried to explain it in terms "Dimension of measurement". This phenomenon can't be explained by the theories put forth so far. The non-linearity of Einstein's equation suggests a possible solution. The oblate spheroid, elongated box, or long cylinder might work as a line segment model. As an immediate analogy with the Schwarzschild solution, the simplified solution would be that, in the co-linear direction, the metric might have a large distortion, while in the width direction the distortion could be small. If this is the case, predictions from the theory will be fairly consistent with the experimental results. The reason is that if a geodesic line crosses a line segment at a grazing angle, the geodesic will be captured in the line, whereas, since the co-linear direction shrinks greatly, the geodesic travels farther in this direction than in the width direction. This gives a simple explanation for a line segment not being displaced: if a line segment is set on the target dot, the line will create a large shrinkage in the co-linear direction, so the geodesic will tend to be straighter in the line segment. This also gives a consistent prediction for the Fraser illusion and an inverse Zöllner illusion. Angle overestimation is also predictable since space shrinks more in the direction of the bisecting line of the acute angle. The Helmholtz square is also predictable since each line simply shrinks.

3.6 References

Coren, S., & Girgus, J.A. (1978). *Seeing is deceiving: The psychology of visual illusions.* Hillsdale, NJ: Lawrence Erlbaum Associates.

Coren, S. (1986). An efferent component in the visual perception of direction and extent. *Psychological Review, 93,* 391-410.

Hubel, D.H., & Wiesel, T.N. (1977). Ferrier lecture; functional architecture of macaque monkey visual cortex. *Proceedings of the Royal Society London, Series B, 198,* 1-59.

Lehmann, G. (1981). *Figurale Wechselwirkungen im Gesichtsfeld: Experimentelle Analysen stochastischer Feldtheorien der subjektiven Figurbildung.* Göttingen: Verlag für Psychologie, Hogrefe.

Livingston, M.S., & Hubel, D.H. (1987). Psychophsical evidence for separate channels for the perception of form, color, movement, and depth. *The Journal of Neuroscience, 7,* 3416-3468.

Misner, C.W., Thorne, K.S., & Wheeler, J.A. (1973). *Gravitation.* New York: Freeman and Co.

Morinaga, S., & Ikeda, H. (1965). Paradox of displacement in geometrical illusion and the problem of dimensions. A contribution to the study of space perception. *Japanese Journal of Psychology* [in Japanese, with an English abstract], *36,* 231-238.

Naito, S. (1991). The gravity lens illusion (submitted).

Obonai, T. (1954). Induction effects in estimates of extent. *Journal of Experimental Psychology, 47,* 57-60.

Robinson, J.O. (1972). *The psychology of visual illusion.* London: Hutchinson & Co.

Schutz, B. (1988). *Geometrical methods of mathematical physics.* Cambridge: Cambridge University Press.

Taylor, M.M. (1962). Geometry of a visual illusion. *Journal of the Optical Society of America, 52,* 565-569.

Uchiyama, M., & Muraki, M. (1961). Field measurement of shape by flickering small spot. *The 25th convention of the Japanese Psychological Society* [in Japanese], p. 74.

von der Heydt, R., Peterhans, E., & Baumgartner, G. (1984). Illusory contours and cortical neuron responses. *Science, 224,* 1260-1262.

Watson, A. (1977). A theory of the visual illusions. *The British Journal of Mathematical and Statistical Psychology, 30,* 43-59.

Watson, A. (1978). A Riemann geometric explanation of the visual illusions and figural after-effects. In E.L.J. Leeuwenberg & H.F.J.M. Buffart (Eds.), *Formal Theories of Visual Perception* (pp. 139-169). New York: John Wiley.

Yokose, Z., & Kawamura, H. (1952). A study of the direction of the field force in shape perception (Report No.1). *Japanese Journal of Psychology* [in Japanese], *23,* 133-143.

Yokose, Z., & Ichikawa, N. (1953). A study of the direction of the field force in shape perception (Report No.2). *Japanese Journal of Psychology* [in Japanese], *23,* 261-274.

4

Analysis of a Model for Source Monitoring

William H. Batchelder[1], **Xiangen Hu**[1], **and David M. Riefer**[2]

ABSTRACT This paper extends and analyzes the Batchelder and Riefer (1990, *Psychological Review*) model for source monitoring. Source monitoring is a popular memory paradigm in which subjects study items from several different sources and later are tested for their ability to recognize the source of each item. The extended model is identifiable, it can handle an arbitrary number of sources, and it enables one to disentangle response bias factors from memory capacities. The model is used to determine the causes of the "picture superiority effect", where items presented as pictures perform better on source monitoring than items presented as printed words. We analyze a new experiment that shows that the picture superiority effect is due to memory capacity factors rather than response bias factors.

4.1 Introduction

Batchelder and Riefer (1990) have developed a multinomial model for the analysis of source monitoring. In a source monitoring experiment, subjects study items from two or more different sources. For example, sources may be items presented as pictures or words, or they may be presented in one of two languages for bilinguals. Once these items have been studied, a recognition memory test is given to measure separately the ability to detect old items from new distractors, and the ability to discriminate the source of detected old items. We conducted a survey of the memory literature and found that over fifty studies fall into this category of source monitoring experiments; however, usually data from these experiments are not analyzed by a substantive mathematical model.

Batchelder and Riefer (1990) developed their model for two sources, and they analyzed several sets of published data with their model. One of the unfortunate features of the model for two sources is its inability to differentiate whether a given data pattern is due to differential ability to discriminate sources, or instead is due to response bias factors. Batchelder and Riefer (1990) suggest that a resolution to this problem can be achieved by extending the model to three or more sources.

---

[1]Department of Cognitive Sciences, University of California, Irvine, CA 92717, USA
[2]Department of Psychology, California State University, San Bernardino, CA 92407, USA
Work on this paper was supported by NSF grant No. BNS-8910552 to Batchelder and Riefer. For copies of computer programs and related papers write to W. H. Batchelder.

The purpose of this paper is to extend and analyze the model for more than two sources. In Section 4.2 and Section 4.3., we extend the model and show that it is identifiable. In Section 4.4., we discuss available software for conducting statistical inference for the model. In Section 4.5., we discuss the nature of response bias factors in the model. In particular, we consider the situation where memory factors are held constant, and we analyze the effects on performance of various response bias strategies. In Section 4.6., we provide an example from the literature where data from a two-source experiment do not permit one to interpret whether a difference is due to memory factors or to response bias factors. In Section 4.7., we describe one of our experiments with three sources that provides a resolution to this problem of interpretation. Thus, we show how researchers can estimate memory factors disentangled from bias factors. Hopefully, future work in source monitoring can profit from applying our model analyses to these problems.

4.2 The General Source Monitoring Model

Suppose there are M sources of old items labeled by S_1, S_2,...,S_M with corresponding response categories R_1, R_2,...,R_M. New distractor items are denoted by S_0 and the corresponding "new" response by R_0. The data for a source monitoring experiment consist of frequency counts of the various responses to each of the $M+1$ stimulus categories. Let Y_{ij} be the frequency of the response R_j to category S_i, $0 \leq i, j \leq M$. The data can be arranged in an $(M+1) \times (M+1)$ table

$$\mathbf{T} = [Y_{ij}]_{M+1,M+1}, \tag{4.1}$$

where S_0 and R_0 are the last row and last column, respectively. The row marginals, denoted by

$$Y_{i\cdot} = \sum_{j=0}^{M} Y_{ij},$$

are fixed by the experimenter. Equation (4.1) can be regarded as an ordinary confusion matrix except for the added category S_0 and its corresponding response R_0, for new items and new responses, respectively.

The model assumes that the responses $\mathbf{Y_i} = \langle Y_{ij} \rangle_{j=0}^{J}$ to a particular source S_i are governed by the multinomial distribution

$$P(\mathbf{Y}_i, \langle p_{ij} \rangle) = Y_{i\cdot}! \prod_{j=0}^{M} \frac{p_{ij}^{Y_{ij}}}{Y_{ij}!}, \tag{4.2}$$

where $p_{ij} = P(R_j|S_i)$. Further, for convenience, it is assumed that the rows of

T are independent; therefore,

$$P(\mathbf{T}, \langle p_{ij} \rangle) = \prod_{i=0}^{M} P(\mathbf{Y}_i, \langle p_{ij} \rangle). \tag{4.3}$$

Thus, the model in (4.2) and (4.3) is a joint multinomial model in the sense of Riefer and Batchelder (1988). Because the row marginals are fixed, there are $M(M+1)$ degrees of freedom in **T**.

Next, we generalize the Batchelder and Riefer (1990) high threshold model for two sources to handle arbitrarily many sources (see Batchelder & Riefer, 1990, for a discussion of the motivation for the two-source model as well as alternative approaches). The source monitoring model assumes that subjects' response probabilities are a function of three classes of parameters: stimulus detection, source discrimination, and response biases.

Stimulus Detection

On test trials, an old item from source S_i is assumed to be either detected as an old item, with probability D_i, or to be undetected, with probability $1 - D_i$, $i = 1, ..., M$.

Source Discrimination

If an item from source S_i is detected as old, then its source is discriminated with probability d_i, and with probability $1 - d_i$ its source is not discriminated. If the source is discriminated, the correct response is made.

Response Biases

There are three types of response biases in the model. First, when an old item from source S_i is detected but not discriminated, with probability $D_i(1 - d_i)$, then it is assigned to response R_j with probability a_j, $j = 1, ..., M$. Second, if an old item from source S_i is undetected, with probability $1 - D_i$, then with probability b, the false alarm probability, it is assumed to have come from some source anyway. Otherwise, with probability $1 - b$, it is assigned to response R_0. Third, false alarms are assigned to response R_j with probability g_j, $j = 1, ..., M$. New items from category S_0 are treated as old, undetected items by the model.

From the foregoing, it is easy to express the probabilities p_{ij} in (4.2) and (4.3) in terms of the parameters as follows:

For old items:

$$P(R_j|S_i) = \begin{cases} D_i\left[d_i + (1-d_i)a_i\right] + (1-D_i)bg_i, & i = j, \\ D_i(1-d_i)a_j + (1-D_i)bg_j, & 0 < j \neq i, \\ (1-D_i)(1-b), & 0 = j, \end{cases} \tag{4.4}$$

for $i = 1, ..., M$. And for new items:

$$P(R_j|S_0) = \begin{cases} bg_j, & 0 < j \leq M, \\ 1-b, & 0 = j. \end{cases} \tag{4.5}$$

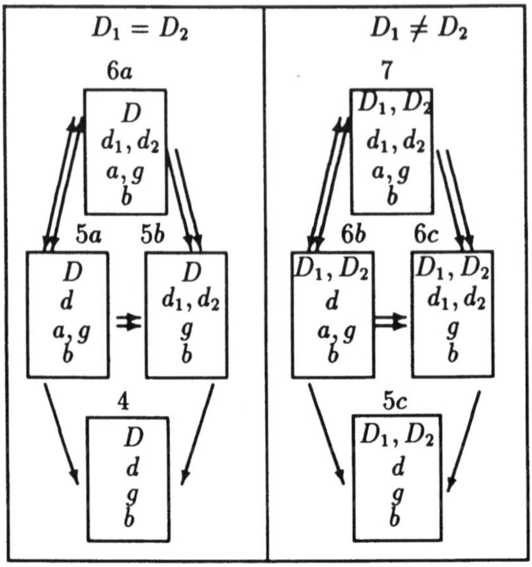

FIGURE 4.1. Nesting relations among eight submodels of the model with $M = 2$ sources. (Two double arrows indicate models that are equivalent. Two single arrows indicate a proper subset relation between models that have the same number of degree of freedom. One single arrow indicates a nested relationship between models that differ by one degree of freedom.)

4.3 Formal Analysis of the Model

An important issue in analyzing any parametric multinomial model is whether or not the model is *identifiable*. By identifiable we mean that there is at most one value of the model parameters consistent with any value of the p_{ij} in (4.2). Identifiability is a necessary condition for most asymptotic parameter estimation techniques (e.g., Read & Cressie, 1988) and it is essential if one wishes to use the model as a measurement tool (Riefer & Batchelder, 1988).

Equations (4.4) and (4.5) involve a total of $4M - 1$ parameters: M detection parameters, D_i; M discrimination parameters, d_i; $M - 1$ old item bias parameters, a_j (since $\sum_{j=1}^{M} a_j = 1$); $M - 1$ guessing biases, g_i; and one false alarm probability, b. For convenience, we hereafter assume that all parameters are in the open unit interval. For two sources there are 7 parameters and only 6 degrees of freedom in $\mathbf{T}$; thus the model is nonidentifiable (cf. end of this section). Batchelder and Riefer (1990) provide a complete analysis of eight nested submodels defined by the presence or absence of the restrictions: $D_1 = D_2$, $d_1 = d_2$, or $a = g$. Figure 4.1 provides a summary of that analysis.

The left and right panels in Figure 4.1 are defined by whether or not the restriction $D_1 = D_2$ holds. There is a type of isomorphism between the two panels. Models 6a and 7 are nonidentifiable even though both models entail restrictions on the p_{ij} of the general joint multinomial model in (4.2) and (4.3). Model 5a and 6a are equivalent (entail the same set of possible p_{ij}) and so are models 7 and 6b; however, models 5a and 6b are both identifiable. Model 5b is a proper subset of model 5a and model 6c is a proper subset of model 6b; however, in neither case is there a reduction in degrees of freedom. These last relationships are unfortunate because they do not permit one to differentiate the cases where $a = g$ from those where $d_1 = d_2$. We will return to this issue in Section 4.6. Finally, model 4 and model 5c both represent a one-dimensional reduction in the parameter space of the models immediately above them in their respective hierarchies.

For $M > 2$ sources, there are fewer parameters in the model than the data in $\mathbf{T}$ provide as shown by

$$M(M+1) - (4M-1) = M^2 - 3M + 1 > 0,$$

for $M = 3, 4, \ldots$. Observation 1 shows that the model is identifiable for $M > 2$.

Observation 1 *For $M > 2$, the full model with $4M-1$ parameters and $M(M+1)$ degrees of freedom is identifiable.*

Proof: Let $\Theta = \langle D_i, d_i, a_i, g_i, b\rangle$ and $\Theta^* = \langle D_i^*, d_i^*, a_i^*, g_i^*, b^*\rangle$, for $i = 1, \ldots, M$, be two sets of parameters for the model with M sources. Assume $p_{ij}(\Theta) = p_{ij}(\Theta^*)$, $0 \le i, j \le M$. The model is identifiable if we can show that $\Theta = \Theta^*$. First, from $p_{00}(\Theta) = p_{00}(\Theta^*)$, we have $b = b^*$. Next from $p_{0j}(\Theta) = p_{0j}(\Theta^*)$ we have $g_j = g_j^*$,for $j = 1, \ldots, M$. From $p_{i0}(\Theta) = p_{i0}(\Theta^*)$, we have $D_i = D_i^*$, for $i = 1, \ldots, M$. When the above results are inserted into $p_{ij}(\Theta) = p_{ij}(\Theta^*)$, for $1 \le i, j \le M$, (4.4) yields

$$d_i + (1 - d_i)a_i = d_i^* + (1 - d_i^*)a_i^* \tag{4.6}$$

and

$$(1 - d_i)a_j = (1 - d_i^*)a_j^*, \tag{4.7}$$

for $i \neq j$. For $M > 2$, (4.7) gives $a_i^*/a_i = a_j^*/a_j$, and thus $a_i = \alpha a_i^*$ for some $0 < \alpha < 1$. Substituting in (4.6) we obtain $d_i = d_i^*$ and thus, back in (4.7), $a_i = a_i^*$. □

From the proof of Observation 1, it is easy to see that even in the case of $M = 2$, the parameters D_1, D_2, b and g are identifiable, and the only problem comes from the tradeoff among d_1, d_2 and a through (4.6) and (4.7).

4.4 Statistical Analysis of the Model

Hu (1990) has developed software for the statistical analysis of the model based on the EM (expectation-maximization) algorithm described in Dempster, Laird and Rubin (1977). Hu and Batchelder (To Appear) adapt the EM algorithm for the class of multinomial processing tree models as defined by Riefer and Batchelder (1988) by making use of the special form of the model equations. That paper provides computationally simple methods suitable for a personal computer for parameter estimation and hypothesis testing based on the likelihood function. Also Hu and Batchelder (To Appear) prove convergence properties of the algorithm for processing tree models that go beyond those known to hold for the exponential class, and they extend the algorithm to cover other members of the Read and Cressie (1988) power divergence family (also see Batchelder, 1991).

4.5 Response Bias in Source Monitoring

Source monitoring is analogous to signal detection (see Green & Swets, 1966) except that the "signal" is one of several sources rather than the usual case of a single source. In fact, if there is a single source, there is no need for a source discrimination process, and our model is related in mathematical form to the standard high-threshold signal detection model of Blackwell (1963). Signal detection models are routinely used to analyze recognition memory paradigms (e.g., see Klatzky, 1975), and our model can be viewed as a natural addition to that area.

In signal detection analyses, one often treats the stimulus variables as fixed by a subject's capacities and the experimental conditions. On the other hand, the bias parameters are assumed to represent a decision process that is under the control of the subject. For example, in yes-no signal-detection, isosensitivity curves plot the false alarm and hit rates on a receiver-operator-characteristic (ROC) curve as functions of the payoff matrix and the signal probability under conditions of constant signal sensitivity, e.g., d'.

In the source monitoring model of (4.4) and 4.5), the detection and discrimination parameters, the D_i and d_i, are analogous to the sensitivity parameters, and the bias parameters are the a_j, g_j, and b. It is theoretically interesting to study the effects of the bias parameters on performance under conditions of fixed D_i and d_i. For example, subjects' may have some metacognitive awareness of their own memory capabilities, e.g., Devolder and Pressley (1989) and Nelson and Narens (1990), and they may use this information to set their response strategy, e.g., Lindsay and Johnson (1991). In this section, we will assume the D_i and d_i are fixed and study performance as a function of the bias parameters.

The simplest way to disentangle memory processes and response bias processes is to assume that the memory parameters determine the "memory state" for an item, and that the bias parameters determine the responses to items in that memory state. In source monitoring, a test item may result in one of several memory states. For example, in the model of (4.4) and (4.5), there are three possible memory states: detected and discriminated, denote by m_1; detected and not discriminated, m_2; and not detected, m_3. More generally, assume that there is a set $\mathcal{M}$ of possible memory states. To simplify, assume the usual experimental condition that all sources, including the new items, appear on the test with equal frequencies.

In order to analyze the response bias process, let u_{ij} be the payoff (real valued utility on some scale) for making response R_j to an item from source S_i. Let $\mathbf{U}_j$ be a random variable denoting the payoff of making response R_j. Our goal is to study the response bias process that maximizes $\mathbf{U}_j$ as a function of the memory state of an item. It is easy to show by Bayes theorem that, for all $m \in \mathcal{M}$,

$$E[\mathbf{U}_j|m] = \sum_{i=0}^{M} u_{ij}P(S_i|m) = \frac{\sum_{i=0}^{M} u_{ij}P(m|S_i)}{\sum_{k=0}^{M} P(m|S_k)}, \tag{4.8}$$

since $P(S_i) = 1/(M+1)$, for all i, by assumption.

First, let us consider the special case of maximizing correct responses regardless of source. In this case, we can set $u_{ii} = 1$ and $u_{ij} = 0$, for $i \neq j$. Then it is easy to see from (4.8) that the R_j that maximizes $E[\mathbf{U}_j|m]$ is given by the j that maximizes $P(m|S_j)$; that is, one should make the response corresponding to the source that most likely results in memory state m. In the case of the model, $P(m_3|S_i)$ is maximized by S_0, since

$$P(m_3|S_i) = \begin{cases} 1 - D_i, & 1 \leq i \leq M, \\ 1, & i = 0, \end{cases} \tag{4.9}$$

so correct responses in the nondetect state are maximized by making response R_0. In the case of the detected state m_2,

$$P(m_2|S_i) = \begin{cases} D_i(1 - d_i), & 1 \leq i \leq M, \\ 0, & i = 0; \end{cases} \tag{4.10}$$

thus correct responses are maximized by responding to the source with the largest value of $D_i(1 - d_i)$.

It is well known that various correct responses and errors may have different payoffs to subjects in recognition memory and signal detection. This fact can lead to behavior that does not maximize the number of correct responses on the memory test. We can simplify the analysis for this situation by assuming "symmetry" of payoffs over sources, e.g., $u_{ii} = A$ and $u_{ij} = B$, for all $1 \leq i \neq j \leq M$. In this case, the payoffs are given by Table 4.1

In Table 4.1, the quantities A, C, D, and E are positive, reflecting the fact that correct responses have positive payoff and errors have negative payoff. However, the quantity B may be either positive or negative depending on the relative values of correctly reporting an old source item as old, but assigning it to the incorrect source category. It is reasonable to assume $-C < B < A$ since correct old responses are more correct than misclassified old responses, and misclassified old responses are more correct than misses.

When these payoffs are analyzed in terms of (4.8), several results can be obtained for the source monitoring model in the next Observation.

Observation 2 *Assume the payoff matrix in Table 4.1, and assume the probabilities of each stimulus category are given by* $P(S_i) = \frac{1}{M+1}$, *for* $i = 0, 1, ..., M$. *Then the following two response bias rules hold:*

TABLE 4.1. Payoffs for various responses in the case of symmetric sources

	Response		
Source	$1 \leq i = j \leq M$	$1 \leq i \neq j \leq M$	$j = 0$
$1 \leq i \leq M$	A	B	$-C$
$i = 0$	$-D$	$-D$	E

Note: The table reports payoffs u_{ij}, of making response j to source i, $0 \leq i, j \leq M$. The quantities A, C, D, and E are positive reals; and B is a real number.

Rule 1 *If $D_i > D_j$, then $g_i < g_j$ results in higher payoff than $g_i > g_j$.*

Rule 2 *If $D_i(1-d_i) > D_j(1-d_j)$, then $a_i > a_j$ results in higher payoff than $a_j > a_i$.*

Proof: If $D_i > D_j$, then from (4.9), $P(m_3|S_i) < P(m_3|S_j)$. Thus from (4.8), (4.9), and Table 4.1,

$$\begin{aligned} \frac{E[\mathbf{U}_i|m_3]}{E[\mathbf{U}_j|m_3]} &= \frac{P(m_3|S_i)A + \sum_{j\neq i} P(m_3|S_j)B - P(m_3|S_0)D}{P(m_3|S_j)A + \sum_{i\neq j} P(m_3|S_i)B - P(m_3|S_0)D} \\ &= \frac{(1-D_i)A + \sum_{j\neq i}(1-D_j)B - D}{(1-D_j)A + \sum_{i\neq j}(1-D_i)B - D} \\ &< 1, \end{aligned} \tag{4.11}$$

and this implies Rule 1. Next suppose $D_i(1-d_i) > D_j(1-d_j)$, then from (4.10), $P(m_2|S_i) > P(m_2|S_j)$. Thus from (4.8), (4.10), and Table 4.1, it is easy to show

$$\frac{E[\mathbf{U}_i|m_2]}{E[\mathbf{U}_j|m_2]} = \frac{D_i(1-d_i)A + \sum_{j\neq i} D_j(1-d_j)B}{D_j(1-d_j)A + \sum_{i\neq j} D_i(1-d_i)B} > 1, \tag{4.12}$$

and this implies Rule 2. □

Rule 1 and 2 are interesting because they can predict a "bias reversal phenomenon" under certain conditions. To see this, suppose the d_i are all equal. Then it is easy to see that if the D_i differ, optimal performance is achieved by reversing the order of the biases, that is, if $g_i < g_j$ then $a_i > a_j$. This result, more than any other, motivates the inclusion of two sets of bias distributions in the source monitoring model.

In Section 4.7, we consider an experiment with three sources. The next observation provides a set of sufficient conditions to conclude the bias reversal given by $g_1 < g_2$ and $a_1 > a_2$.

TABLE 4.2. Frequency counts in Durso and Johnson (1980)

		Response			Totals
		P	W	N	
	P	742	41	57	840
Source	W	159	620	61	840
	N	50	75	715	840

Note: The entries in Table 4.2 are the frequency counts of responses to the three types of items. *P* items were presented as pictures, *W* items were presented as words, and *N* were new distractors. These data are from Durso and Johnson's Function condition.

Observation 3 *Suppose $M = 3$ in (4.1). Then if $0 < b < 1, p_{01} < p_{02}$, and $p_{31} > p_{32}$, it follows that $g_1 < g_2$ and $a_1 > a_2$.*

Proof: From $p_{01} < p_{02}$, $bg_1 < bg_2$; thus $g_1 < g_2$. On the other hand, if $p_{31} > p_{32}$, then $p_{31} - p_{32} = D_3(1 - d_3)(a_1 - a_2) + (1 - D_3)b(g_1 - g_2)$. Since $g_1 - g_2 < 0$, it follows that $a_1 > a_2$. □

The three hypotheses in Observation 3 are easy to test statistically. Let $P_{ij} = Y_{ij}/Y_{i\cdot}$. First, if $P_{00} < 1$, then $b > 0$. Second, consider the hypotheses, H_0: $p_{01} = p_{02}$ versus H_1: $p_{01} < p_{02}$. A one-tail test of H_0 can be accomplished by a test of equality of the correlated proportions P_{01} and P_{02}. Finally, the hypotheses H_0: $p_{31} = p_{32}$ versus the alternative H_1: $p_{31} > p_{32}$ can be tested on correlated proportions in the same way as the other two hypotheses. These two tests are independent by the assumptions in (4.2) and (4.3).

4.6 An Empirical Example

Durso and Johnson (1980) and Foley, Durso, Wilder, and Friedman (1991) have studied source monitoring for items that were presented either as pictures or words. One set of the data derived from Durso and Johnson (1980) is presented in Table 4.2.

Table 4.2 presents data in the form of (4.1) for an $M = 2$ source experiment. The authors concluded that there was a "picture superiority effect". This conclusion is supported because the frequency of "picture" responses to words, $Y_{WP} = 159$, is significantly larger than the frequency of "word" responses to pictures, $Y_{PW} = 41$. Durso and Johnson (1980) also concluded that the picture superiority effect was due to memory factors rather than bias factors because the proportion of "picture" responses to distractors, $Y_{NP} = 50$, was significantly smaller than the "word" responses to distractors, $Y_{NW} = 75$. However, they did not analyze their data with a mathematical model.

TABLE 4.3. Parameter estimates and goodness-of-fit statistics for the Durso and Johnson (1980) data

Model 5b	Model 5a
$d_1 = 0.92$	$d_1 = 0.75$
$d_2 = 0.50$	$d_2 = 0.75$
$a = 0.40$	$a = 0.82$
$g = 0.40$	$g = 0.40$
$G^2(1) = 0.14159$	$G^2(1) = 0.14159$

We applied our model for two sources to their data. A statistical test failed to reject the hypothesis $D_1 = D_2$ ($p > 0.70$), since the number of "new" responses to pictures and words (57 versus 61) was practically identical. This result implies that the proper model for the data is in the left panel of Figure 4.1. Model 5*b*, with the source discrimination parameters $d_1 \neq d_2$, and and the response bias parameters $a = g$, expresses the hypothesis that the picture superiority effect is due to memory discrimination differences. Model 5*a*, with $d_1 = d_2$ and $a \neq g$, represents the hypothesis that the source discrimination parameters are equal, and thus, bias differences account for the effect. We applied these two models to the data, and the maximum likelihood parameter estimates are exhibited in Table 4.3. Also in Table 4.3 are the values of G^2, the logarithm of the likelihood ratio goodness-of-fit statistic, which is asymptoticaly distributed as $\chi^2(1)$.

The result in Table 4.3 shows that both models fit the data quite well. In fact, the fits are identical, which is not surprising in light of the fact that model 5*b* is nested in model 5*a*, with the same number of parameters; thus, the two models are indistinguishable for many data sets. The conclusion is that there is an epistemological impasse in interpretating the Durso and Johnson (1980) data, i.e., the model analysis leaves open whether or not the picture superiority effect is a memory phenomenon or a response bias phenomenon.

4.7 A Solution

It is possible to resolve the question of the origin of the picture superiority effect by conducting an experiment with three sources. Observation 1 shows that the model for $M = 3$ sources is identifiable, so unique recovery of the parameters is possible. We conducted an experiment similar to Durso and Johnson's (1980), in which subjects memorized items that were presented visually as pictures or words. However, we also added a third source of items that were presented auditorilly.

4.7.1 METHOD

Twenty-eight undergraduate volunteers from the University of California at Irvine were presented with a list of items for memorization. Subjects were run in four groups, with seven subjects in each group. The stimuli used in the experiment were 48 items taken from Snodgrass and Vanderwart (1980). Half of the 48 items were unrelated to each other (noncategorized items) while the remaining 24 consisted of six categories of four items each (categorized items). Subjects within each of the groups were presented with 36 items randomly chosen from this pool of 48 stimuli, under the constraint that 18 noncategorized items were selected along with three items from each of the six categories. The twelve nonchosen items for each group constituted the new distractors on a later recognition test.

Subjects were presented with the 36 list items in a random order, one at a time on a slide projector at a rate of five seconds per item. Each item appeared in one of three formats: picture, visual word, or spoken word. The twelve picture items appeared on slides containing a line drawing from Snodgrass and Vanderwart (1980). For the twelve visual words, subjects saw the corresponding English word for each item. For the twelve spoken words, subjects saw a blank slide for five seconds while a tape recorder presented the word spoken in a male voice. Presentation of the lists was counterbalanced across the four groups of subjects so that each of the 48 items in the stimulus pool appeared equally often as picture, visual word, spoken word, or new distractor.

Following list presentation, subjects engaged in a distractor task for three minutes, in which they attempted to detect slight differences between pairs of almost-identical cartoon drawings. These drawings were unrelated to the pictures presented in the earlier list. After this, subjects were tested for their memory of the list items. Because the spoken words were the added source in this experiment, it was decided to present items in this format for the memory test in order to minimize bias effects for the other two sources. A tape recorder auditorilly presented the 48 items from the original stimulus pool, in a randomly-determined order and at a presentation rate of ten seconds per item. For each item, subjects indicated in writing whether they thought that item was presented as a picture, visual word, spoken word, or was a new item.

4.7.2 RESULTS

Table 4.4 presents the 4×4 table of response frequencies for the four groups combined. To determine the fit of the model to the data, we computed the G^2 goodness-of-fit statistic separately for each of the four groups of subjects run in the experiment. None of these G^2 values exceeded statistical significance, nor did the combined G^2 value summed across the four groups, $G^2(4) = 7.56, p > 0.1$. However, G^2 computed for the combined data in Table 4.4 did exceed statistical significance, $G^2(1) = 6.67, p < 0.01$. It is not surprising whenever the fit of

TABLE 4.4. Frequency counts from the experiment

		Response				Totals
		P	VW	SW	N	
	P	279	14	24	19	336
Source	VW	38	182	51	63	334
items	SW	18	48	217	53	336
	N	13	23	37	263	336

Note: P=Picture, VW=Visual Word, SW=spoken word, N=New Distractor. The total for visual words reflects two non-responses.

TABLE 4.5. Parameter estimates from the model

Pictures	Visual Words	Spoken Words
$D_1 = 0.93$	$D_2 = 0.76$	$D_3 = 0.80$
$d_1 = 0.86$	$d_2 = 0.48$	$d_3 = 0.66$
$a_1 = 0.23$	$a_2 = 0.41$	$a_3 = 0.37$
$g_1 = 0.18$	$g_2 = 0.32$	$g_3 = 0.51$
	$b = 0.22$	

Note: D_i = probability of detecting Source S_i items, d_i = probability of discriminating the source of detected S_i items, a_i = probability of giving response R_i to detected but nondiscriminated items, b = probability assigning an old source response to undetected items, g_i = probability of giving response R_i to nondetected items.

a model to aggregate data fails even though the model successfully fits data from individual subjects or groups. For one thing, such fits are based on larger samples which elevates the power of the test. Also any individual differences in the separate groups may combine to yield significance at the aggregate level. Riefer and Batchelder (1991) discuss the robustness of multinomial models under conditions where data are combined across subjects.

We replicated Durso and Johnson's (1980) empirical finding that the proportion of source confusions for visual words was higher than for pictures (0.33 vs. 0.12, respectively). However, as we indicated earlier, this empirical result in itself does not reveal whether this difference is due to source discrimination differences or to response biases. Fortunately, the multinomial model provides a method for settling this issue. Table 4.5 presents the parameter estimates computed from the data in Table 4.4. For convenience of presentation, we have computed the parameter values across all 28 subjects in the experiment, as opposed to computing the values for each of the four groups separately. The overall pattern of results is the same in each case.

Pictures were detected significantly better than visual words ($D_1 = 0.93$ vs. $D_2 = 0.76$), $G^2(1) = 27.05, p < 0.01$. This result is consistent with the standard picture superiority effect. In addition, pictures were also better on their source discriminability ($d_1 = 0.86$ vs. $d_2 = 0.46$), $G^2(1) = 35.87, p < 0.01$. This is an important finding because it confirms Durso and Johnson's (1980) conclusions regarding the cause of the superior source memory for pictures over words.

However, the model reveals that response biases were also a factor in the experiment. Because $D_1 > D_2$, Rule 1 predicts $g_1 < g_2$. This prediction is confirmed for this experiment, although the difference between g_1 and g_2 was only marginally significant, $G^2(1) = 2.76, p < 0.10$. In addition, $D_2(1 - d_2) > D_1(1 - d_1)$, so Rule 2 predicts $a_2 > a_1$. This prediction is also confirmed, with the value of a_2 significantly larger than a_1, $G^2(1) = 6.11, p < 0.05$.

The results of this experiment nicely illustrate how the general source-monitoring model can be used by researchers in their analysis of data. The experiment showed that pictures were superior to words for both recognition memory as well as source memory. But the model revealed that these differences were due not only to differences in item detection and source discrimination, but also to differences in the various response biases as well. By separately measuring each of these individual processes, the model provides a clearer picture of the overall results than what is usually afforded by standard empirical statistics alone.

4.8 Alternative Models

A reasonable concern with the analysis based on our source monitoring model is that alternative multinomial models consistent with the assumptions of (4.2) and (4.3) may yield different interpretations of the data. In fact our high threshold assumption that new items are never "detected" as old has been discredited in favor of signal detection theory as a model of yes-no signal-detection and also, though less clearly, as a model of yes-no recognition memory (see Green & Swets, 1966; Laming, 1973; and Macmillan & Creelman, 1991). The negative evidence comes from such bases as comparing conditions under the assumption of isosensitivity discussed in Section 4.5. and noting that the ROC curves predicted from the high threshold model do not correspond to those observed in the experiments. In fact Macmillan and Creelman (1991, Chapter 10) describe how a model based on signal detection theory can be constructed for a paradigm like ours, based on simultaneous detection and recognition. Unfortunately, the signal detection approach requires more parameters than degrees of freedom in a single condition of a two or three source experiment. Further, our three source model does have the advantage of testability, since there are twelve degrees of freedom and eleven parameters. So we would argue that if the goodness-of-fit index is acceptable, then it is a useful measurement tool for a single condition of a source monitoring experiment. Nevertheless, we did examine two generalizations of our model by introducing a new parameter D_0. In the first version, D_0 is the probability that a new item is detected as new, and in the second version, D_0' is the probability that a new item is detected as old. For these versions, (4.4) remains the same, however (4.5) is modified. For version 1, (4.5) becomes

$$P(R_j|S_0) = \begin{cases} (1 - D_0)bg_j, & 0 < j \leq M \\ D_0 + (1 - D_0)(1 - b), & 0 = j, \end{cases} \tag{4.13}$$

and for the second version,

$$P(R_j|S_0) = \begin{cases} D_0'a_j + (1 - D_0')bg_j, & 0 < j \leq M \\ (1 - D_0')(1 - b), & 0 = j. \end{cases} \tag{4.14}$$

The first version is related to the double high-threshold theory discussed in Macmillan and Creelman (1991, Chapter 4) and it yields an identifiable model for $M > 2$; however, the sample variance of the estimators of D_0 and some other parameters are often unacceptably large. In particular, for the three source data in Table 4.4, only marginal improvement in the fit is achieved with version 1, and further the relationship between the D_i and the d_i are the same as found with the eleven parameter model. So more work is needed to see if version 1 provides a useful alternative for analyzing data from three sources.

The second version, which is a discrete state weakening of the high threshold assumption, yields a nonidentifiable model. However, for any given value of D_0', the model is identifiable. For the data in Table 4.4, the equivalent best fitting versions of this model yield $D_0' \in (0.00, 0.165)$ and the other parameter estimates that constituted our explanation of the picture superiority effect support the same conclusions we drew in Section 4.6. For $D_0' > 0.165$, the fit worsens.

Consequently, we conclude that the analysis of source monitoring data with our model is acceptably robust under reasonable variations in the assumptions about the responses to new items. So until a better model is derived, we recommend that it be used to supplement other empirical analyses of source monitoring data.

4.9 References

Batchelder, W.H., & Riefer, D. M. (1990). Multinomial processing models of source monitoring. *Psychological Review, 97,* 548-564.

Batchelder, W.H. (1991). Getting wise about minimum distance measures. Review of *goodness-of-fit statistics for discrete multivariate data.* By Read, T. R. C., & Cressie, N. A. C. *Journal of Mathematical Psychology, 35,* 267-273.

Blackwell, H. R. (1963). Neural theories of simple visual discrimination. *Journal of the Optical Society of America, 53,* 129-160.

Dempster, A. P., Laird, N. M., & Rubin, D. B. (1977). Maximum likelihood from incomplete data via the EM algorithm. *Journal of the Royal Statistical Society, Series B, 39,* 1-38.

Devolder, A. P., & Pressley, M. (1989). Metamemory across the adult lifespan. *Canadian Psychology, 30,* 578-587.

Durso, F.T., & Johnson, M.K. (1980). The effect of orienting tasks on recognition, recall, and modality confusion of pictures and words. *Journal of Verbal Learning and Verbal Behavior, 19,* 416-429.

Foley, M.A., Durso, F.T., Wilder, A., & Friedman, R. (1991). Developmental comparisons of explicit versus implicit imagery and reality monitoring. *Journal of Experimental Child Psychology, 51,* 1-13.

Green, D.M., & Swets, J.A. (1966). *Signal detection theory and psychophysics.* New York: Wiley.

Hu, X. (1990). Statistical inference program for multinomial source monitoring models. Version 1.0. University of California, Irvine (avilable upon request).

Hu, X., & Batchelder, W.H. (To Appear). The statistical analysis of multinomial processing tree models with the EM algorithm. *Psychometrika.*

Klatzky, R.L. (1975). *Human memory.* San Francisco: Freeman.

Laming, D. (1973), *Mathematical psychology.* New York: Academic Press.

Lindsay, S.D., & Johnson, M.K. (1991). Recognition memory and source monitoring. *Bulletin of the Psychonomic Society, 29,* 203-205.

Macmillan, N.A., & Creelman, C.D. (1991). *Detection theory: A users guide.* New York: Combridge University Press.

Nelson, T.O., & Narens, L. (1990). Metamemory, a theoretical framework and new findings. In G.H. Bower (Ed.) *The psychology of learning and motivation* (Vol. 26, pp. 125-173). New York: Academic Press.

Read, T.R.C., & Cressie, N.A.C. (1988). *Goodness-of-fit statistics for discrete multivariate data.* New York: Springer-Verlag.

Riefer, D.M. , & Batchelder, W.H. (1988). Multinomial modeling and the measurement of cognitive processes. *Psychological Review, 95,* 318-339.

Riefer, D.M., & Batchelder, W.H. (1991). Statistical inference for multinomial processing tree models. In J.-P. Doignon & J.-Cl. Falmagne (Eds.), *Mathematical psychology: Current developments* (pp. 313-316). Berlin: Springer-Verlag.

Snodgrass, J.G., & Vanderwart, M. (1980). A standardized set of 260 pictures: Norms for name agreement, image agreement, familiarity, and visual complexity. *Journal of Experimental Psychology: Human Learning and Memory, 6,* 174-215.

5

Feature Representations in Connectionist Systems

John Bankart[1], Philip T. Smith[1], Mark Bishop[2], and Paul Minchinton[2]

ABSTRACT This paper has two goals: to demonstrate the importance of the feature representation chosen for a connectionist model, and to examine the properties of a particular model, the back-propagation algorithm, in conditions intended to simulate the 'graceful degradation' encountered in human ageing. In a concept learning simulation, the number of zero or near-zero values of the input features and in the responses of the hidden units was found to influence speed of learning, strength of response to a prototype, and performance with distorted input (the latter two being inversely related). Degradation of the network prior to learning enhanced prototype performance but disrupted distortion performance. In the light of these results we discuss the design of efficient learning algorithms and the potentiality of these networks as models for human ageing.

5.1 Feature representations

Features are widely used in linguistics, scaling and psychological modelling to express similarities and dissimilarities between entities. For example, MAN and WOMAN might be said to share the feature +HUMAN but are differentiated on the feature MALE: MAN is +MALE, WOMAN is −MALE. Feminists (and others) will note an asymmetry in this formulation: MALE is both the name of the feature and one of the values the feature takes. This is a characteristic of a pair of feature values which are said to exhibit a marked/unmarked opposition: +MALE is the unmarked (default) value, −MALE the marked (exceptional) value. Such a formulation was introduced into linguistics in the 1930s by Trubetzkoy (1939, 1969) and has been developed by Greenberg (1966) and Chomsky & Halle (1968) in linguistics, and Clark (1969) in psychology. Not all feature pairs have this structure: Trubetzkoy calls the marked/unmarked opposition privative, but he also distinguishes two further kinds of opposition: (1) equipollent oppositions (where the opposing feature values have equal weight, e.g., red v. green); (2) gradual oppositions (where the distinction is the degree to which some property is manifest along some dimension, e.g., 60-watt v. 100-watt light bulbs).

---

[1]University of Reading, Department of Psychology, Earley Gate, Whiteknights, Reading RG6 2AL, England

[2]University of Reading, Department of Cybernetics, Whiteknights, Reading RG6 2AX, England

This research was carried out while the first author held an SERC studentship. We gratefully acknowledge the help of two anonymous referees.

In numerical terms it seems natural to code privative features by 0 (absence) and 1 (presence), equipollent features by +1 and −1 (that is, equal but opposite) and gradual features by a and b (with $a, b > 0$). In scaling it is rare to find privative features, with zero often being taken as an arbitrary point along a continuum, with no special theoretical significance, though work on asymmetries between subjects' judgments of similarity and dissimilarity suggests that modelling in terms of presence/absence of particular features can be fruitful (e.g., Tversky, 1977).

5.2 Connectionist models

Feature representations in connectionist models are not arbitrary, and sometimes they can have quite striking effects which are unlikely to be met in humans. For example, consider the auto-associator (see McClelland & Rumelhart, 1988, for a full discussion, including references to earlier work). This is a device capable of unsupervised learning which extracts prototypes from sequences of exemplars. It has the property that, given any specific pattern, it will respond equally strongly to this pattern and to the 'photographic negative' of the pattern (one obtained by changing the sign of all feature values). This is because the auto-associator is effectively performing a principal components analysis, and such an analysis remains essentially unchanged if we change the sign of the components. If we train such a system to recognize YOUNG WOMAN with a feature description +HUMAN −MALE −OLD it will respond equally strongly to the pattern −HUMAN +MALE +OLD: such performance in an organism intent on selecting a partner for a candlelit dinner, for example, is clearly unacceptable.

Restricting the autoassociator to features with non-negative values circumvents this difficulty. We shall show that there is no such simple solution for the back-propagation algorithm and, more generally, for connectionist systems which contain hidden units and which use a version of the delta rule to learn. Again the reader is referred to McClelland & Rumelhart (1988) for background.

The backpropagation algorithm can be defined by a series of equations. We shall concentrate on one of these equations, the generalized delta rule,

$$\Delta w_{ij}^{(n+1)} = \beta(\delta_{pi} o_{pj}) + \alpha \Delta w_{ij}^{(n)}. \tag{5.1}$$

This equation specifies the change in the strength of the connection (weight) $\Delta w_{ij}^{(n+1)}$ between source unit j and destination unit i on trial $n+1$, as a function of an error signal δ_{pi} at unit i when pattern p is presented, the output o_{pj} of unit j, and the change in weight on the previous trial $\Delta w_{ij}^{(n)}$, where α and β are constants. Note that the equation has the satisfactory property that, if δ_{pi} is small, the first term on the right hand side of the equation is small, and hence $w_{ij}^{(n+1)}$ will be small (i.e., if the error signal is small, indicating good performance, the weights do not need to be changed much). However, (5.1) has the unattractive property that the first term on the right hand side will also be small if o_{pj} is small, even if the error signal is quite large (i.e., the weights are not changed much in response to an error signal if they are associated with a low output). In practice, the algorithm starts with small random weights before learning begins, and the observation in the previous sentence implies that the random weight associated with a zero output will remain unchanged throughout learning, making learning less efficient.

This point we illustrate with reference to a concept learning simulation we have carried out. Table 5.1 shows 10 patterns (p) consisting of 8 input features and 6 output features. A network of 8 input units, 6 output units, and 4 hidden units was trained with these patterns, using a back propagation algorithm (the algorithm provided by the McClelland & Rumelhart (1988) package). Learning rate β in (5.1) was set at 0.5, Momentum α in (5.1) was set at 0.9. At the start of each simulation, weights and biases were randomly assigned values in the range -0.5 to $+0.5$ (biases are terms added to the net inputs to the hidden units and the output units, and act rather like thresholds). The 10 patterns were presented in a random order (this constitutes an epoch of training): if the criterion for successful training (to be discussed below) was not reached, a further epoch of training was initiated (with a different random order for the presentation of patterns). This process continued until criterion was reached. The learning algorithm was implemented after the presentation of each pattern. All the details of the algorithm we have just outlined are the default values of the McClelland & Rumelhart package, and we have no reason to believe the results we have obtained are the result of unusual or inappropriate parameter values.

Inspection of Table 5.1 will show that this is a concept learning task: the 10 inputs are being mapped on to just 2 distinct outputs. Moreover, the first 5 features of the input describe (non-presented) prototypical forms, in relation to which the actual inputs are imperfect exemplars, differing from the prototype on one feature. The prototypes are 5 zeros for the first feature, or 5 ones. The reader may find it helpful to think of the first 5 features as discriminating between birds and mammals, with 5 zeros being a prototypical bird, and 5 ones a prototypical mammal: the first feature might represent capacity for flight, in which case pattern $\boldsymbol{a}$ is a bird which differs from prototype on flight capacity, for instance, an ostrich, and pattern $\boldsymbol{f}$ is a mammal which differs from prototype on flight capacity, for instance, a bat. The values of the remaining 3 features of the input were assigned randomly (but fixed throughout the simulations): they are irrelevant for the discrimination and the algorithm should learn to ignore them. The output features are arbitrary: our experience with such simulations suggests the particular values are not important, and these were kept constant throughout the simulations.

Goodness of learning in these systems is measured by pattern sum of squares (pss). Given a target output of t_{pi} for unit i in response to pattern $\boldsymbol{p}$, when the

TABLE 5.1. Patterns used in training

p	Input								Output					
a	**1**	**0**	**0**	**0**	**0**	**0**	**1**	**1**	**0**	**0**	**0**	**1**	**1**	**0**
b	0	1	0	0	0	0	1	1	0	0	0	1	1	0
c	0	0	1	0	0	1	0	0	0	0	0	1	1	0
d	0	0	0	1	0	1	0	1	0	0	0	1	1	0
e	0	0	0	0	1	1	1	1	0	0	0	1	1	0
f	0	1	1	1	1	1	1	1	0	1	1	0	1	1
g	1	0	1	1	1	0	1	1	0	1	1	0	1	1
h	1	1	0	1	1	1	1	0	0	1	1	0	1	1
i	**1**	**1**	**1**	**0**	**1**	**0**	**1**	**1**	**0**	**1**	**1**	**0**	**1**	**1**
j	1	1	1	1	0	1	0	1	0	1	1	0	1	1

actual output is o_{pi},

$$pss(p) = \sum_i (o_{pi} - t_{pi})^2. \tag{5.2}$$

Good performance is indicated by small pss.

A system such as this should display at least three properties: first, performance on the (non-presented) prototype is at least as good as on the corresponding exemplars. If P is the prototype of pattern $\boldsymbol{p}$ (constructed by replacing any of the first five features of $\boldsymbol{p}$ that do not match prototype by the prototypical value), then we expect $pss(P) \leq pss(p)$. Second, performance on minor distortions of a pattern should be good: if $\boldsymbol{p_s}$ is an input pattern formed by changing one feature of pattern $\boldsymbol{p}$, define $dss(p)$, distortion sum of squares for pattern $\boldsymbol{p}$, by

$$dss(p) = \sum_s pss(p_s)/n, \tag{5.3}$$

where n is the number of distorted patterns. We expect dss to be small. Third, the system should exhibit graceful degradation; in particular, random deletion of a small number of connections between units prior to training (that is, setting some weights to zero and keeping them at zero during training) should not greatly alter performance. We deleted Deg % of the connections randomly, subject to the restriction that every unit retained at least one input and one output connection. Deg took the values 0, 12.5, 25 and 37.5. There are several other ways that degradation could be implemented, of which the removal of neurons, not just the connections between them, is the most obvious and plausible. Removal of entire units in our very small-scale networks would have catastrophic effects: for example, removal of the first two input units would mean that patterns a and b could never be discriminated. In larger networks with more redundancy we speculate that removal of units or their interconnections would produce qualitatively similar results, but until large networks are implemented

TABLE 5.2. Median performance

Deg	$pss(\mathbf{a})$	$pss(\mathbf{i})$	$pss(\mathbf{A})$	$pss(\mathbf{I})$	$dss(\mathbf{a})$	$dss(\mathbf{i})$
0	0.0027	0.0041	0.0011	0.0027	0.1418	0.0217
12.5	0.0030	0.0039	0.0010	0.0023	0.1613	0.0332
25	0.0034	0.0038	0.0010	0.0018	0.3120	0.1304
37.5	0.0043	0.0051	0.0008	0.0019	0.5017	0.5083

A is the prototype of a, and I the prototype of i

with a variety of forms of degradation our proposal should be regarded as an arbitrary initial step.

Training continued until a criterion of the sum of the pattern sums of squares for all the individual items was less than or equal to 0.04. Training to a criterion, rather than testing after a fixed number of epochs, is used to ensure that the performance of the system with prototypes and distorted patterns is uncontaminated by incomplete learning of individual patterns. Number of epochs to criterion increased with *Deg*: for $Deg = 0$, median epochs to criterion was 17 (range 9 to 49); for $Deg = 37.5$, median epochs to criterion was 45.5 (range 24 to 56). When criterion was reached, training was discontinued and the system tested with the original patterns, the prototypes and distorted versions of patterns $\boldsymbol{a}$ and $\boldsymbol{i}$ (appearing in bold in Table 5.1). Patterns $\boldsymbol{a}$ and $\boldsymbol{i}$ were chosen because they contained predominantly zeros and ones, respectively (further simulations confirm that it is only the proportions of zeros and ones in the input patterns that are crucial, the proportions in the output patterns are irrelevant). The choice of patterns $\boldsymbol{a}$ and $\boldsymbol{i}$ is to some extent arbitrary, and essentially the same results can be obtained with any of the other patterns. Table 5.2 shows the results of simulations, using two different training sets and two different networks having the same amounts of degradation. The second training set was constructed by interchanging all the zeros and ones in Table 5.1: in both training sets "pattern $\boldsymbol{a}$" is used to refer to the pattern composed predominantly of zeros and "pattern $\boldsymbol{i}$" is composed predominantly of ones. The table shows medians based on the results of 10 simulations for undegraded networks and 20 simulations for the degraded networks.

The table enables us to make the following points. (1) There is a prototype effect: $pss(\boldsymbol{A}) < pss(\boldsymbol{a})$ and $pss(\boldsymbol{I}) < pss(\boldsymbol{i})$. This effect is very reliable and is present in 99.3% of the individual cases. (2) $pss(\boldsymbol{A}) < pss(\boldsymbol{I})$. This effect is reliable and is present in 78.6% of the individual cases. (3) Performance with distorted input is good for $Deg = 0$ and 12.5 with pattern $\boldsymbol{i}$, but performance is markedly worse for pattern $\boldsymbol{a}$: $dss(\boldsymbol{a}) > dss(\boldsymbol{i})$ in 96.7% of individual cases. For $Deg = 25$ the effect is weaker; and it is absent for $Deg = 37.5$. For these particular patterns, a dss of 0.15 represents the upper limit of acceptability, since for dss larger than this the obtained and target outputs are likely to have at least one mismatch even when the obtained outputs are rounded to the nearest whole number. (4) At each level of degradation, $pss(\boldsymbol{A})$ and $dss(\boldsymbol{a})$ are negatively correlated, as are $pss(\boldsymbol{I})$ and $dss(\boldsymbol{i})$. That is, networks that are good for prototypes are bad for resistance to distortion, and *vice versa*. (5) The effect of degradation is to preserve or even enhance the prototype effect, but to impair resistance to distortion.

We have examined the effects of using Trubetzkoy's gradual features for this problem, by replacing the values of 0 and 1 in Table 5.1 by 0.4 and 0.6. It is quite clear that this is not the solution to the problem. Not only is learning slower, but the prototype effect disappears (present on only 25% of occasions in 30 simulations).

Using Trubetzkoy's equipollent features, for instance, replacing 0 and 1 by -1 and 1, is a more viable alternative, since there are now no zeros among the input patterns. We have not implemented detailed simulations yet, but there is one reason why this cannot be the best solution to the problem: although the modeller has control of the input, and can ensure that no features with near-zero values are input, similar control cannot be exerted over the hidden units, where, if the net input to a particular hidden unit turns out to be near-zero for a particular pattern, inefficiency of learning will take place. This is illustrated with a fine-grain analysis of some of our simulations. Consider an undegraded network with 4 hidden units being trained with the patterns of Table 5.1. For patterns $\boldsymbol{a}$ and $\boldsymbol{i}$ we can ask how many of the hidden units, after training, respond with activations near 1 (> 0.75) and near zero (< 0.25). Let $\mathit{diff}(\boldsymbol{a})$ be the difference between the number of near-one units and the number of near-zero hidden units responding to pattern $\boldsymbol{a}$; let $\mathit{diff}(\boldsymbol{i})$ be similarly defined. Potentially these *diff* functions can vary between -4 and $+4$, though in the 10 simulations we have carried out the range was from -2 to $+2$. Now it turns out that $\mathit{diff}(\boldsymbol{a})$ is negatively correlated with $pss(\boldsymbol{a})$ ($r = -0.70, P < 0.05$) and with $pss(\boldsymbol{A})$ ($r = -0.53, P = 0.11$), and $\mathit{diff}(\boldsymbol{i})$ is negatively correlated with $pss(\boldsymbol{i})$ ($r = -0.89, P < 0.001$) and $pss(\boldsymbol{I})$ ($r = -0.92, P < 0.001$). Moreover, speed of learning is related to these functions: $\mathit{diff}(\boldsymbol{a})$ is positively correlated with the number of trials to reach criterion ($r = 0.68, P < 0.05$), and $\mathit{diff}(\boldsymbol{i})$ is negatively correlated ($r = -0.75, P < 0.05$). A specific example illustrates this last point: on one simulation when it turned out that all of the hidden units responded with near-zero values to pattern $\boldsymbol{i}$, the algorithm took 33 epochs to reach criterion; on another simulation when 2 hidden units responded with near-one values and the other 2 units responded with near-zero values, the algorithm took only 13 epochs to reach criterion. A referee has pointed out that it may be a mistake to equate efficiency with speed of learning of one set of patterns: the network may be required to learn other sets of patterns and having hidden units responding with near-zero values to the first set of patterns may permit them to be more effective in learning other sets of patterns. More extensive simulation would be needed to establish this point.

A final way, that we have examined, of coding binary oppositions is by pairs of features, $[f_1, f_2]$. One value of a basic opposition is coded $[+f_1, -f_2]$, the other value is coded $[-f_1, +f_2]$ (this is like coding RED as [red, not green] and GREEN as [not red, green]. The attraction of this proposal, when zeros and ones are being used as feature values, is that there are equal numbers of zeros and ones for each input pattern, so asymmetries of the sort we have been discussing will not arise. The unattractive aspects of this proposal are that, in doubling the number of input features, we put additional demands on the processing capacity of the system and, as we have remarked above, we cannot exert control over the coding of the hidden units: if any such units respond with near zero values to a particular pattern, learning will be less efficient and, as we have verified, prototype effects are reduced.

5.3 Discussion

5.3.1 Implications for Ageing Research

The primary interest here is in how performance changes as the network is degraded. Our main result is that degrading usually disrupts resistance to distortion but slightly enhances response to a prototype. Informal observation suggests old people are less able than the young to cope with changes in previously well-learnt patterns (e.g., changes in size or shape of coins or paper currency). A similar phenomenon is seen in extreme form in Alzheimer's Disease, with sufferers increasingly less able to cope with seemingly quite trivial changes in their domestic environment. In contrast, old people's handling of prototypical information, as reflected in their performance on a variety of tasks involving semantic memory, semantic priming for instance, remains constant (Howard, 1988) or in some cases actually seems slightly enhanced in comparison with younger controls (Burke & Harrold, 1988, Table 6.1). This is consistent with the slight enhancement of prototype effects with increasing degradation in our model, where the enhancement is sometimes so slight as to be indistinguishable in practice from constancy. Thus our model seems to be able to deal with several phenomena reported in the ageing literature, and we have been able to show that these phenomena derive in a simple way from the degrading of a learning network. It should be pointed out that our results – strictly speaking – apply only to learning of new material after degradation, not to the maintenance or modification of old material learned prior to degradation. We think, however, very similar results will be observed in this latter case, since, as connections are lost or become unreliable, the network will be obliged to re-learn the old material: the only difference from the *tabula rasa* learning we have studied is that the network is able to start the relearning process from a location in weight space closer to an appropriate solution. It should also be pointed out that these degradation results apply to all the patterns we have studied (e.g., whether the stimulus is composed of predominantly zeros or ones), and so these issues are orthogonal to those we have raised in respect of feature descriptions and the back propagation algorithm.

5.3.2 IMPLICATIONS FOR THE DESIGN OF EFFICIENT LEARNING ALGORITHMS

Our results have shown that a system using the delta rule and having inputs that can take zero or near-zero values changes its properties as a function of the number of zeros in the input (the results we report are not confined to the patterns shown in Table 5.1, but have been confirmed on a wide variety of patterns, numbers of features, and numbers of zeros).

Our final result, that how well a prototype is learnt and the number of epochs needed to achieve satisfactory learning are dependent on the number of hidden units that take near-zero values, indicates that there is an important source of variance in these algorithms that cannot be removed simply by avoiding zero values at input. There are two possible reactions to this observation: one is to abandon the delta rule or any similar rule that involves multiplicative combinations of error score and activation. The other reaction is to take these results at face value and try to explain some of the variance we observe in human learning as a consequence of asymmetries in coding within systems using delta rules or the like. But we would need to know much more about how the brain encodes information before this might be a viable approach.

5.4 References

Burke, D. M., & Harrold, R. M. (1988). Automatic and effortful processes in old age: Experimental and naturalistic processes. In L. Light & D.M. Burke (Eds.) *Language, memory and aging* (pp. 100–116). Cambridge: Cambridge University Press.

Chomsky, N., & Halle, M. (1968). *The sound pattern of English.* New York: Harper & Row.

Clark, H. H. (1969). Linguistic processes in deductive reasoning. *Psychological Review, 76,* 387–404.

Greenberg, J. (1966). *Language universals.* The Hague: Mouton.

Howard, D.V. (1988) Aging and memory activation: The priming of semantic and episodic memories. In L. Light & D.M. Burke (Eds.) *Language, Memory and Aging* (pp. 77–79). Cambridge: Cambridge University Press.

McClelland, J. L., & Rumelhart, D. E. (1988). *Explorations in parallel distributed processing.* Cambridge, MA: MIT Press.

Trubetzkoy, N. S. (1939/1969). *Grundzüge der Phonologie [Principles of phonology].* Los Angeles, CA: University of California Press.

Tversky, A. (1977). Features of similarity. *Psychological Review, 84,* 327–352.

6

Finite Markov Learning Models for Knowledge Structures

Jean-Claude Falmagne[1]

ABSTRACT Some finite Markov learning models are described, which explain the transitions, over time, between the various states of a knowledge structure, from the empty state to the full domain of the structure. Any one of these models sets constraints on the probability distribution on the collection of knowledge states, thereby reducing considerably the number of parameters that need to be estimated. Such models are intended for use at an early stage in the analysis of data. The resulting knowledge structure should be regarded as tentative, and can be refined by further analysis with more demanding models.

The number of states in empirical knowledge structures tends to be quite large. For example, in a recent experiment involving 50 items, the number of states estimated by the QUERY routine, applied to five human experts, ranged from several hundreds to several thousands (Kambouri et al., 1991; for an introduction to knowledge structures, see Doignon and Falmagne, 1985, or Falmagne et al., 1990; the QUERY routine is described by Koppen in this volume). This presents a problem in practical applications, since it means that a prohibitively large number of parameters – e.g. the probabilities of all these states in the relevant population – may have to be estimated from the empirical frequencies of the response patterns.

More precisely, let $(Q, \mathcal{K})$ be a knowledge structure, with $|Q| = n$. Here, Q is a set of items called the *domain* of the structure, which is supposed finite, and $\mathcal{K}$ is a distinguished family of subsets of Q representing the *knowledge states*. It is assumed that $\mathcal{K}$ always contains the empty state $\emptyset$ and the domain Q. Occasionally, I shall refer to $\mathcal{K}$ itself as the knowledge structure; since $\cup\mathcal{K} = Q$, no ambiguity can arise. A general probabilistic model for such a structure postulates the existence of a probability distribution $p : K \mapsto p(K)$ on $\mathcal{K}$:

$$p(K) \geq 0, \qquad \forall K \in \mathcal{K};$$

$$\textstyle\sum_{K \in \mathcal{K}} \; p(K) = 1.$$

Thus, $p(K)$ may be regarded as the probability of finding a subject, arbitrarily chosen from the reference population, in state K. It is convenient to represent

---

[1]This work is supported by NSF Grant IRI 8919068 to Jean-Claude Falmagne at the University of California, Irvine, CA 92717, USA

The author is grateful to Dina Blok, Stefan Fries, Frank Gilson, Kamakshi Lakshminarayan, Mike Regenwetter, Josh Stern and especially Jean-Paul Doignon, for their reactions to previous drafts of this material. The reactions of the referees have also been useful.

the complete list of (correct or incorrect) responses made by a subject to the n items of the test by the subset $R \subseteq Q$, where R contains all the correct responses. Any $R \subseteq Q$ is referred to as a ***response pattern***. The probability $P(R)$ of any response pattern R can then be obtained from the state probabilities by means of the formula

$$P(R) = \sum_{K \in \mathcal{K}} \rho(R, K) p(K), \tag{6.1}$$

where $\rho(R, K)$ denotes the conditional probability of the response pattern R, given the state $K \in \mathcal{K}$. These conditional probabilities can be modelled by some form or other of the 'local independence' condition of psychometric theory (Lord and Novick, 1974). Such a model will be specified in a moment.

Practical applications encounter difficulties in that the number of probabilities $p(K)$ to be estimated is typically very large. Even with substantial data sets, reliable estimates may be hard to obtain (see, however, Villano, 1991).

This paper discusses a number of Markovian learning models which set constraints on the probabilities $p(K)$ of the states, or on the probabilities of the learning paths, and thereby reduce the number of parameters to a manageable level. I stress that these models are meant to be used at a preliminary stage of analysis, as a first pass at the construction of the knowledge structure. More realistic – but also more demanding – models can be found in Falmagne (1989, in press).

6.1 Basic concepts and an example

I begin by modelling the conditional probabilities $\rho(R, K)$. As in Falmagne (1989, in press) and Falmagne et al. (1990), suppose that, for each item $q \in Q$, there is a parameter β_q, $0 \le \beta_q \le 1$, representing the probability of a careless error in response if this item is contained in the current knowledge state. There is also a collection of parameters η_q representing the probability of a lucky guess for a response to an item $q \in Q$ not contained in the current state. It is assumed that, given the state, the responses are mutually independent events. This means that the parameters β_q and η_q specify the conditional probability $\rho(R, K)$ of (6.1), of observing a response pattern R, given a knowledge state K, according to the following formula:

$$\rho(R, K) = \Big[\prod_{q \in K \setminus R} \beta_q\Big]\Big[\prod_{q \in K \cap R} (1 - \beta_q)\Big]\Big[\prod_{q \in R \setminus K} \eta_q\Big]\Big[\prod_{q \in \overline{R \cup K}} (1 - \eta_q)\Big], \tag{6.2}$$

in which the complement $\overline{R \cup K}$ within the last pair of brackets is to be taken with respect to Q. For example, with $Q = \{1, 2, 3, 4\}$, a subject in the knowledge state $\{2, 3, 4\}$ will give the response pattern $\{1, 3\}$ with probability

$$\rho(\{1,3\}, \{2,3,4\}) = \eta_1 \beta_2 (1 - \beta_3) \beta_4 . \tag{6.3}$$

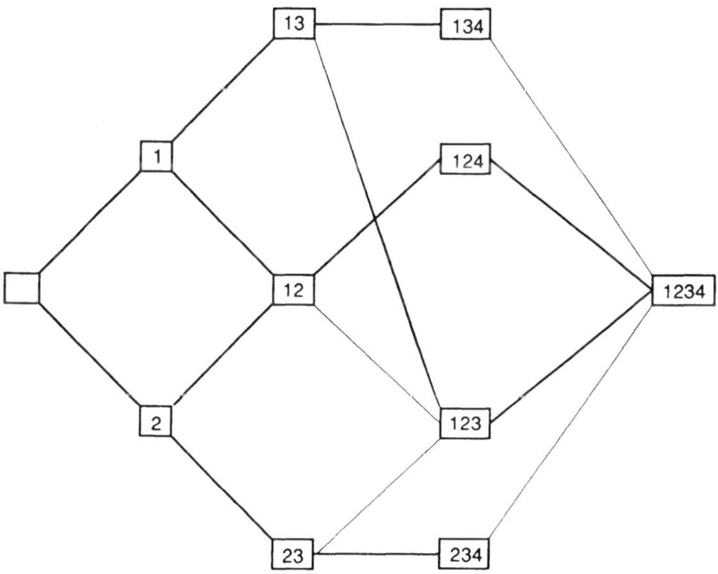

FIGURE 6.1. Transitions between states in the knowledge structure $\mathcal{F}$ of (6.4)

Formula (6.2) may seem formidable. In some situations, however – with open responses, for example –, it may be assumed that some or all of the guessing probabilities η_q are zero; this results in a substantial simplification of the formula.

All the models in this paper will be described in the framework of one particular knowledge structure:

$$\mathcal{F} = \{\emptyset, \{1\}, \{2\}, \{1,2\}, \{1,3\}, \{2,3\}, \{1,2,3\}, \{1,2,4\}, \{1,3,4\}, \{2,3,4\}, \{1,2,3,4\}\}. \tag{6.4}$$

This structure is also represented in the diagram of Figure 6.1, which displays the possible transitions between states, from the empty state $\emptyset$ to the domain $Q = \{1,2,3,4\}$ of the structure. I next recall some basic definitions (cf. Doignon and Falmagne, 1985, Falmagne et al., 1990):

Definition 1 *A knowledge structure* $(Q, \mathcal{K})$ *is called a knowledge space if the family* $\mathcal{K}$ *is closed under union. Any maximal*[2] *chain of states is called a learning path. Any maximal chain of subsets of* Q*, all of which are states, is a gradation. A knowledge structure in which every learning path is a gradation is said to be well graded. Suppose that a knowledge structure* $(Q, \mathcal{K})$ *satisfies the condition:*

[2]A chain $\mathcal{C}$ of subsets (resp. states) is called *maximal* if, whenever $\mathcal{C} \cup \{K\}$ is a chain of subsets (resp. states), then $K \in \mathcal{C}$.

for all $x, y \in Q$,

$$(\forall K \in \mathcal{K}, x \in K \Longleftrightarrow y \in K) \Rightarrow x = y\,;$$

then $(Q, \mathcal{K})$ *is called discriminating. It is easily verified that any well graded knowledge structure is discriminating.*

As made clear by Figure 6.1, the knowledge structure $\mathcal{F}$ specified by (6.1) is a discriminating knowledge space, with eight gradations.

In this paper, consideration is restricted to well graded knowledge structures. All the models discussed here are based on the following basic idea:

> *The probability that, at the time of the test, a subject is in a particular state K of the structure is expressed as the probability that this subject*
>
> *(i) has successively mastered all the items in the state K, and*
>
> *(ii) has failed to master any item immediately accessible from K.*

For any state K, the set of all items *accessible from* K will be denoted by

$$S(K) = \{q | q \notin K \text{ and } (K + \{q\}) \in \mathcal{K}\}\,. \tag{6.5}$$

For example, in $\mathcal{F}$, $S\{1,2\} = \{3,4\}$.

6.2 Model 1

The first model also makes strong independence assumptions regarding the learning process. It introduces one parameter g_q, $0 \leq g_q \leq 1$, for each item q in Q. It is assumed that, for any state K, all the events in the two classes: '*mastering any item* $q \in K$' and '*failing to master any item* $q \in S(K)$', are pairwise independent in the sense that

$$p(K) = \begin{cases} \prod_{r \in S(K)} (1 - g_r) & \text{if } K = \emptyset \\ \prod_{q \in K} g_q & \text{if } K = Q \\ \prod_{q \in K} g_q \prod_{r \in S(K)} (1 - g_r) & \text{if } K \neq \emptyset, Q. \end{cases} \tag{6.6}$$

In this model the state probabilities are specified in terms of only $n = |Q|$ parameters, irrespective of the number of states. The probabilities of the states in the standard example are listed in Table 6.1. In this way the probabilities of the eleven states are expressed in terms of just 4 parameters. It is easy to verify that these probabilities add up to one. In general, however, the quantities $p(K)$ defined by (6.6) do not necessarily specify a probability distribution. To see that this is so, consider the knowledge structure:

$$\mathcal{H} = \{\emptyset, \{1\}, \{2\}, \{3\}, \{1,2\}, \{1,3\}, \{1,2,3\}\}$$

TABLE 6.1. Probabilities of the states of the structure $\mathcal{F}$ in Model 1

$\emptyset$	$(1-g_1)(1-g_2)$
$\{1\}$	$g_1(1-g_2)(1-g_3)$
$\{2\}$	$g_2(1-g_1)(1-g_3)$
$\{1,3\}$	$g_1g_3(1-g_4)(1-g_2)$
$\{1,2\}$	$g_1g_2(1-g_4)(1-g_3)$
$\{2,3\}$	$g_2g_3(1-g_1)(1-g_4)$
$\{1,3,4\}$	$g_1g_3g_4(1-g_2)$
$\{1,2,4\}$	$g_1g_2g_4(1-g_3)$
$\{1,2,3\}$	$g_1g_2g_3(1-g_4)$
$\{2,3,4\}$	$g_2g_3g_4(1-g_1)$
$\{1,2,3,4\}$	$g_1g_2g_3g_4$.

with domain $\{1,2,3\}$. Define $g_q = \frac{1}{2}$ for all $q \in \{1,2,3\}$. Equation (6.6) gives

$$p(\emptyset) = p\{1\} = p\{1,2\} = p\{1,3\} = p\{1,2,3\} = \frac{1}{8},$$

and

$$p\{2\} = p\{3\} = \frac{1}{4},$$

yielding $\sum_{K\in\mathcal{H}} p(K) = \frac{9}{8}$. Notice that this knowledge structure is not a knowledge space: $\{2\}, \{3\} \in \mathcal{H}$, but $\{2,3\} \notin \mathcal{H}$. Conditions will be investigated under which (6.6) defines a genuine probability distribution on a well graded knowledge structure $\mathcal{K}$.

Some intuition can be gained from examining the situation in the case of the knowledge structure $\mathcal{F}$ of (6.4) (cf. Figure 6.1 and Table 6.1). The sum of the state probabilities is obtained from the following formula, which is displayed in a staggered style, to facilitate the parsing:

$$\begin{aligned}
1 = (1-g_1)(1-g_2) & \\
&+g_1(1-g_2)(1-g_3)\\
&+g_2(1-g_1)(1-g_3)\\
&\qquad+g_1g_3(1-g_2)(1-g_4)\\
&\qquad+g_1g_2(1-g_3)(1-g_4)\\
&\qquad+g_2g_3(1-g_1)(1-g_4)\\
&\qquad\qquad+g_1g_3g_4(1-g_2)\\
&\qquad\qquad+g_1g_2g_4(1-g_3)\\
&\qquad\qquad+g_1g_2g_3(1-g_4)\\
&\qquad\qquad+g_2g_3g_4(1-g_1)\\
&\qquad\qquad\qquad+g_1g_2g_3g_4
\end{aligned}$$

Multiplying the factors in each of the state probabilities gives, in the right-hand member, an expanded expression in which there is a single term 1 (coming from the probability $(1 - g_1)(1 - g_2)$ of the empty state in the first line of the display), and all the other terms are monomials of the form $g_i, g_ig_j, \ldots, g_1g_2g_3g_4$, with a positive or negative sign. The sum is equal to one because each of these monomials arises with a positive sign exactly as often as with a negative sign. Notice that, scanning the expression from right to left, the first time that a given monomial $g_i \ldots g_j$ appears, it arises from the probability of the state $\{i, \ldots, j\}$ and with a positive sign. It may also appear in the column of terms immediately on the left with a negative sign. It may even appear in the next column on the left with a positive sign again, etc. For example, $g_1g_3g_4$ appears first as a result of the multiplication of $g_1g_3g_4(1-g_2) = g_1g_3g_4 - g_1g_2g_3g_4$, which is the probability of the state $\{1,3,4\}$. There is also the term $-g_1g_3g_4$ appearing as the result of the multiplication of $g_1g_3(1-g_2)(1-g_4)$. These two occurences cancel, and there is no other occurence of this monomial in the formula.

This example is suggestive of a more general relation. In the next definition this idea is expressed in combinatorial terms.

Definition 2 *Let $(Q, \mathcal{K})$ be a finite well graded knowledge structure. A state J is an access (state) to a state K if*

$$J \subseteq K \subseteq J \cup S(J).$$

Thus, any state K has at least one access, namely, itself; and any non empty state has at least two accesses. An access J to K is called a m-access if $|K \setminus J| = m$. In this case J is said to be an access of rank m to K. An m-access J to a state K is called odd if m is odd; otherwise, J is said to be even. Take a state K in some arbitrary finite knowledge structure $\mathcal{K}$, and let m be the largest positive integer such that K has some m-access J. Then J is said to be an initial access to K.

In the knowledge structure $\mathcal{F}$,

$$\{1,2\} \subseteq \{1,2,4\} \subseteq \{1,2\} \cup S(\{1,2\}) = \{1,2,3,4\}.$$

The state $\{1,2,4\}$ has two accesses, itself and $\{1,2\}$. The state $\{1,2,3\}$ has 6 accesses. The domain $\{1,2,3,4\}$ has 8 accesses, and 3 of those are initial, namely, $\{1,2\}$, $\{1,3\}$ and $\{2,3\}$. Consider some non empty state $K = \{i, \ldots, j\} \in \mathcal{F}$. In the expression of the sum of the state probabilities, each even access to K yields the monomial $g_i \ldots g_j$, and each odd access the monomial $-g_i \ldots g_j$. In this example it is easy to verify that each state has the same number of even and odd accesses. This ensures that the state probabilities add up to one. In general:

Theorem 1 *If every non empty state of a well graded, finite knowledge structure has the same number of even and odd accesses, then the function p defined by (6.6) specifies a genuine probability distribution on the structure.*

The proof is omitted; it would do little more than paraphrase the argument above. The condition in this theorem is not very appealing. However, there is also the following, less obvious, result:

Theorem 2 *In a well graded, finite knowledge space, every non empty state has the same number of even and odd accesses. Consequently, the function p defined by (6.6) always defines a genuine probability distribution on the space.*

Before providing a proof, I consider a special case. Suppose that some state K of a well graded knowledge space has a unique initial access J, and let m be the rank of this access. Thus, K has $|J|+m$ elements. This means that there are $\binom{m}{1}$ states of the form $J \cup \{q\}$, with $q \in K \setminus J$. Since the knowledge structure is closed under union, each one of these $\binom{m}{1}$ states is an access of rank $m-1$ to K. By a similar argument, K has $\binom{m}{2}$ accesses of rank $m-2$, $\binom{m}{3}$ accesses of rank $m-3$, etc. From the standard binomial identity

$$\binom{m}{0} - \binom{m}{1} + \binom{m}{2} - \ldots \pm \binom{m}{m} = 0 \tag{6.7}$$

it follows that K necessarily has the same number of even and odd accesses.

The proof given below generalizes this argument for the case in which a state can have more than one initial access.

Proof of Theorem 2: Consider the general case where a state K has some initial access(es) of rank m. Note that any two initial accesses must have the same number of elements. Indeed, if J, L are two initial m-accesses, it follows by definition that $J, L \subseteq K$, with $|K \setminus J| = |K \setminus L| = m$. In fact, the set of all initial accesses of K can be described quite precisely.

Let $\mathcal{F}_j(K)$ be the collection of all the accesses of rank j of K, and let

$$T(K) = \cup\mathcal{F}_m(K) \setminus \cap\mathcal{F}_m(K).$$

Lemma 1 *There is a nonnegative integer $r \leq |T(K)| = k$, such that K has exactly $\binom{k}{r}$ initial accesses. Specifically, a set J is an initial access of K if and only if for some $J' \subseteq T(K)$, with $|J'| = r$*

$$J = \cap\mathcal{F}_m(K) + J'.$$

In other words, J is an initial access of K if and only if J is the union of $\cap\mathcal{F}_m(K)$ with one of the $\binom{k}{r}$ subsets of size r of $T(K)$.

Proof: Take some initial access J of K and define $r = |J \setminus \cap\mathcal{F}_m(K)|$. Write

$$\mathcal{T}_r(K) = \{A \subseteq T(K) \,|\, |A| = r\},$$

(so that, $\mathcal{T}_r(K)$ is the family of all subsets of size r of $T(K)$.) Take some arbitrary set $B' \in \mathcal{T}_r(K)$ and consider $B = \cap\mathcal{F}_m(K) + B'$. (Note that the union must be disjoint.) If B is not an initial access of K, then all the $A' \in \mathcal{T}_r(K)$ such that $A' + \cap\mathcal{F}_m(K)$ is an initial access of K have at least one point in common, say q. But then $q \in \left[\cap \mathcal{F}_m(K)\right] \cap T(K)$, and this generates a contradiction. □

For the remainder of this proof, intuition will be guided by an example. Envisage a set K containing 10 elements, with $\binom{7}{5} = 21$ initial accesses of rank $10 - 5 = 5$. In the notation of the Lemma, $k = 7$, $r = 5$, and each of the $\binom{7}{5}$ initial accesses is missing $2 + 3 = 5$ elements from K. An access of rank 4 of K may be formed by taking some initial access $J \in \mathcal{F}_5(K)$ and adding one element to it. This element may come either from $K \setminus \cup\mathcal{F}_5(K)$ or from $\cup\mathcal{F}_5(K) \setminus J$. This means that there are

$$\binom{7}{5}\binom{3}{1} + \binom{7}{6}$$

accesses of rank 4. The accesses of rank 3 contain 7 elements; there are exactly

$$\binom{7}{5}\binom{3}{2} + \binom{7}{6}\binom{3}{1} + \binom{7}{7}$$

such accesses. These accesses may be formed in 3 ways: either by picking one of the $\binom{7}{5}$ initial accesses and adding 2 elements, taken from $K \setminus \cup\mathcal{F}_5(K)$; or by picking one of the $\binom{7}{6}$ subsets of size 6 of $T(K)$ and adding one element, taken from $K \setminus \cup\mathcal{F}_5(K)$; or by $T(K)$ itself.

The accesses of rank 2, 1 and 0 are obtained similarly. There are respectively,

$$\binom{7}{5}\binom{3}{3} + \binom{7}{6}\binom{3}{2} + \binom{7}{7}\binom{3}{1} \qquad \text{accesses of rank 2,}$$

$$\binom{7}{6}\binom{3}{3} + \binom{7}{7}\binom{3}{2} \qquad \text{accesses of rank 1,}$$

$$\binom{7}{7}\binom{3}{3} \qquad \text{access of rank 0.}$$

The theorem holds in this particular case, in view of the following computation in which the accesses of odd ranks are associated with positive signs:

$$\binom{7}{5} - \left[\binom{7}{5}\binom{3}{1} + \binom{7}{6}\right] + \left[\binom{7}{5}\binom{3}{2} + \binom{7}{6}\binom{3}{1} + \binom{7}{7}\right]$$

$$-\left[\binom{7}{5}\binom{3}{3} + \binom{7}{6}\binom{3}{2} + \binom{7}{7}\binom{3}{1}\right] + \left[\binom{7}{6}\binom{3}{3} + \binom{7}{7}\binom{3}{2}\right] - \binom{7}{7}\binom{3}{3}$$

$$= \binom{7}{5}\left[\binom{3}{0} - \binom{3}{1} + \binom{3}{2} - \binom{3}{3}\right]$$

$$-\binom{7}{6}\left[\binom{3}{0} - \binom{3}{1} + \binom{3}{2} - \binom{3}{3}\right]$$

$$+\binom{7}{7}\left[\binom{3}{0} - \binom{3}{1} + \binom{3}{2} - \binom{3}{3}\right]$$

$$= 0$$

by the binomial identity (6.7), applied to each term. Implicit in the last equation is a partition of the collection of all accesses to K into $\binom{7}{5} + \binom{7}{6} + \binom{7}{7} = 29$ classes, with each of these classes containing the same number of accesses, i.e. $\binom{3}{0}+\binom{3}{1}+\binom{3}{2}+\binom{3}{3} = 8$. The theorem holds because, in each of these 29 classes, the number of even and odd accesses is the same, namely, $\binom{3}{0} + \binom{3}{2} = \binom{3}{1} + \binom{3}{3} = 4$. This example points to an argument for the general case.

Any access to K can be obtained by the following 3-stage procedure:

1. First, take a subset J' of size r of $T(K)$. Form the set

$$J = \cap\mathcal{F}_m(K) + J'.$$

2. Next, take a subset J'' of $\cup\mathcal{F}_m \setminus J$ and form the set $J + J''$.

3. Finally, take some elements from $K \setminus \cup\mathcal{F}_m(K)$ and add them to $J + J''$.

As in the example, there is a partition on the set of accesses to K, the classes of which are specified by the choice of the pair (J', J'') in steps 1 and 2. Denote by $C(J', J'')$ a class of that partition. The accesses in each class $C(J', J'')$ are obtained by adding elements from $K \setminus \cup\mathcal{F}_m(K)$ to $\cap\mathcal{F}_m(K) + J' + J''$. These accesses can be odd or even. Actually, within each class $C(J', J'')$, the number of odd accesses is the same as the number of even accesses. Write $s = |K\setminus\cup\mathcal{F}_m(K)|$. (Thus, $s = 10 - 7 = 3$ in the above example.) It can be seen that $1 = \binom{s}{0}$ access can be formed by adding 0 elements to $\cap\mathcal{F}_m(K) + J' + J''$, $s = \binom{s}{1}$ accesses can be formed by adding one element to $\cap\mathcal{F}_m(K) + J' + J''$, etc. In general, there are $\binom{s}{l}$ accesses formed by adding l elements to $\cap\mathcal{F}_m(K) + J' + J''$. The equality of the numbers of odd and even accesses results from the identity (6.7). Since this holds for any class $C(J', J'')$ of the partition, the result follows. □

Objections can certainly be made to the assumption that the probability of mastering an item q does not depend on the current state K of the subject, provided that the item is learnable from that state, i.e. $q \in S(K)$. This seems very strong. The independence assumptions are also difficult to accept. It can be elaborated by assuming that the probability of mastering an item may depend upon past events, for example, upon the last item learned. This idea will not be developed here, however. A different kind of model is considered in the next section, in which a knowledge structure is regarded as the state space of a Markov chain describing the learning process.

6.3 Model 2

We assume that learning takes place in discrete steps. On any given step, at most one item is mastered. In the example above, a transition from state $\emptyset$ to state $\{1\}$ or to state $\{2\}$ may occur on step one or later, if neither of the two states, $\{1\}$ or $\{2\}$, has yet been achieved. The probabilities of such transitions are specified by parameters. It is assumed that these probabilities do not depend upon past events. Thus, the probability of a transition from state K to state $K+\{r\}$, with $r \in S(K)$, is equal to g_r. A sample of subjects tested at a given time is assumed to have accomplished some number m of steps. This number is a parameter which has to be estimated from the data. If m is large, the probabilities of states containing many items will also be large.

This model is a Markov chain having as a state space the knowledge structure $\mathcal{K}$, and transition probabilities defined by a matrix M. The states of the Markov chain coincide with the knowledge states. Such a matrix for the standard example is given below (cf. Figure 6.1 and Table 6.2). For simplicity, any knowledge state is represented by a string listing its elements. Also, since the matrix is quite large, the abbreviations

$$\overline{g_q} = 1 - g_q \quad \text{and} \quad \overline{g_{qr}} = 1 - g_q - g_r$$

are adopted.

The process begins with a vector v_o specifying the initial probabilities of the states. (Since the states of the Markov chains are confounded with the knowledge states, no clash of terminology can arise.) The probabilities after one step are thus given by the vector

$$v_1 = v_o\, M.$$

If, as may often seem sensible, the subjects are assumed to start the learning process in state $\emptyset$, the initial probability vector will take the form

$$v_0 = \underbrace{(1,0,\ldots,0)}_{|\mathcal{K}|\text{ terms}}.$$

If this assumption is incorporated in the present example, the probabilities of

TABLE 6.2. Transition matrix M of the Markov chain in Model 2, for our standard example

	$\emptyset$	1	2	12	13	23	134	124	123	234	Q
$\emptyset$	$\overline{g_{12}}$	g_1	g_2	0	0	0	0	0	0	0	0
1	0	$\overline{g_{23}}$	0	g_2	g_3	0	0	0	0	0	0
2	0	0	$\overline{g_{13}}$	g_1	g_3	0	0	0	0	0	0
12	0	0	0	$\overline{g_{34}}$	0	0	0	g_4	g_3	0	0
13	0	0	0	0	$\overline{g_{24}}$	0	g_4	0	g_2	0	0
23	0	0	0	0	0	$\overline{g_{14}}$	0	0	g_1	g_4	0
134	0	0	0	0	0	0	$\overline{g_2}$	0	0	0	g_2
124	0	0	0	0	0	0	0	$\overline{g_3}$	0	0	g_3
123	0	0	0	0	0	0	0	0	$\overline{g_4}$	0	g_4
234	0	0	0	0	0	0	0	0	0	$\overline{g_1}$	g_1
Q	0	0	0	0	0	0	0	0	0	0	1

the states after the first and the second step are

$$v_1 = (\overline{g_{12}}, g_1, g_2, 0, 0, 0, 0, 0, 0, 0, 0),$$

$$v_2 = (\overline{g_{12}}^2, \overline{g_{12}}g_1 + g_1\overline{g_{23}}, \overline{g_{12}}g_2 + g_2\overline{g_{13}}, g_1g_2, g_1g_3, g_2g_3, 0, 0, 0, 0, 0),$$

the first rows of M and $M^{(2)}$ respectively.

In general, the state probabilities after step k may be obtained from

$$v_k = v_o \, M^{(k)}.$$

Notice that the number of parameters, including the parameters β_q and η_q entering in the specification of the function ρ in (6.2), cannot exceed $3n$.

To some extent this model is vulnerable to the same kind of criticisms as those addressed to Model 1, namely, that the probabilities of learning new items do not depend upon past events. It is sufficient that an item r can be learned from the current state K, that is, $r \in S(K)$. However, there is an easy adaptation of this model which is discussed next.

6.4 Modcl 3

The main concepts of Model 2 are explained, except that the last item learned affects the probabilities of learning new items. This model also admits representation by a Markov chain, but it is now necessary to keep track of the last item learned. In other words, except for the empty set, the states of the Markov chain take the form of a pair (K, q), with

$$q \in R(K) = \{s \in \mathcal{K} | (K \setminus \{s\}) \in \mathcal{K}\}.$$

To avoid confusion, the states of this Markov chain will be referred to as *Markov-states.* The probability of a transition from Markov-state (K, q), with $q \in K$, to some Markov-state $(K+\{r\}, r)$ – this corresponds to a transition from knowledge state K to a knowledge state $K + \{r\}$ – is g_{qr}. This transition probability g_{qr} depends upon the state K, but only through q and r. For every Markov-state (K, q), the probabilities g_{qr} satisfy the constraint

$$\sum_{r \in S(K)} g_{qr} \leq 1 .$$

Set the probability of remaining in Markov-state (K, q) equal to $1 - \sum_{r \in S(K)} g_{qr}$. The probabilities of transition from Markov-state (K, q) to any other Markov-state (K', r), with $K' \neq K + \{r\}$, are zero. Needless to say, the empty set is also a Markov-state. The probability of transition from that state to state $(\{q\}, q)$ is g_{0q}.

In the example there are 23 Markov-states; these can be read from Figure 6.2. The full transition table of the Markov chain is too large for convenient display. I shall only give some example values. For simplicity, abbreviate the notation of the Markov-state (K, q) by listing the elements of K as a string, terminating with q. Thus,

$$(\{1,2,3\},2) \qquad \text{is abbreviated as} \qquad 132.$$

(The order of 1 and 3 in the string is arbitrary.)

Writing $P_{\sigma,\sigma'}$ for the probability of transition from the Markov-state σ to the Markov-state σ' gives, for instance,

$$P_{\emptyset,\emptyset} = 1 - g_{01} - g_{02}, \qquad P_{12,123} = g_{23},$$

$$P_{12,23} = 0, \qquad P_{23,234} = g_{34},$$

$$P_{234,2341} = g_{41}, \qquad \text{etc.}$$

6.5 Remarks on testing

Each one of these models specifies – up to the values of the response parameters β_q and η_q and of the parameters g – the probability of all the patterns of reponses in a particular situation. As an illustration, consider Model 1 applied to the standard example. To simplify matters, suppose that open responses have been required from the subjects, so that all the η_q may be set to zero. In this case a response pattern R can be elicited from a state K only if $R \subseteq K$. Equation (6.1) simplifies to

$$P(R) = \sum_{K \supseteq R, K \in \mathcal{K}} \rho(R, K) p(K) . \tag{6.8}$$

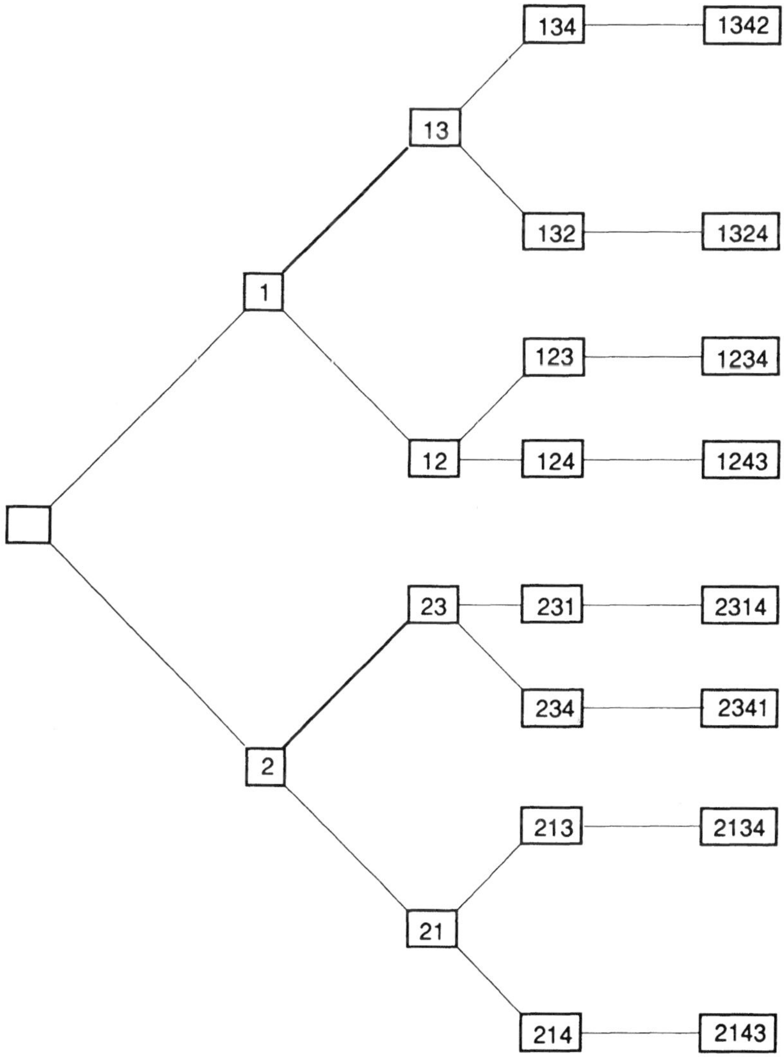

FIGURE 6.2. Transitions diagram in Model 3, for the standard example

For the knowledge structure $\mathcal{F}$ of (6.4), it follows from (6.8) and (6.2) that

$$\begin{aligned} P\{1,2,3\} &= \rho(\{1,2,3\},\{1,2,3\})p\{1,2,3\} + \rho(\{1,2,3\},\{1,2,3,4\}p\{1,2,3,4\} \\ &= (1-\beta_1)(1-\beta_2)(1-\beta_3)g_1g_2g_3(1-g_4) \\ &\qquad +(1-\beta_1)(1-\beta_2)(1-\beta_3)\beta_4g_1g_2g_3g_4 \,. \end{aligned}$$

These models can therefore be evaluated by standard statistical methods. The parameters could be estimated by maximizing the likelihood function, using an optimization routine. The models could then be tested by likelihood ratio or other techniques.

Models 2 and 3 are also capable of predicting the results of a sequence of tests. Suppose that the subjects in a sample have been tested twice (before and after some training period, for example). Each subject has thereby provided a pair (R_1, R_2) of response patterns, where R_1 and R_2 stand for the response patterns observed in the first and second test respectively. The predictions concern the joint probability $P(R_1, R_2)$ of each pair $(R_1, R_2) \in 2^Q \times 2^Q$. The calculation outlined below indicates how these predictions could be obtained in the case of either of the two models.

Write K_1 and K_2 for the knowledge states at the time of tests 1 and 2, and $P(K_1, K_2)$ for the corresponding joint probability of these states. Assuming that

$$P(R_1, R_2 | K_1, K_2) = \rho(R_1, K_1)\rho(R_2, K_2),$$

it follows, with obvious justifications,

$$\begin{aligned} P(R_1, R_2) &= \textstyle\sum_{(K_1,K_2)\in\mathcal{K}\times\mathcal{K}} P(R_1, R_2|K_1, K_2)P(K_1, K_2) \\ &= \textstyle\sum_{(K_1,K_2)\in\mathcal{K}\times\mathcal{K}} \rho(R_1, K_1)\rho(R_2, K_2)P(K_2|K_1)P(K_1)\,. \end{aligned} \tag{6.9}$$

For both models, explicit expressions can be derived for each of the quantities entering in (6.9). These expressions depend, of course, on all the parameters in the models. In particular, the numbers of steps of the Markov chain should be treated as free parameters. In Model 2, for example, $P(K_1)$ would be the probability that the Markov chain is in state K_1 after exactly k steps[3]. Similarly, $P(K_2|K_1)$ would be the probability that the Markov chain moves from K_1 to K_2 in exactly m steps, where both k and m have to be estimated from the data. Obviously, analogous derivations can be worked out for situations in which more than two successive tests have been performed.

Some objections to the models discussed in this paper have already been raised. Their most blatant weakness is probably the absence of parameters representing individual subjects. Nevertheless, the application of these models may

[3] We recall that, in Model 2, the knowledge states are confounded with the states of the Markov chain.

prove to be both informative and cost-effective as a first-pass analysis of complex data. Notice, in particular, that it is possible to apply Model 3 while assuming that *all* subsets of Q are feasible knowledge states. In the case of an 8-item test applied once, there are 2^8 possible response patterns. This gives 255 degrees of freedom in the data, greatly exceeding the number of parameters in Model 3: 7×8 parameters $g_{q,q'}$, and 16 response parameters β_q and η_q, that is, 72 parameters overall. Application of this model may reveal that the estimates of some of the parameters $g_{q,q'}$ are negligible, pointing to a knowledge structure with some smaller number of knowledge states. The next step would then be to reanalyze the data using more realistic representations, such as those proposed by the author elsewhere (Falmagne, 1989, in press).

6.6 References

Doignon, J.-P., & Falmagne, J.-Cl. (1985). Spaces for the assessment of knowledge. *International Journal of Man-Machine Studies, 23,* 175–196.

Falmagne, J.-Cl. (1989). A latent trait theory via stochastic learning theory for a knowledge space. *Psychometrika, 53,* 283–303.

Falmagne, J.-Cl. (in press). Stochastic learning paths in knowledge structures. *Journal of Mathematical Psychology.*

Falmagne, J.-Cl., Koppen, M., Villano, M., Doignon, J.-P., & Johannesen, L. (1990). Introduction to knowledge spaces: How to build, test and search them. *Psychological Review, 97 (2),* 201–224.

Kambouri, M., Koppen, M., Villano, M., & Falmagne, J.-Cl. Knowledge assessment: Tapping human expertise by the QUERY routine. *Technical Report MBS 91-20, Irvine Research Unit in Mathematical Behavioral Sciences.* Irvine: University of California. (Submitted for publication.)

Lord, F.M., & Novick, M.R. (1974). *Statistical theories of mental tests scores.* (2nd. ed.) Reading, MA: Addison Wesley.

Villano, M. (1991). *Computerized knowledge assessment: Building the knowledge structure and calibrating the assessment routine.* Doctoral dissertation. New York: New York University.

7

Stochastic Learning Paths – Estimation and Simulation

J.-Cl. Falmagne[1] and K. Lakshminarayan[1]

ABSTRACT The focus of this paper is a stochastic theory (due to Falmagne, in press) capable of describing in detail the progress of students learning a particular field. This theory provides explicit predictions for the joint probabilities of observing, at times $t_1 < t_2 < \ldots < t_n$, the successive response patterns $R_1, R_2, \ldots, R_n$. The main concern of this paper is the feasibility of parameter estimation. The difficulties are that the number of parameters is large (the data to be explained are rich and intricate), and also that no closed form or approximation is currently available for a particular form of an "n-dimensional" beta integral playing a central role in the predictions. It is nevertheless shown, through simulated data, that very accurate estimates of the parameters can be obtained in exemplary, representative cases. The estimation procedure uses a combination of standard approximations and numerical integration. These results indicate that the model is applicable in realistic situations.

This paper is devoted to a stochastic theory providing a detailed description of the progress of students through some educational material. We suppose that students sampled from a specified population have been tested repeatedly, at times $t_1 < t_2 < \ldots < t_n$, using the same set Q of questions or problems. Each of the problems is assumed to have a unique correct response. The test result of a student at a given time can be represented by the subset $R \subseteq Q$ of all the items correctly responded to. For any $t > 0$, we denote by $\boldsymbol{R}_t$ a random variable specifying the response pattern observed at time t. Thus, $\boldsymbol{R}_t$ takes its values in 2^Q. The main purpose of this theory is to predict the joint probabilities

$$\mathbb{P}(\boldsymbol{R}_{t_1} = R_1, \boldsymbol{R}_{t_2} = R_2, \ldots, \boldsymbol{R}_{t_n} = R_n) \tag{7.1}$$

of observing the response patterns $R_1, R_2, \ldots, R_n$ at times $t_1 < t_2 < \ldots < t_n$. The basic idea is that a succession of observable response patterns can be explained by the transitions of a subject through the states of a knowledge structure in the sense of Doignon and Falmagne (1985; for a review of this work, see Falmagne, Koppen, Villano, Doignon and Johanessen, 1990). Only the main features of the stochastic theory are presented here. (The detailed derivations have been given in Falmagne, in press.) In particular, no proofs will be found in this paper. Our main purpose is to discuss a number of technical issues related

---

[1]University of California, Irvine, CA 92717, USA
Funding for this work has been provided by NSF Grant IRI 8919068 to Jean-Claude Falmagne at the University of California, Irvine. The authors are grateful to the editors and to one referee for their comments, and also to Bill Batchelder and Josh Stern for their reactions to previous presentations of this material. They also thank the UCI office of academic computing for the use of the Convex C240.

to parameter estimation and to simulation. We begin by briefly recalling some of the basic concepts of knowledge space theory.

7.1 Basic concepts

Definition 1 *A pair* $(Q, \mathcal{K})$ *is a knowledge structure if* Q *is a set of questions or items, and* $\mathcal{K}$ *is a family of subsets of* Q *containing the empty set* $\emptyset$ *and the set* Q *itself. Since* $\cup\mathcal{K} = Q$, *we will occasionally, without ambiguity, refer to the family* $\mathcal{K}$ *as the knowledge structure. The set* Q *is the domain of the knowledge structure. The elements of* $\mathcal{K}$ *are called (knowledge) states. In this paper, only finite domains will be considered. A knowledge structure* $\mathcal{K}$ *which is closed under union* ($K, K' \in \mathcal{K}$ *implies* $K \cup K' \in \mathcal{K}$) *is a (knowledge) space. A learning path is a maximal chain of states, that is, a collection* ν *of states ordered by inclusion, containing* $\emptyset$ *and* Q, *and such that if* $K \subset K''$ *are two immediately successive states in the chain* ν, *then there is no state* K' *such that* $K \subset K' \subset K''$. *A learning path in which any two immediately successive states* $K \subset K'$ *always differ by exactly one question* ($|K|+1 = |K'|$) *is called a gradation. A knowledge structure in which all learning paths are gradations is said to be well graded.*

An example of a knowledge structure for the domain

$$Q = \{1, 2, 3, 4, 5\} \tag{7.2}$$

is the collection of subsets

$$\begin{aligned}\mathcal{F} = \ & \{\emptyset, \{5\}, \{1,5\}, \{2,5\}, \{4,5\}, \{1,2,5\}, \{1,4,5\}, \{3,4,5\}, \\ & \{1,2,4,5\}, \{1,3,4,5\}, \{1,2,3,4,5\}\}\,.\end{aligned} \tag{7.3}$$

This example will be used throughout the paper, and is pictured in Figure 7.1.

Notice that $\mathcal{F}$ is not a knowledge space: $\{2,5\}$ and $\{4,5\}$ are states, while $\{2,4,5\}$ is not. The knowledge structure $\mathcal{F}$ is well graded, with seven gradations. With obvious abbreviations, these gradations are

$$54312, \quad 54132, \quad 54123, \quad 51432, \quad 51423, \quad 51243, \quad 52143\,.$$

(Thus, 54312 represents the gradation $\emptyset \subset \{5\} \subset \{4,5\} \subset \{3,4,5\} \subset \{1,3,4,5\} \subset Q$.) Scanning Figure 7.1 from left to right, we consider the learning history of a subject, sampled from a specified population and learning at rate $\lambda > 0$, this being the value of some random variable $\boldsymbol{L}$ associated with the population. (The density of that random variable is represented by the graph in the upper left corner of the figure.) We also suppose the learning path of the subject to be the gradation 54312 indicated by the arrows linking the boxes representing the states on the figure. The choice of this particular learning path is random, and governed by a probability distribution assigning a probability p_ν to each learning path ν.

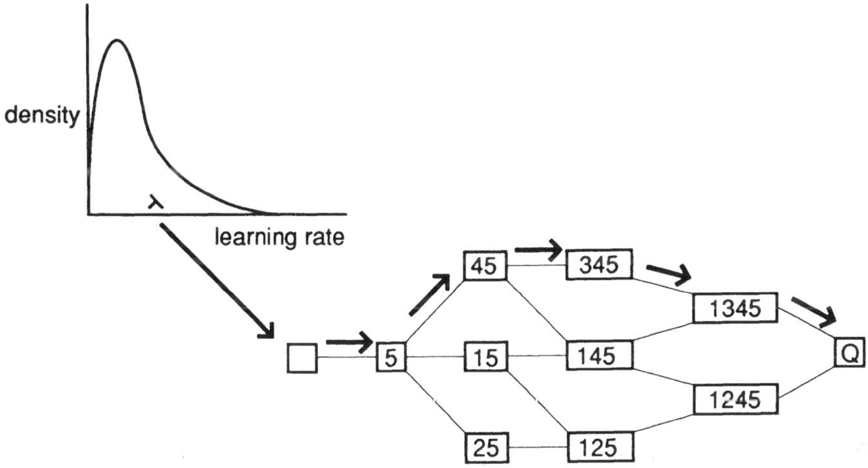

FIGURE 7.1. The knowledge structure $\mathcal{F}$ of (6.3), the learning rate density, and the learning path 54312

At first, the subject is naive and begins the learning process in state $\emptyset$. The second state encountered is $\{5\}$, and we suppose that the time required to master item 5 is a random variable $\boldsymbol{T}_{5,\lambda}$, depending on the item and on the learning rate of the subject. The second question to be mastered is 4, which requires a random time $\boldsymbol{T}_{4,\lambda}$. Then come, successively 3, 1 and 2, with their associated random times $\boldsymbol{T}_{i,\lambda}$, $i = 1, 2, 3$. The probability that this subject, following this particular path, is in state $\{4, 5\}$ at some time t of a test is

$$\begin{aligned} &\mathbb{P}(\boldsymbol{T}_{5,\lambda} + \boldsymbol{T}_{4,\lambda} \leq t < \boldsymbol{T}_{5,\lambda} + \boldsymbol{T}_{4,\lambda} + \boldsymbol{T}_{3,\lambda}) \\ &= \mathbb{P}(\boldsymbol{T}_{5,\lambda} + \boldsymbol{T}_{4,\lambda} \leq t) - \mathbb{P}(\boldsymbol{T}_{5,\lambda} + \boldsymbol{T}_{4,\lambda} + \boldsymbol{T}_{3,\lambda} \leq t)\,. \end{aligned} \tag{7.4}$$

Distributional assumptions are required to derive explicit predictions for the joint probabilities of (7.1). We suppose that the learning rate random variable $\boldsymbol{L}$ has a gamma distribution, with density f and parameters $\alpha > 0$ and $\xi > 0$; that is:

$$f(\lambda) = \begin{cases} \frac{\xi^\alpha}{\Gamma(\alpha)}\lambda^{\alpha-1}e^{-\xi\lambda} & \text{for } \lambda > 0, \\ 0 & \text{for } \lambda \leq 0, \end{cases} \tag{7.5}$$

with $\Gamma(\alpha) = \int_0^\infty y^{\alpha-1}\, e^{-y}\, dy$. Thus, Γ is the standard gamma function, and $E(\boldsymbol{L}) = \frac{\alpha}{\xi}$ and $Var(\boldsymbol{L}) = \frac{\alpha}{\xi^2}$. We also suppose that each of the learning time random variables has a gamma distribution, with respective parameters μ_i ($i = 1, \ldots, 5$) and λ. Thus, $E(\boldsymbol{T}_{i,\lambda}) = \frac{\mu_i}{\lambda}$ and $Var(\boldsymbol{T}_{i,\lambda}) = \frac{\mu_i}{\lambda^2}$. All these random variables are assumed to be independent. In view of the additivity property of gamma random variables, the right-hand member of 7.4 becomes

$$\int_0^t \frac{\lambda^{\mu_5+\mu_4} e^{-\lambda\tau} \tau^{\mu_5+\mu_4-1}}{\Gamma(\mu_5+\mu_4)} d\tau - \int_0^t \frac{\lambda^{\mu_5+\mu_4+\mu_3} e^{-\lambda\tau} \tau^{\mu_5+\mu_4+\mu_3-1}}{\Gamma(\mu_5+\mu_4+\mu_3)} d\tau . \tag{7.6}$$

To obtain the probability that *any* subject following gradation 54312 (whatever the learning rate) be in state $\{4,5\}$ at time t, we have to average over λ in (7.6).

With $x, y, t, \xi > 0$ and $\theta = \frac{t}{t+\xi}$, we write

$$\begin{aligned} B(x,y) &= \frac{\Gamma(x)\Gamma(y)}{\Gamma(x+y)} , \\ I_\theta(x,y) &= \frac{1}{B(x,y)} \int_0^\theta \tau^{x-1}(1-\tau)^{y-1}\, d\tau = \frac{1}{B(x,y)} \int_0^t \frac{\tau^{x-1}\xi^y}{(\tau+\xi)^{x+y}}\, d\tau \end{aligned} \tag{7.7}$$

as usual, for the *beta* function, and for the *incomplete beta function ratio*, respectively (Johnson and Kotz, 1970). After multiplying in (7.6) by the density of $\boldsymbol{L}$ in (7.5), and integrating over λ, we obtain by a standard result (Johnson and Kotz, 1970, vol. 1, p. 195),

$$I_{\frac{t}{t+\xi}}(\mu_5+\mu_4, \alpha) - I_{\frac{t}{t+\xi}}(\mu_5+\mu_4+\mu_3, \alpha) . \tag{7.8}$$

The probability that any subject is in state $\{4,5\}$ at time t (whatever path is taken) is the weighted sum of three such probabilities, since three learning paths go through state $\{4,5\}$, the weights being the probabilities p_ν of the paths ν. The probabilities of the subject being in any other state at time t can be derived by similar computations. As an example, take any state $K \neq \emptyset, Q$. Denoting by K^ν the state immediately following K along learning path ν and writing $\mathcal{C}(K)$ for the set of all learning paths ν going through K, and $\boldsymbol{K}_t$ for the state of the subject at time t, we have, with $\mu_K = \sum_{i\in K} \mu_i$,

$$\mathbb{P}(\boldsymbol{K}_t = K) = \sum_{\nu\in\mathcal{C}(K)} p_\nu [I_{\frac{t}{t+\xi}}(\mu_K, \alpha) - I_{\frac{t}{t+\xi}}(\mu_{K^\nu}, \alpha)] . \tag{7.9}$$

The probability of a particular response pattern R at time t is obtained from

$$\mathbb{P}(\boldsymbol{R}_t = R) = \sum_{K\in\mathcal{F}} \mathbb{P}(\boldsymbol{R}_t = R | \boldsymbol{K}_t = K) \mathbb{P}(\boldsymbol{K}_t = K) . \tag{7.10}$$

In this equation, the state probabilities are computed as in Example (7.9) above, and the conditional probabilities are defined by parameters, in the spirit of the 'local independence' condition of psychometric theory (Lord & Novick, 1974). Specifically, for each question i, $\beta_i \in [0,1[$ is the probability of a careless error in responding to some question i belonging to the subject's state. Similarly,

$\eta_i \in [0,1[$ is the probability of a correct response to a question not contained in the subject's state. Conditional on the subject's state, the responses to the questions are independent events. We thus have, for instance,

$$\mathbb{P}(\boldsymbol{R}_t = \{1,3\} | \boldsymbol{K}_t = \{3,4,5\}) = \eta_1(1-\eta_2)(1-\beta_3)\beta_4\beta_5 .$$

The derivations in the general case of n tests, as in (7.1), are similar, but a more systematic development is required. After reviewing our notation in the next section, the stochastic theory will be specified axiomatically for the case of an arbitrary number of questions. A slight generalization will be introduced in this formulation. In the example knowledge structure $\mathcal{F}$ of (7.3), all the learning paths were gradations: the subject was assumed to master the items successively, each new item giving rise to a new knowledge state. This assumption is abandoned in the axioms given below, which allow for the possibility that some items might be learned 'as a package', in order to reach the next state along the learning path followed. Further remarks can be found after the axioms.

Glossary

$(Q,\mathcal{K})$	a knowledge structure;
$\boldsymbol{K}_t = K \in \mathcal{K}$	specifies that K is the state at time $t \geq 0$;
$\boldsymbol{R}_t = R \in 2^Q$	specifies the set R of correct responses given at time $t \geq 0$;
$\boldsymbol{T}_{i,\lambda}$	time taken for student with learning rate λ to master question i;
μ_i	difficulty of question $i \in Q$;
$\boldsymbol{L} = \lambda$	the learning rate is equal to some number $\lambda \geq 0$;
$\mathcal{C}$	the collection of all learning paths;
$\mathcal{C}(K, K', \ldots)$	the collection of all learning paths containing $K, K', \ldots$;
$\boldsymbol{C} = \nu \in \mathcal{C}$	the learning path of the subject is ν;
p_ν	the probability of learning path $\nu \in \mathcal{C}$: $\mathbb{P}(\boldsymbol{C} = \nu) = p_\nu$;
β_i	probability of a careless error to question $i \in Q$;
η_i	probability of a lucky guess on item i.

Axioms

[A] **Learning Ability.** The random variable $\boldsymbol{L}$ measuring the learning rate has a gamma distribution with parameters $\alpha > 0$ and $\xi > 0$, and a density function f specified by (7.5). We have thus $E(\boldsymbol{L}) = \frac{\alpha}{\xi}$ and $Var(\boldsymbol{L}) = \frac{\alpha}{\xi^2}$.

[I] **Independence of Learning Path and Learning Rate.** The choice of a learning path is governed by a random variable $\boldsymbol{C}$, independent of the learning rate. That is, for all $\lambda \in \mathbb{R}$ and $\nu \in \mathcal{C}$,

$$\mathbb{P}(\boldsymbol{L} \leq \lambda, \boldsymbol{C} = \nu) = \mathbb{P}(\boldsymbol{L} \leq \lambda)\,\mathbb{P}(\boldsymbol{C} = \nu) .$$

[R] **Response Rule** (*Local independence.*) There is a function $r : \mathcal{K} \times 2^Q \mapsto [0,1]$ such that, for all $\nu \in \mathcal{C}$, all positive integers n, all real numbers λ and $t_n > t_{n-1} > \ldots > t_1 \geq 0$, and any event $\mathcal{E}$ determined only by

$$((\boldsymbol{K}_{t_1}, \boldsymbol{R}_{t_1}), (\boldsymbol{K}_{t_2}, \boldsymbol{R}_{t_2}), \ldots, (\boldsymbol{K}_{t_{n-1}}, \boldsymbol{R}_{t_{n-1}})),$$

we have

$$\begin{aligned} \mathbb{P}(\boldsymbol{R}_{t_n} = R_n | \boldsymbol{K}_{t_n} = K_n, \mathbf{E}, \boldsymbol{L} = \lambda, \boldsymbol{C} = \nu) &= \mathbb{P}(\boldsymbol{R}_{t_n} = R_n | \boldsymbol{K}_{t_n} = K_n) \\ &= r(K_n, R_n), \end{aligned}$$

with, moreover, for any $(K, R) \in \mathcal{K} \times 2^Q$,

$$r(K,R) = \left[\prod_{i \in K \setminus R} \beta_i\right]\left[\prod_{i \in K \cap R} (1-\beta_i)\right]\left[\prod_{i \in R \setminus K} \eta_i\right]\left[\prod_{i \in \overline{R \cup K}} (1-\eta_i)\right],$$

in which the complement $\overline{R \cup K}$ in the last factor is taken with respect to the domain Q.

[B] **Beginning State.** For all $\nu \in \mathcal{C}$ and $\lambda \in \mathbb{R}$,

$$\mathbb{P}(\boldsymbol{K}_0 = \emptyset | \boldsymbol{L} = \lambda, \boldsymbol{C} = \nu) = 1.$$

[L] **Learning Rule.** There is a function $\ell : \mathcal{K} \times \mathcal{K} \times \mathbb{R}^2 \times \mathcal{C} \mapsto [0,1]$, such that, for all $\nu \in \mathcal{C}$, all positive integers n, all real numbers λ and $t_{n+1} > t_n > \ldots > t_1 \geq 0$, and any event $\mathcal{E}$ determined only by

$$((\boldsymbol{K}_{t_1}, \boldsymbol{R}_{t_1}), (\boldsymbol{K}_{t_2}, \boldsymbol{R}_{t_2}), \ldots, (\boldsymbol{K}_{t_{n-1}}, \boldsymbol{R}_{t_{n-1}})),$$

we have

$$\begin{aligned} &\mathbb{P}(\boldsymbol{K}_{t_{n+1}} = K_{n+1} | \boldsymbol{K}_{t_n} = K_n, \boldsymbol{R}_n = R_n, \mathcal{E}, \boldsymbol{L} = \lambda, \boldsymbol{C} = \nu) \\ &= \mathbb{P}(\boldsymbol{K}_{t_{n+1}} = K_{n+1} | \boldsymbol{K}_{t_n} = K_n, \boldsymbol{L} = \lambda, \boldsymbol{C} = \nu) \\ &= \ell(K_n, K_{n+1}, t_{n+1} - t_n, \lambda, \nu). \end{aligned}$$

Moreover, the function ℓ is assumed to satisfy the following two conditions: For any $K, K' \in \mathcal{K}$, $\nu, \nu' \in \mathcal{C}$, $\delta > 0$ and any $\lambda \in \mathbb{R}$,

$$\ell(K, K', \delta, \lambda, \nu) = 0 \quad \text{if} \quad \nu \notin \mathcal{C}(K, K'); \qquad (L1)$$

$$\ell(K, K', \delta, \lambda, \nu) = \ell(K, K', \delta, \lambda, \nu') \quad \text{if} \quad \nu, \nu' \in \mathcal{C}(K, K'). \quad (L2)$$

[T] **Learning Times.** We assume that, for a subject with learning rate λ, the time required for the mastery of item i (this item being accessible from that subject's current state) is a random variable $\boldsymbol{T}_{i,\lambda}$, which has a gamma distribution, with parameters μ_i and λ. Thus, $E(\boldsymbol{T}_{i,\lambda}) = \frac{\mu_i}{\lambda}$ and $Var(\boldsymbol{T}_{i,\lambda}) = \frac{\mu_i}{\lambda^2}$. All these random variables (for all $i \in Q$ and $\lambda > 0$) are assumed to be independent. The items are mastered successively.

This means that the total time to master items $i, i', \ldots$ is the sum $\boldsymbol{T}_{i,\lambda} + \boldsymbol{T}_{i',\lambda} + \ldots$. Formally, for any positive real numbers δ and λ, any learning path $\nu \in \mathcal{C}$, and any two states $K, K' \in \nu$, we have

$$\ell(K, K', \delta, \lambda, \nu) =$$

$$= \begin{cases} \mathbb{P}(\sum_{i \in K^\nu \setminus K} \boldsymbol{T}_{i,\lambda} > \delta) & \text{if } K = K' \neq Q \\ \mathbb{P}(\sum_{i \in K' \setminus K} \boldsymbol{T}_{i,\lambda} \leq \delta) - \mathbb{P}(\sum_{i \in K'^\nu \setminus K} \boldsymbol{T}_{i,\lambda} \leq \delta) & \text{if } K \subset K' \neq Q \\ \mathbb{P}(\sum_{i \in K' \setminus K} \boldsymbol{T}_{i,\lambda} \leq \delta) & \text{if } K \subset K' = Q \\ 1 & \text{if } K = K' = Q \\ 0 & \text{in all other cases} \end{cases}$$

$$= \begin{cases} \displaystyle\int_\delta^\infty \frac{\lambda^{\mu_{K^\nu \setminus K}} e^{-\lambda\tau} \tau^{\mu_{K^\nu \setminus K} - 1}}{\Gamma(\mu_{K^\nu \setminus K})}\, d\tau & \text{if } K = K' \neq Q \\ \displaystyle\int_0^\delta \left[\frac{\lambda^{\mu_{K' \setminus K}} e^{-\lambda\tau} \tau^{\mu_{K' \setminus K} - 1}}{\Gamma(\mu_{K' \setminus K})} - \frac{\lambda^{\mu_{K'^\nu \setminus K}} e^{-\lambda\tau} \tau^{\mu_{K'^\nu \setminus K} - 1}}{\Gamma(\mu_{K'^\nu \setminus K})} \right] d\tau & \text{if } K \subset K' \neq Q \\ \displaystyle\int_0^\delta \frac{\lambda^{\mu_{K' \setminus K}} e^{-\lambda\tau} \tau^{\mu_{K' \setminus K} - 1}}{\Gamma(\mu_{K' \setminus K})}\, d\tau & \text{if } K \subset K' = Q \\ 1 & \text{if } K = K' = Q \\ 0 & \text{in all other cases.} \end{cases}$$

Remarks

1) The reader may be reluctant to accept the notion that the learning path followed by a subject is governed by a probability distribution p, with each of the p_ν being a parameter to be estimated from the data (subject only to the constraint that the p_ν sum to 1). In principle, there may be many learning paths, rendering the number of such parameters too large for practical applications of the theory. In fact, in turns out that, at least in some cases, the number of learning paths is actually quite small. Morover, the matter can be simplified by Markovian assumptions regarding the transitions between states (c.f. Falmagne,

in press). The principle of these assumptions is that the probability of a transition from any state K to a state $K \cup \{q\}$ should depend only on q and K. The probability of a given path would then be the product of successive transition probabilities. Note that the model in which the probabilities are directly attached to the path is the most general case. More specific assumptions along these lines would also be considered, which could further reduce the number of parameters required to specify the probabilities of the learning paths. We return to this issue in the discussion.

2) The independence axiom [I] is debatable. There has to be some relation between the learning rate and the school chosen by the student. In turn, the school will exert a strong influence on the learning path. However, this axiom renders the theory manageable and should be abandoned only for good reasons. Hopefully, the predictions will turn out to be robust in this respect.

3) Regarding the response rule Axiom [R], in many cases, the design of the items may preclude educated (or lucky) guesses of the correct responses, justifying the assumption that the parameters η_i are all zero.

4) Axioms [B], [L] and [T] taken together specify the stochastic process $((\boldsymbol{K}_t, \boldsymbol{R}_t)_{t\geq 0}, \nu, \lambda)$ describing the progress of a particular subject equipped with a learning rate λ, along learning path ν. These axioms have been discussed at length earlier in this paper. Axiom [B] initiates the process. It may be objected that one does not necessarily know when learning begins. However, in such a case the time t_1 of the first test may be taken to be a parameter to be estimated from the data. Thus, the data itself provide an estimate of the beginning of learning. This remark can be generalized. In some situations (for example, when some intensive training has been given to the subjects), it will make sense to consider some of the times t_i as parameters. A large value estimated to such a parameter would be indicative of an efficient training period.

5) The assumption of gamma distributions for the learning times are simply convenient. Moreover, there is some indication that the predictions may be robust with respect to distributional assumptions (Falmagne, 1989). The assumption of independence between the random variables $\boldsymbol{T}_{i,\lambda}$ is not particularly significant; whatever dependence there may be in nature can be accommodated by the common parameter λ which enters into the specification of these random variables.

6) Applications of this theory in the case of a single test ($n = 1$) can be found in Falmagne et al. (1990), and Taagepera et al. (submitted for publication).

The remainder of this paper is concerned with practical issues of parameter estimation. We only consider two successive tests, and the derivations from the axioms given in the next section deal only with this case. General results, proofs, and further remarks on the axioms can be found in Falmagne (in press).

7.2 Predictions for the case of two successive tests: $n = 2$ in (6.1)

We begin with a result relating the responses precisely to the states. (For proofs, see Falmagne, in press.)

Theorem 1 *For any real numbers $t \geq 0$, $\delta > 0$, and any $R, R' \subseteq Q$,*

$$\mathbb{P}(\boldsymbol{R}_t = R, \boldsymbol{R}_{t+\delta} = R') = \sum_{(K,K') \in \mathcal{K} \times \mathcal{K}} r(R,K)\, r(R',K')\, \mathbb{P}(\boldsymbol{K}_t = K, \boldsymbol{K}_{t+\delta} = K').$$

The values of the function r are specified by Axiom [R] in terms of the parameters β_i and η_i. Thus, the joint probabilities of the response patterns at times t, $t+\delta$ will be obtained from the joint probabilities $\mathbb{P}(\boldsymbol{K}_t = K, \boldsymbol{K}_{t+\delta} = K')$ of all the states at those times. We adopt the following notation for the distribution function of the gamma random variable with parameters μ_S and λ:

$$H_{S,\lambda}(\delta) = \int_0^\delta \frac{\lambda^{\mu_S} e^{-\lambda\tau} \tau^{\mu_S - 1}}{\Gamma(\mu_S)}\, d\tau, \tag{7.11}$$

for any non empty $S \subseteq Q$ and any real numbers $\mu_S, \lambda, \delta \geq 0$. By convention, we shall also write:

$$H_{\emptyset,\lambda}(\delta) = 1, \qquad \text{and} \qquad H_{Q^\nu \setminus S,\lambda}(\delta) = 0. \tag{7.12}$$

The general result for the joint probabilities of the states in the $n = 2$ case is then as follows:

Theorem 2 *Writing f as in (7.5) for the gamma density function of the random variable $\boldsymbol{L}$, there are, for states $\emptyset \subseteq K \subseteq K' \subseteq Q$, and any real numbers $t \geq 0$, $\delta > 0$,*

$$\mathbb{P}(\boldsymbol{K}_t = K, \boldsymbol{K}_{t+\delta} = K')$$

$$= \sum_{\nu \in \mathcal{C}(K,K')} p_\nu \left\{ \int_0^\infty [H_{K,\lambda}(t) - H_{K^\nu,\lambda}(t)][H_{K'\setminus K,\lambda}(\delta) - H_{K'^\nu\setminus K,\lambda}(\delta)]\, f(\lambda)\, d\lambda \right\}. \tag{7.13}$$

The first bracketed factor under the integral specifies, for a subject with learning rate λ, the probability of reaching state K but not going beyond it (i.e., to K^ν), in t units of time. The second bracketed expression similarly concerns a transition from K to K' in δ units of time. The integral and the summation compute the averages over learning rates and learning paths, respectively and successively. Analyzing the condition $\emptyset \subseteq K \subseteq K' \subseteq Q$ of this theorem yields 7 different types of situations, depending on which, if any, of the 3 inclusions are strict. On the other hand, multiplying the two bracketed factors in (7.13) gives a sum of 4 different expressions, each one of them containing an integral of a product of two gamma distribution functions H. Thus, there are 28 cases overall. They fall in four categories. Two of them are trivial:

(i) Each of the two distribution functions H takes value 1. One of the 28 cases corresponds to this description, that with $K = K' = \emptyset$,

$$\int_0^\infty H_{K,\lambda}(t)H_{K'\setminus K,\lambda}(\delta)\, f(\lambda)\, d\lambda = \int_0^\infty f(\lambda)\, d\lambda = 1 .$$

(ii) One of the two functions H vanishes, which implies that the integral also vanishes. This happens in 7 of the 28 cases.

The two remaining categories are more difficult.

(iii) One of the two functions H takes the value 1, and the other one does not vanish. Eight of the 28 cases fall in this category. Integration yields the incomplete beta function ratio. We have, for instance,

$$\begin{aligned}\int_0^\infty H_{\emptyset,\lambda}(t)H_{\emptyset^\nu\setminus\emptyset,\lambda}(\delta)f(\lambda)\, d\lambda &= \int_0^\infty H_{\emptyset^\nu,\lambda}(\delta)f(\lambda)\, d\lambda \\ &= \int_0^\infty\int_0^\delta \frac{\lambda^{\mu_{\emptyset^\nu}} e^{-\lambda\tau}\tau^{\mu_{\emptyset^\nu}-1}}{\Gamma(\mu_{\emptyset^\nu})}\cdot\frac{\xi^\alpha\lambda^{\alpha-1}e^{-\xi\lambda}}{\Gamma(\alpha)}\, d\tau\, d\lambda \\ &= I_{\frac{\delta}{\delta+\xi}}(\mu_{\emptyset^\nu}, \alpha) .\end{aligned}$$

(iv) In the remaining 12 cases, neither of the two functions H takes the value 1 or 0. For example,

$$\int_0^\infty H_{K,\lambda}(t)H_{K'\setminus K,\lambda}(\delta)f(\lambda)\, d\lambda \tag{7.14}$$

$$= \int_0^\infty\left[\int_0^t \frac{\lambda^{\mu_K}e^{-\lambda\tau}\tau^{\mu_K-1}}{\Gamma(\mu_K)}\, d\tau \int_0^\delta \frac{\lambda^{\mu_{K'\setminus K}}e^{-\lambda v}v^{\mu_{K'\setminus K}-1}}{\Gamma(\mu_{K'\setminus K})}\, dv\right]\frac{\xi^\alpha\lambda^{\alpha-1}e^{-\xi\lambda}}{\Gamma(\alpha)}\, d\lambda \tag{7.15}$$

$$= \frac{1}{\Gamma(x)\Gamma(y)\Gamma(\alpha)}\int_0^t\int_0^\delta \tau^{x-1}v^{y-1}\xi^\alpha\left[\int_0^\infty e^{-\lambda(\tau+v+\xi)}\lambda^{x+y+\alpha-1}\, d\lambda\right] d\tau\, dv . \tag{7.16}$$

The last equation results from setting $x = \mu_K$ and $y = \mu_{K'\setminus K}$. This integral depends upon the parameters x, y and α, and upon the ratios $\frac{t}{\xi}$ and $\frac{\delta}{\xi}$. To see this, we first apply the transformation $\lambda \mapsto \frac{\lambda}{\tau+\upsilon+\xi}$ and integrate over λ. The right-hand member of (7.16) becomes

$$\frac{\Gamma(x+y+\alpha)}{\Gamma(x)\Gamma(y)\Gamma(\alpha)} \int_0^t \int_0^\delta \frac{\tau^{x-1}\upsilon^{y-1}\xi^\alpha}{(\tau+\upsilon+\xi)^{x+y+\alpha}} \, d\tau \, d\upsilon \,, \tag{7.17}$$

yielding a 2-dimensional form of the incomplete beta function ratio, cf. (7.7). Next, we apply the two transformations $\tau \mapsto \frac{\tau}{\tau+\xi} = z$ and $\upsilon \mapsto \frac{\upsilon}{\upsilon+\xi} = u$ and finally obtain, after rearranging (the details are omitted),

$$\frac{\Gamma(x+y+\alpha)}{\Gamma(x)\Gamma(y)\Gamma(\alpha)} \int_0^{\frac{t}{t+\xi}} \int_0^{\frac{\delta}{\delta+\xi}} \frac{z^{x-1}u^{y-1}(1-z)^{y+\alpha-1}(1-u)^{x+\alpha-1}}{(1+zu)^{x+y+\alpha}} \, du \, dz \,. \tag{7.18}$$

Similar results are obtained for the 11 remaining cases in this category.

The integrals of Category (iii) do not create any difficulty, since very good approximations are available for the incomplete beta function ratio, which enters in many scientific applications. We used the approximation of the International Mathematical and Statistical Library (IMSL) which is available on the Convex C240 computer where our analysis was performed. Another approximation is given in Johnson & Kotz (1970). Notice in passing that the application of this theory in the case of a single test ($n = 1$) only involves the integrals of Category (iii). Such an application is described in Falmagne et al. (1990). The situation is less favorable in the case of the integrals of Category (iv), which must be evaluated in the case $n = 2$. As far as we know, no approximation is currently available for the integrals (7.17)–(7.18). In some applications, these integrals may be computed numerically. However, this is not practical for the purpose of parameter estimation, which typically requires several thousand iterated computations of all the predictions. In the next section, we describe a procedure using a combination of numerical integration and approximation which, while computationally intensive, appears to be feasible in practice.

The appropriate data for such an analysis consists of the frequencies of all the pairs of response patterns obtained from a sample of subjects, each of which has been tested at times t and $t + \delta$. We denote by $N(R, R')$, with $R, R' \subseteq Q$, the number of subjects who have produced the two patterns of responses R and R' at times t and $t + \delta$, respectively. The total number of subjects tested is thus

$$N = \sum_{R,R' \subseteq Q} N(R, R') \,.$$

The predictions concern the set of frequencies $N(R, R')$.

7.3 Parameter estimation

The parameters may be estimated by maximizing the likelihood function

$\prod_{R,R'\subseteq Q} \mathbb{P}(\boldsymbol{R}_t = R, \boldsymbol{R}_{t+\delta} = R')^{N(R,R')}$

with

$$\mathbb{P}(\boldsymbol{R}_t = R, \boldsymbol{R}_{t+\delta} = R') = \sum_{(K,K')\in\mathcal{K}\times\mathcal{K}} r(R,K)\,r(R',K')\mathbb{P}(\boldsymbol{K}_t = K, \boldsymbol{K}_{t+\delta} = K'), \tag{7.19}$$

(by Theorem 1), with respect to the various parameters of the theory; these are the response parameters β_i and η_i, the parameters α and ξ of the distribution of the learning rates, the item difficulty parameters μ_i, and the probabilities p_ν of the learning paths. In some cases, it is also possible to treat the times t and $t+\delta$ as parameters. For example, the time t elapsed since the beginning of learning may be difficult to assess accurately. The function r is specified by Axiom [R]. To simplify the computation of the integrals of Case (iv) entering in the expression of the joint probabilities of the states, we use the Wilson-Hilferty approximation to the gamma integral (Wilson and Hilferty, 1931; Johnson and Kotz, 1970, Vol. 1, p. 176). Denoting by Φ the distribution function of the standard normal random variable, we obtain, after simplification, from (7.14):

$$\int_0^\infty H_{K,\lambda}(t)H_{K'\setminus K,\lambda}(\delta)f(\lambda)\,d\lambda$$

$$\approx \int_0^\infty \Phi\left\{[(\frac{\lambda t}{\mu_K})^{\frac{1}{3}} - 1 + (9\mu_K)^{-1}]\sqrt[3]{\mu_K}\right\} \tag{7.20}$$

$$\cdot\Phi\left\{[(\frac{\delta\lambda}{\mu_{K\setminus K'}})^{\frac{1}{3}} - 1 + (9\mu_{K\setminus K'})^{-1}]\sqrt[3]{\mu_{K\setminus K'}}\right\}\frac{\xi^\alpha\lambda^{\alpha-1}e^{-\xi\lambda}}{\Gamma(\alpha)}\,d\lambda\,.$$

For the prediction of the observed frequencies, the integrals of Case (iv) have only to be computed a small number of times (for 5 items, for example, "small" means a few dozen), and this can be achieved by numerical methods. Excellent approximations exist for the Gaussian integral. Thus, once the parameters have been estimated, the data can be predicted without difficulty by numerical integration. For the estimation phase, however, a different course must be taken. The following 2-phase procedure has been investigated for this purpose.

Phase 1. Simultaneous maximization. The marginal frequencies

$$N_1(R) = \sum_{R'\subseteq Q} N(R,R'), \qquad N_2(R') = \sum_{R\subseteq Q} N(R,R')$$

are computed. These marginal frequencies are treated as if they had been generated by two independent samples of subjects, from the same population, each sample containing the same number

$$N = \sum_{R,R' \subseteq Q} N(R, R') = \sum_{R \subseteq Q} N_1(R) = \sum_{R \subseteq Q} N_2(R)$$

of subjects. It is assumed that the parameters for these two samples have identical values, except possibly for

$$\theta_1 = \frac{t}{t+\xi}, \qquad \theta_2 = \frac{\delta}{\delta+\xi}. \tag{7.21}$$

A maximization of the sum of the two log-likelihood functions is performed, namely

$$\left[\log \prod_{R \subseteq Q} \mathbb{P}(\mathbf{R}_t = R)^{N_1(R)}\right] + \left[\log \prod_{R' \subseteq Q} \mathbb{P}(\mathbf{R}_{t+\delta} = R')^{N_2(R')}\right]. \tag{7.22}$$

(The exact maximization technique used is explained later.) This yields a preliminary set of values for all the parameters. As pointed out in the previous section, we are here approximating the two-test situation with two single-test situations, and the expression of the likelihood functions in (7.22) only contains integrals of Cases (i), (ii) or (iii). (We recall that these integrals are either trivial, or involve the incomplete beta function ratio.) As will appear in the simulation study described below, most of the estimation work is actually performed during Phase 1. The parameter values resulting from the maximization in Phase 1 are close to those used for the simulation. However, substantial further improvements can be made.

Phase 2. Joint maximization. Using these preliminary estimates as starting values, the logarithm of the likelihood function in (7.19) is maximized, using the normal approximations for the beta integrals, as in (7.20). The resulting final values of the parameters can then be used for prediction.

The maximization technique used was a modification of a conjugate gradient search method of Powell (1964). The modification was due to Brent (1973). The actual program used was Praxis (Brent, 1973; Gegenfurtner, Submitted for publication).

The computational hardware used for both phases was the UCI-Convex C240 computer. The computational time for phase 1 ranged from 2000 to 30000 CPU seconds depending on the starting values of the parameters. We started at different values to ensure that the final estimated parameters in Phase 1 did not change depending on the starting point in the parameter space. Prior to the introduction of Phase 1, the model could not be used because the computational time needed was prohibitive.

7.4 Simulation studies

The parameter values in the first column of Table 7.1 were used to perform a Monte-Carlo simulation of the data of 20,000 fictitious students, for the knowledge structure $\mathcal{F}$ defined in (7.3) and pictured in Figure 7.1. Since we were not sure beforehand of the computational time involved, we set to probability 0 all but 3 of the seven possible gradations in the knowledge structure in Figure 7.1. These were chosen quite arbitrarily as were the probabilities of gradations used to simulate the data. The three gradations were $\nu_1 = 51432$, $\nu_2 = 54312$, $\nu_3 = 52143$. This simulation provided 20,000 test results distributed in $(2^5)^2 = 1024$ possible patterns of responses. (Actually, not all the response patterns were represented in the data; see below.) The estimation procedure described in the last section gave the results contained in Table 7.1. The first column of the table contains the parameter values chosen for simulating the data. The second column lists the estimated parameter values at the end of Phase 1 of the procedure. These estimates are based on the maximization of the sum of the logarithms of the two marginal likelihood functions, cf. (7.22). It is clear that Phase 1 was very efficient: with a few noticeable exceptions, the parameter values in column 2 of Table 7.1 are close to those used for the simulation. The exceptions are: the value of η_5, which is overestimated, and the values of the μ_i's, β_3 and α, which are all underestimated. The underestimation of both the μ_i parameters and α makes some sense. A lower learning rate may be compensated by a lower level of difficulty of the items. When a single test is made, it may be difficult for the theory to differentiate between an increase of the difficulty of all the items, and a decrease of the learning rate in the population. This difficulty seems to disappear in the case of two tests: in the third column of the Table 7.1, the values of practically all the estimates have improved, and most dramatically those of the α and μ_i parameters. We note in passing that the chi-square log-likelihood ratio statistic was also computed, using the values of the parameters obtained in Phase 2. The likelihood function was a smooth function of the parameters of the model. The denominator of the likelihood ratio was the standard maximum-likelihood index for the multinomial model. The numerator was the maximum-likelihood value of the stochastic learning paths theory (using the parameter values listed in the last column of Table 7.1). The value of the chi-square was 614 for 574 degrees of freedom. (Given the values of parameters β_i and η_i, all 1024 possible response patterns may, in principle, appear in any one simulation. In fact, only 594 response patterns were represented in the simulated data. The 574 degrees of freedom resulted from grouping some of the response categories, and subtracting 20 degrees of freedom for the $21 - 1$ estimated parameters. The criterion for grouping was set at an estimated value of 5 per cell.) This value is not significant. More importantly, the parameter values are very close to the target values at the end of Phase 2.

TABLE 7.1. Results of the simulation. The first column contains the parameter values used for the simulation of the data. The values resulting from the simultaneous maximization (Phase 1) are in the second column. The third column contains the values obtained from the joint maximization (Phase 2). The starred values are explained in the remark.

	Estimated values	
Parameters	Sim. fit	Joint fit
$\beta_1 = .05$	.051384	.048232
$\beta_2 = .05$	.064101	.052990
$\beta_3 = .05$	.027550	.051946
$\beta_4 = .05$	.058596	.049970
$\beta_5 = .05$	.052723	.049091
$\eta_1 = .25$	.264890	.246034
$\eta_2 = .25$	.256932	.254318
$\eta_3 = .25$	.243144	.257355
$\eta_4 = .25$	.255865	.250405
$\eta_5 = .25$	.431079	.285079
$\mu_1 = 3.0$	1.965390	3.070257
$\mu_2 = 3.5$	1.696523	3.526136
$\mu_3 = 3.0$	2.441855	3.019277
$\mu_4 = 3.5$	2.516962	3.989693
$\mu_5 = 3.0$	1.794383	3.172711
$\alpha = 2.5$	1.516883	2.342488
$\theta_1 = .66667$	0.625994*	0.688067
$\theta_2 = .80$	0.816148*	0.820376
$\nu_1 = .178179$	0.180966	0.175729
$\nu_2 = .000006$	0.000000	0.000000
$\nu_3 = .824271$	0.819034	0.824271

Remark. The parameters θ_1, θ_2 with estimated values in column 2 of Table 7.1 are those specified by the equations in (7.21) in terms of ξ, t, and δ. These values were obtained from a reparametrization. Indeed, a simultaneous application of the theory to the data of two single tests at times t and $t + \delta$ yields estimates of the two parameters,

$$\theta_1' = \frac{t}{t + \xi}, \qquad \theta_2' = \frac{t + \delta}{t + \delta + \xi},$$

since the second test takes place $t+\delta$ units of time after the beginning of learning. We thus have $\theta_1' = \theta_1$. With $v(x) = (\frac{1}{x} - 1)^{-1}$, $0 < x < 1$, it follows that

$$v(\theta_2) = v(\theta_2') - v(\theta_1).$$

TABLE 7.2. Results of the simulation for the knowledge structure $\mathcal{H}$ of (6.23). The first column contains the parameter values used for the simulation of the data. The second column contains the values obtained from the joint maximization (Phase 2).

Parameters	Estimates
$\beta_1 = .05$	.048691
$\beta_2 = .05$	.048361
$\beta_3 = .05$	.051062
$\eta_1 = .00$	.000000
$\eta_2 = .00$	.000000
$\eta_3 = .00$	.000000
$\mu_1 = 3.0$	3.142417
$\mu_2 = 3.5$	3.497725
$\mu_3 = 3.0$	2.889061
$\alpha = 2.5$	2.772243
$\theta_1 = 0.6667$	0.646706
$\theta_2 = .80$	0.776167
$\nu_1 = .75$	0.747726
$\nu_2 = .25$	0.252274

The estimate of θ_2 in the second column of Table 7.1 was obtained from the estimates of $\theta_1' = \theta_1$ and θ_2' by

$$\theta_2 = v^{-1}[v(\theta_2') - v(\theta_1)]\,.$$

Another, smaller knowledge structure

$$\mathcal{H} = \{\emptyset, \{1\}, \{1,2\}, \{1,3\}, \{1,2,3\}\} \tag{7.23}$$

was subjected to the same treatment. The results of 20,000 subjects were simulated, using the parameter values in the first column of Table 7.2.

In view of the expected smaller number of computations of the relevant integral, Phase 1 of the procedure was omitted; (that is, the parameter values were estimated by brute-force numerical integration.) The resulting estimated values are contained in the second column of Table 7.2. On computing the log-likelihood ratio statistic with the estimated values of the parameters we get a chi-square value of 44, for 29 degrees of freedom. (There are $(2^3)^2 = 64$ possible response patterns. Many response patterns did not appear in the simulated data. This was to be expected in view of the knowledge structure and the parameter values used in the simulation.)

7.5 Discussion

These results suggest that realistic applications of the theory are feasible. At this stage, however, such applications remain computationally expensive, and good approximations for the 2-dimensional or n-dimensional beta type integral (7.15), (7.16), (7.17), (7.18) are highly desirable.

Some issues were raised by the referees, concerning the scope of this theory, which we now address. The following two-pronged argument can be made that this theory is practicable only in situations in which the number of questions is very small: 10 questions at most, maybe only 5 or 6, with a maximum of two tests. To illustrate the difficulty, we use an example proposed by one of the referees. Consider a situation in which a 10-item test has been applied three times to the same sample of subjects. The number of response categories is $2^{30} > 10^9$. One can ask – as does the referee – how many subjects will be required if the computations are to be reasonably determinate? The second problem concerns the number of parameters, which may be prohibitive, even if the number of learning paths is only a small fraction of the 10! possible ones.

An appropriate discussion of these objections requires an explanation of our position with respect to the entire enterprise. We believe that repeated, comprehensive testing of representative samples of the student population is potentially very informative. The results would constitute basic educational data, which should be collected, analyzed and understood as best as we know how. If there are any laments to be expressed, they should be directed to the complexity of such data, not to the theories purporting to explain them. It is quite true that, as indicated by the example of the referee, a head-on application of the theory, in the exact form presented here, to a case involving a sizable number of questions and several repetitions, would encounter serious difficulties. This, however, is not an intrinsic limitation of the theory: *any* theory for such data would meet the same kind of difficulties. Actually, essentially the same objections can be raised with any of the standard psychometric models. Typically, the number of parameters of a 'latent trait' psychometric model tends to be smaller than that of the stochastic learning path theory described here. However, this greater simplicity of the psychometric models may come at the expense of their realism. In fact, the particular case of the stochastic learning path theory in which there is only one path strongly resembles a one-dimensional 'latent trait' model. (Notice in this connection that a learning path is essentially a Guttman scale. If one wishes, the stochastic learning path theory may be regarded as a stochastic, multidimensional elaboration of the concept of a Guttman scale.)

With these considerations in mind, let us address the objections above concerning the feasibility of large scale applications. Consider the case of a 50 item test applied 3 times, at times t_1, t_2 and t_3, and suppose that a credible knowledge structure is available. (Techniques for developing such a structure have been developed; c.f. Falmagne et al., 1990; Koppen & Doignon, 1990; Kambouri et al., 1991). Let us first take the issue of the number of parameters, especially the possibly staggeringly large number of learning paths. In the presentation given here, each learning path is assigned a probability. The only constraint on these probabilities is that they must add up to one. Since, in this situation, one should expect thousands of learning paths, the number of parameters is overwhelming. Notice, however, that this is only the most general case. Constraints on the collection of learning path probabilities can be explored, which might reduce the number of parameters considerably. For instance, the succession of states could have a Markovian character. To illustrate this idea, consider the case of the knowledge structure $\mathcal{F}$ of Figure 7.1. The probability of a transition from $\emptyset$ to $\{5\}$ would be one, which we would write as $p_{\emptyset,\{5\}} = 1$. There are three states immediately following state $\{5\}$, namely $\{4,5\}$, $\{1,5\}$ and $\{2,5\}$. Introduce three transition probabilities which, with obvious notation, satisfy

$$p_{\{5\},\{4,5\}} + p_{\{5\},\{1,5\}} + p_{\{5\},\{2,5\}} = 1 .$$

In general, writing p_{K,K^ν} for the probability of a transition from state K to state K^ν (the state immediately following K in the learning path ν), requires that

$$\sum_{\nu \in \mathcal{C}(K)} p_{K,K^\nu} = 1$$

for every state K. The probability of a particular learning path ν is then obtained from the product

$$p_\nu = p_{\emptyset,\emptyset^\nu} \cdot p_{\emptyset^\nu,(\emptyset^\nu)^\nu} \cdot \ldots \cdot p_{{}^\nu Q,Q}. \tag{7.24}$$

in which ${}^\nu Q$ stands for the state immediately preceding Q on the learning path ν. The quantities p_ν, defined by (7.24), clearly satisfy $\sum_{\nu \in \mathcal{C}} p_\nu = 1$. In the case of the example of Figure 7.1, the number of (independent) parameters specifying the probabilities of the learning paths is reduced to 5 (from the original $7-1=6$). This may appear a negligible reduction. Examination of a number of examples suggests, however, that the reduction becomes much more substantial as the number of items grows. Besides, much stronger constraints can be investigated, the more extreme of which would involve the assumption that the probability of a transition from state K to state K^ν would depend only upon the items in $K^\nu \setminus K$. Suppose, for example, that to each item i is attached a scale value $v(i) > 0$, and define

$$p_{K,K^\nu} = \frac{\sum_{i \in (K^\nu \setminus K)} v(i)}{\sum_{\nu' \in \mathcal{C}(K)} \sum_{j \in (K^{\nu'} \setminus K)} v(j)} .$$

In this case, since the $v(K)$ are only defined up to a multiplication by a positive constant, the number of parameters needed to specify the learning path probabilities is reduced to $|Q| - 1$. Other, less demanding assumptions of the same kind could be studied, which could keep the number of such parameters under control.

The other prong of the criticism concerns the very large number of response categories. Since this number far exceeds the number of students tested, many of these categories will be empty. There is no harm in grouping all these categories into one cell, in such a manner that the expected number of responses falling into that cell satisfies the appropriate criterion for the convergence of the likelihood ratio statistic[2]. Such grouping is standard procedure, and should not create any qualms.

In any event, these arguments should suggest that it may be practicable to apply the stochastic learning theory to non-trivial parts of the data, involving sets of 6-7 questions, and two of the three presentations of the test. Success in a sufficiently large number of such applications would grant some plausibility to the assumption that the theory holds for the full data. Hopefully, the estimated values of some of the parameters would be, to a reasonable extent, invariant with the subset of items and with the chosen pair of presentations of the test, so that the contours of the overall picture would emerge. One can also be more ambitious and engage in a gradual construction of the full picture, by piecing together the results obtained for the parts. Obviously, we are considering here a very substantial enterprise, the details of which are far from clear at this stage. However, it is one that may be well worth the trouble.

The theory discussed in this paper offers a detailed description of a complex educational process, together with a collection of useful measures concerning various aspects of the student's performance, such as the difficulty of the items, the times required to master them, the likelihood of learning sequences, the probability of the various knowledge states, etc. Such measures can be used for the implementation of efficient individual knowledge assessment routines, with a potentially useful impact on the educational progress of the students.

7.6 References

Brent, R.P. (1974). *Algorithms for minimization without derivatives* (2nd ed.). Englewood Cliffs: Prentice Hall.

Doignon, J.-P., & Falmagne, J.-Cl. (1985). Spaces for the assessment of knowledge. *International Journal of Man-Machine Studies, 23,* 175–196.

Falmagne, J.-Cl. (1989). A latent trait theory via stochastic learning theory for a knowledge space. *Psychometrika, 53,* 283–303.

Falmagne, J.-Cl. (in press) Stochastic learning paths in a knowledge structure. *Journal of Mathematical Psychology.*

Falmagne, J.-Cl., Koppen, M., Villano, M., Doignon, J.-P., & Johannesen, L. (1990). Introduction to knowledge spaces: How to build, test, and search them. *Psychological Review, 97,* 201–224.

[2]The criterion used here (that the expected cell size should exceed 5) is conservative. For large, sparse data matrices, a criterion of approximately 1 may be appropriate (see Fienberg, 1981).

Fienberg, S.E. (1981). *The analysis of cross-classified categorical data* (2nd ed.). Cambridge, MA: MIT Press.

Gegenfurtner, K. *PRAXIS: Brent's algorithm for functions minimization.* (Submitted for publication).

Johnson, N.L., & Kotz, S. (1970). *Distributions in statistics. Continuous univariate distributions, I, II.* New York: Houghton Mifflin.

Kambouri, M., Koppen, M., Villano, M., & Falmagne, J.-Cl. (1991). Knowledge assessment: Tapping human expertise by the QUERY routine. *Technical report MBS 91-20.* Irvine: Irvine Research Unit in Mathematical Behavioral Sciences. (Submitted for publication)

Koppen, M., & Doignon, J.-P. (1990). How to build a knowledge space by querying an expert. *Journal of Mathematical Psychology, 34,* 311–331.

Koppen, M. (in press). Extracting human expertise for constructing knowledges spaces: An algorithm. *Journal of Mathematical Psychology.*

Lord, F.M., & Novick, M.R. (1974). *Statistical theories of mental tests scores* (2nd ed.) Reading, MA: Addison Wesley.

Powell, M.J.D. (1964). An efficient method for finding the minimum of a function in several variables without calculating derivatives. *Computer Journal, 7,* 155–162.

Taagepera, M., Potter, F. Miller, G.E., & Lakshminarayan, K. *Defining the critical learning path from knowledge state analysis of the density concept.* (Submitted for publication).

Wilson, E.B., & Hilferty, M.M. (1931). The distribution of chi-square. *Proceedings of the National Academy of Sciences, Washington, 17,* 684-688.

8

Knowledge Spaces and Skill Assignments

Jean-Paul Doignon[1]

ABSTRACT The concept of a knowledge space is at the heart of a descriptive model of knowledge in a given body of information. Another model explains the observed knowledge of individuals by latent skills. We here reconcile these two underlying approaches by showing that each finite knowledge space can be generated from a skill assignment that is minimal and unique up to an isomorphism. Some more technicalities are required in the infinite case. Part of the results reformulate theorems from the theory of Galois lattices of relations.

8.1 Introduction

A knowledge structure is a fairly simple and purely descriptive model of knowledge in a body of information. The elementary pieces of this body are called questions or problems; at any time, any person or student knows a definite subset of this set of questions. Thus, a collection of specific subsets is obtained, consisting of all possible states of knowledge with respect to that body of information. This set-theoretic approach to knowledge representation was conceived with the aim of designing automatic procedures for the assessment of knowledge (Doignon and Falmagne, 1985; Falmagne and Doignon, 1988a, 1988b). In order to derive results about the model, various assumptions have been considered on the collection of knowledge states. For instance, stability under union requires that any union of knowledge states is again a knowledge state. This axiom defines knowledge spaces. It will play an important rôle here.

Following another approach (see, e.g., Marshall, 1981), knowledge is explained by skills: a student masters all the problems for which he or she possesses the required skills. The link between skills and problems can be formalized in several ways. Both a disjunctive model and a conjunctive model will be discussed. The first model assumes that a problem is mastered as soon as at least one of its assigned skills is acquired by the student. The conjunctive model, on the contrary, states that all of the assigned skills are necessary for the mastery of a problem.

Not too much will be said about the pros and cons of the two approaches based respectively on knowledge states and on skill assignments. As previously said, the first approach is descriptive, while the second aims at discovering hidden

---

[1]Université Libre de Bruxelles, c.p. 216, Bd du Triomphe, B-1050 Bruxelles, Belgium

Our work in this area is partially supported by NSF grant IRI 8919068 to Jean-Claude Falmagne at the University of California, Irvine.

factors. The goal here is, rather, to reconcile the two approaches in some way, by showing that skills can always be assigned in a manner which generates the observed family of knowledge states. Even further, each finite knowledge space corresponds in this way to a unique *minimal* skill assignment. Some formal definitions will be needed before the results can be stated rigorously.

Not all of these results are completely new. Skills were mentioned by Falmagne et al. (1990). However, only preliminary results were reported. It will be shown here that a skill assignment can be produced for each given knowledge space and, moreover, that in many circumstances there is a canonical skill assignment (uniquely determined up to an isomorphism). As any set of questions may be allowed (either finite or infinite), some technicalities have to be taken care of. On the other hand, it will turn out that part of the results is, in fact, a reformulation of known theorems on Galois lattices of relations. Because of the quite distinct motivation, it is worth stating precisely the links between knowledge structures and skill assignments.

8.2 Background

Some definitions and results from Doignon and Falmagne (1985) are repeated here for convenience.

Definition 1 *A knowledge structure $(Q, \mathcal{K})$ consists of a nonempty set Q of questions or problems, together with a collection $\mathcal{K}$ of subsets of Q, called knowledge states; the empty set and the whole set Q are assumed to be knowledge states, that is $\emptyset \in \mathcal{K}$ and $Q \in \mathcal{K}$.*

The set Q may be finite or infinite. In some papers, it is not required that the empty set and/or the whole set Q are members of $\mathcal{K}$; these assumptions avoid complications in the results which follow.

Notation 1 *For q, a question in a knowledge structure $(Q, \mathcal{K})$, let*

$$\mathcal{K}_q = \{K \in \mathcal{K} \mid q \in K\}. \tag{8.1}$$

Definition 2 *A knowledge structure $(Q, \mathcal{K})$ is a knowledge space when the family $\mathcal{K}$ is stable under union; that is, the union of any subfamily of $\mathcal{K}$ belongs to $\mathcal{K}$. Formally,*

$$\text{if } \mathcal{L} \subseteq \mathcal{K}, \text{ then } \cup\mathcal{L} \in \mathcal{K}. \tag{8.2}$$

Several motives can be given for Definition 2. Knowledge spaces arise in close connection with surmise systems, introduced as a variant of the AND/OR graphs from artificial intelligence (see Doignon and Falmagne, 1985). Stability under union (or, dually, under intersection) allows the design of an automated, efficient procedure to uncover possible knowledge states by querying experts (Koppen and Doignon, 1990, or Müller, 1989). Moreover, the dual axiom of stability under intersection holds for several classes of mathematical structures, see, e.g., Birkhoff (1967).

Definition 3 *A closure space is a pair $(X, \mathcal{C})$, where X is a nonempty set and $\mathcal{C}$ a family of subsets of X that is stable under intersection,*

$$\text{if } \mathcal{D} \subseteq \mathcal{C}, \text{ then } \cap\mathcal{D} \in \mathcal{C}. \tag{8.3}$$

Moreover, it is always assumed here that $\emptyset \in \mathcal{C}$. Note that $X \in \mathcal{C}$ is a consequence of Equation 8.3: the intersection of an empty subfamily of subsets of X equals X.

To each knowledge space $(Q, \mathcal{K})$, there corresponds biuniquely the closure space $(Q, \mathcal{C})$ with

$$\mathcal{C} = \{Q \setminus K \mid K \in \mathcal{K}\}. \tag{8.4}$$

Consequently, there is a duality between the theory of knowledge spaces and the theory of closure spaces.

Definition 4 *A spanning subfamily of a knowledge structure $(Q, \mathcal{K})$ is a subfamily $\mathcal{S}$ of $\mathcal{K}$ such that $\mathcal{K}$ consists exactly of all unions of members of $\mathcal{S}$. Then $\mathcal{S}$ is said to span $\mathcal{K}$. A base is a minimal spanning subfamily.*

In this last definition, 'minimal' is to be understood with respect to family inclusion: there is no strict subfamily of the base that is also a spanning subfamily of $\mathcal{K}$. A base $\mathcal{B}$ never contains $\emptyset$, since $\emptyset$ is the union of the empty subfamily of $\mathcal{B}$.

Proposition 1 *If a knowledge structure admits a spanning subfamily, then it is a knowledge space. Each base of a knowledge space is included in each spanning subfamily of this space. Hence, if a knowledge space has a base, this base is unique.*

Notice the existence of knowledge spaces having no base in our sense. To obtain such a space, take the family of all open subsets of the real line. Of course, any finite knowledge space (that is, a space formed by a finite number of questions) necessarily has a base.

8.3 Skill assignments; the disjunctive model

Let us start by formalizing the idea of skills attached to questions.

Definition 5 *Let Q be a set of questions. A skill assignment on Q is formed by a nonempty set S of skills, and a mapping τ from Q to the family of all nonempty subsets of S. This skill assignment is denoted as (Q, S, τ), and essentially consists of a mapping $\tau : Q \to 2^S \setminus \{\emptyset\}$. The skills belonging to $\tau(q)$ are said to be assigned to q.*

The disjunctive model of knowledge in presence of a skill assignment is defined next, followed by an example and a first, easy result. Remember the interpretation underlying the next Definition 6: A student in possession of a subset T of skills masters exactly those questions that have at least one assigned skill in the subset T.

Definition 6 *Given a skill assignment (Q, S, τ), the knowledge state K delineated by a subset T of S is*

$$K = \{q \in Q \mid \tau(q) \cap T \neq \emptyset\}. \tag{8.5}$$

Example 1 Let $Q = \{a, b, c, d\}$, and $S = \{s, t, u\}$. Define $\tau : Q \to 2^S$ by

$$\tau(a) = \{s, t\}, \tag{8.6}$$
$$\tau(b) = \{t, u\}, \tag{8.7}$$
$$\tau(c) = \{s\}, \tag{8.8}$$
$$\tau(d) = \{s, u\}. \tag{8.9}$$

Then,

$$\emptyset \quad \text{delineates} \quad \emptyset, \tag{8.10}$$
$$\{s\} \qquad \{a, c, d\}, \tag{8.11}$$
$$\{t\} \qquad \{a, b\}, \tag{8.12}$$
$$\{u\} \qquad \{b, d\}, \tag{8.13}$$
$$\{s, t\} \qquad \{a, b, c, d\}, \tag{8.14}$$
$$\{t, u\} \qquad \{a, b, d\}, \tag{8.15}$$
$$\{s, u\} \qquad \{a, b, c, d\}, \tag{8.16}$$
$$\{s, t, u\} \qquad \{a, b, c, d\}. \tag{8.17}$$

In Example 1, the union of any two delineated knowledge states is again a delineated knowledge state. This is an instance of a general fact.

Proposition 2 *Let (Q, S, τ) be a skill assignment. The family of all the knowledge states delineated by subsets of S forms a knowledge space on Q.*

Proof: The following assertion is easily seen to hold. If, for each index i in a set I, the knowledge state K_i is delineated by the subset T_i of S, then $\cup_{i \in I} K_i$ is delineated by $\cup_{i \in I} T_i$. Moreover, the empty subset of S always delineates the empty subset of Q, and the whole set S delineates Q. □

Definition 7 *In the situation covered by Proposition 2, the resulting knowledge space is said to be delineated by the skill assignment (Q, S, τ).*

Now, can *any* knowledge space be obtained in this way? The answer is "Yes", as announced in Falmagne et al. (1990).

Proposition 3 *For any knowledge space $(Q, \mathcal{K})$, there exists a skill assignment (Q, S, τ) that delineates $(Q, \mathcal{K})$.*

Proof: Introduce, somewhat artificially, a skill assignment τ on Q by taking $\mathcal{K}$ to be the set of skills, and letting

$$\tau(q) = \mathcal{K}_q \tag{8.18}$$

for $q \in Q$ (thus deciding that the skills assigned to question q are all the knowledge states containing q. Notice that $\tau(q) \neq \emptyset$ follows from $Q \in \mathcal{K}_q$). It will be shown that $(Q, \mathcal{K})$ is delineated by $(Q, \mathcal{K}, \tau)$. First, if $\mathcal{L} \subseteq \mathcal{K}$, the knowledge state K delineated by $\mathcal{L}$, according to Definition 6, is

$$K = \{q \in Q \mid \tau(q) \cap \mathcal{L} \neq \emptyset\}, \tag{8.19}$$

that is,

$$\begin{aligned} K &= \{q \in Q \mid \mathcal{K}_q \cap \mathcal{L} \neq \emptyset\} && (8.20)\\ &= \{q \in Q \mid \exists L \in \mathcal{L} : q \in L\} && (8.21)\\ &= \cup \mathcal{L}. && (8.22) \end{aligned}$$

Since $(Q, \mathcal{K})$ is a knowledge space, $K \in \mathcal{K}$. On the other hand, if $K \in \mathcal{K}$, it can be shown that K is delineated by some subset of $\mathcal{K}$. Take $\{K\}$ to be this subset; it delineates the knowledge state L with

$$L = \{q \in Q \mid \tau(q) \cap \{K\} \neq \emptyset\}, \tag{8.23}$$

or

$$\begin{aligned} L &= \{q \in Q \mid K \in \mathcal{K}_q\} && (8.24)\\ &= K. && (8.25) \end{aligned}$$

In conclusion, the given knowledge space $(Q, \mathcal{K})$ is delineated by the skill asignment $(Q, \mathcal{K}, \tau)$ that was constructed. □

There are many skill assignments that delineate the same knowledge space. To see this, give other 'labels' to the skills, and use the corresponding assignment. In the next Section, this idea is formalized. It is also shown that in many circumstances there is a canonical skill assignment for a given knowledge space.

8.4 Minimal skill assignments

If the names of the skills are changed, but the assignment otherwise remains the same, the skill assignment obtained is not structurally new. Rather, it is isomorphic to the previous one.

Definition 8 *Two skill assignments (Q, S, τ) and (Q, S', τ') on the same set Q of questions are isomorphic when there exists an isomorphism from the first skill assignment onto the second one, that is, a bijective mapping $f : S \to S'$ which, for each q in Q, transforms $\tau(q)$ into $\tau'(q)$:*

$$f(\tau(q)) = \tau'(q), \tag{8.26}$$

or in an even more compact writing,

$$f \circ \tau = \tau' \tag{8.27}$$

(on the left of the composition symbol $\circ$, the same letter f is used to denote the action of f on 2^S).

Example 2 Let $Q = \{a, b, c, d\}$, and $S' = \{1, 2, 3\}$. Define $\tau' : Q \to 2^{S'}$ by

$$\begin{aligned} \tau'(a) &= \{1,2\}, & (8.28)\\ \tau'(b) &= \{1,3\}, & (8.29)\\ \tau'(c) &= \{2\}, & (8.30)\\ \tau'(d) &= \{2,3\}. & (8.31) \end{aligned}$$

The skill assignment from Example 1 is isomorphic to the present one: the mapping $f : \{s, t, u\} \to \{1, 2, 3\}$ sending s onto 2, t onto 1, and u onto 3, is an isomorphism from (Q, S, τ) onto (Q, S', τ').

It should be clear that two isomorphic skill assignments delineate exactly the same knowledge space.

Among all the skill assignments that delineate a given knowledge space, some are minimal in the sense that, after deletion of any skill, they no longer delineate the same space. A precise definition follows. If the knowledge space is finite, the next Proposition 4 implies that these minimal skill assignments are exactly those having the least possible number of skills, and that they are all isomorphic to the one another.

Definition 9 *A skill assignment (Q', S', τ') is a restriction of a skill assignment (Q, S, τ) if $Q' = Q$, $S' \subseteq S$, and $\tau'(q) = \tau(q) \cap S'$ for each q in Q. If, moreover, $S' \subset S$, then (Q', S', τ') is a strict restriction of (Q, S, τ). A skill assignment (Q, S°, τ°) is minimal if none of its strict restrictions delineates the same knowledge space as (Q, S°, τ°) does.*

Some knowledge spaces cannot be delineated by any minimal skill assignment. They are all infinite; an example is obtained by taking the family of all open subsets of the real line.

Proposition 4 *A knowledge space admits a base (as in Definition 4) iff it is delineated by some minimal skill assignment. When $\mathcal{B}$ is the base of the knowledge space $(Q, \mathcal{K})$, and (Q, S°, τ°) is any minimal skill assignment delineating $(Q, \mathcal{K})$, the following holds: $|S^\circ| = |\mathcal{B}|$. In fact, any two minimal skill assignments delineating $(Q, \mathcal{K})$ are isomorphic. Furthermore, any skill assignment delineating $(Q, \mathcal{K})$ admits a minimal restriction delineating $(Q, \mathcal{K})$.*

Proof: First consider any skill assignment (Q, S, τ) and its delineated knowledge space $(Q, \mathcal{K})$. For $s \in S$, denote by $K(s)$ the state from $\mathcal{K}$ delineated by $\{s\}$. Then any subset T of S delineates $\cup_{t \in T} K(t)$. Therefore, $\mathcal{A} = \{K(s) \mid s \in S\}$ spans $\mathcal{K}$. Now pick, for each $M \in \mathcal{A}$, one single element s in S for which $K(s) = M$, and denote by S' the subset of S formed by these s. As, clearly, $\mathcal{A} = \{K(s) \mid s \in S'\}$, each state K from $\mathcal{K}$ is delineated by some subset T of S'. For each q in Q, set $\tau'(q) - \tau(q) \cap S'$. The given skill assignment (Q, S, τ) may thus be restricted, without bringing any change to the delineated space $(Q, \mathcal{K})$, to the skill assignment (Q, S', τ').

Assume now that $(Q, \mathcal{K})$ admits the base $\mathcal{B}$. By Proposition 1, $\mathcal{B} \subseteq \mathcal{A}$. Hence, setting $S^\circ = \{s \in S' \mid K(s) \in \mathcal{B}\}$ gives a subset S° of S satisfying $|S^\circ| = |\mathcal{B}|$. Similarly, as for (Q, S', τ'), a skill assignment (Q, S°, τ°) delineating $\mathcal{K}$ is obtained by taking $\tau^\circ(q) = \tau(q) \cap S^\circ$. The conclusion is that any minimal skill assignment (Q, S^*, τ^*) delineating $\mathcal{K}$ satisfies $|S^*| = |\mathcal{B}|$, and also that any skill assignment (Q, S, τ) delineating $(Q, \mathcal{K})$ has a minimal restriction, namely, (Q, S°, τ°).

A further look at the minimal skill assignment (Q, S°, τ°) just constructed shows that it is determined from the space $\mathcal{K}$ it delineates, up to the names of the skills forming S°. In fact, there is a bijective mapping

$$S^\circ \to \mathcal{B} : s \to K(s), \tag{8.32}$$

and, moreover, for $q \in Q$ and $s \in S^\circ$,

$$s \in \tau^\circ(q) \iff q \in K(s), \tag{8.33}$$

or equivalently,

$$\tau^\circ(q) = \{s \in S^\circ \mid q \in K(s)\}. \tag{8.34}$$

Hence, all the minimal skill assignments delineating $(Q, \mathcal{K})$ are isomorphic to each other.

Finally, similar arguments can be used to show that any knowledge space $(Q, \mathcal{K})$ delineated by a minimal skill assignment (Q, S°, τ°) has the base $\{K(s) \mid s \in S^\circ\}$. □

8.5 The conjunctive model and Galois lattices

The conjunctive model is obtained by changing the way a subset of skills delineates a knowledge state: in this case, all the skills assigned to a question are required for the mastery of this question.

Definition 10 *Given a skill assignment (Q, S, τ), any subset T of S delineates, according to the conjunctive model, the knowledge state*

$$\{q \in Q \mid \tau(q) \subseteq T\}. \tag{8.35}$$

Proposition 5 *Given a skill assignment (Q, S, τ), the family $\mathcal{C}$ of all knowledge states delineated according to the conjunctive model is stable under intersection.*

Definition 11 *Keeping the same notations as in the preceding Proposition 5, the closure space $(Q, \mathcal{C})$ is said to be delineated, according to the conjunctive model, by the skill assignment (Q, S, τ).*

Proposition 5 and Definition 11 are dual to Proposition 2 and Definition 7. Moreover, Proposition 5 is a well-known result in the theory of Galois lattices of relations (cf. Barbut and Monjardet, 1970; Birkhoff, 1967; or Wille, 1987). To see that this is so, define a relation R between sets Q and S by setting, for $q \in Q$ and $s \in S$,

$$q\,R\,s \iff s \in \tau(q). \tag{8.36}$$

Then the knowledge states delineated in Q, according to the conjunctive model, are nothing other than the closed sets constituting the Galois lattice of the relation R; they form a closure space on Q. It is also known that each closure space can be obtained in this way. This property is the dual of Proposition 3.

8.6 Skill multiassignments

According to Proposition 3, any knowledge space is delineated by at least one skill assignment. In order to obtain a similar result for knowledge structures, it is necessary to generalize skill assignments. More than one subset of skills can now be assigned to a given question; each such subset will be called a competency. To master the question, a student has to completely possess at least one of the relevant competences.

Definition 12 *A skill multiassignment on a set Q of questions consists of a set S of skills, and a mapping μ from Q to the collection of all nonempty families of nonempty subsets of S. This skill assignment is denoted as $(Q, S; \mu)$. Hence μ is some mapping $\mu : Q \to 2^{2^S \setminus \{\emptyset\}} \setminus \{\emptyset\}$. The knowledge structure delineated by $(Q, S; \mu)$ is obtained by collecting all subsets K of Q that satisfy, for some subset T of S:*

$$K = \{q \in Q \mid \exists C \in \mu(q) : C \subseteq T\}. \tag{8.37}$$

With the same notations, the knowledge state K is delineated by the subset T of skills.

Notice that S delineates Q (because $\mu(q) \neq \emptyset$ for each question q), and that the empty subset of S delineates the empty subset of Q (because $C \neq \emptyset$ for each competency C).

Proposition 6 *Each knowledge structure $(Q, \mathcal{K})$ is delineated by at least one skill multiassignment.*

Proof: Define a skill multiassignment by setting $S = \mathcal{K}$, and for $q \in Q$,

$$\mu(q) = \{\mathcal{K} \setminus \{M\} \mid M \in \mathcal{K}_q\}; \tag{8.38}$$

that is, given question q, introduce for each knowledge state M containing q a competency $\mathcal{K} \setminus \{M\}$ for q (this competency is nonempty because it contains the empty set). Let $(Q, \mathcal{L})$ be the knowledge structure delineated by the skill multiassignment $(Q, S; \mu)$; it will be shown that $\mathcal{K} = \mathcal{L}$.

First, let $K \in \mathcal{K}$. Then consider the subset $\mathcal{K} \setminus \{K\}$ of $\mathcal{K}$, and compute the state L from $\mathcal{L}$ it delineates,

$$\begin{aligned} L &= \{q \in Q \mid \exists M \in \mathcal{K}_q : \mathcal{K} \setminus \{M\} \subseteq \mathcal{K} \setminus \{K\}\} && (8.39) \\ &= \{q \in Q \mid K \in \mathcal{K}_q\} && (8.40) \\ &= K. && (8.41) \end{aligned}$$

Hence, $K = L \in \mathcal{L}$.

Second, let P in $\mathcal{L}$ be delineated by the subset T of skills. From

$$\begin{aligned} P &= \{q \in Q \mid \exists M \in \mathcal{K}_q : \mathcal{K} \setminus \{M\} \subseteq T\} && (8.42) \\ &= \begin{cases} Q & \text{if } T = \mathcal{K}, \\ K & \text{if } T = \mathcal{K} \setminus \{K\} \text{ for some } K \in \mathcal{K}, \\ \emptyset & \text{otherwise,} \end{cases} && (8.43) \end{aligned}$$

it can be seen that P belongs to $\mathcal{K}$. Thus, $\mathcal{K}$ is indeed delineated by the skill multiassignment $(Q, S; \mu)$. □

8.7 Discussion

The question of how to 'explain' knowledge states from non-observable skills was often raised by the audience at talks on knowledge-space theory. Only a few of the possible answers have been studied in this paper. An anonymous referee pointed out that one approach would simply be to consider that any question from Q constitutes by itself a skill. Technically, this leads to another formalization of skills (since according to Definition 7, such a skill assignment delineates the trivial knowledge space $(Q, 2^Q)$).

But this other approach does not truly reflect the motivation behind the original question. A knowledge state in the present sense forms the observable side of what a person knows in a certain domain. At least theoretically, it can be uncovered by testing the mastery of each of the questions (or, in the case that previous information is available on the family of all possible states, by testing well-selected questions). On the other hand, the search for skills is motivated by parsimony: a limited number of acquisitions by the subject should offer a complete explanation of his or her observable state. Thus, only a small amount of hidden and simple information accounts for complex, observable behaviour (whose direct encoding can be more demanding). Proposition 4 provides, I consider, an adequate formal casting of this point of view.

The fact that skills can be combined in different ways (e.g., according to the disjunctive or the conjunctive model) raises another question: What are the best assumptions to be made about the family of knowledge states? Stability under union was preferred in Doignon and Falmagne (1985). The argument was based on a mathematical result: this stability characterizes the family of states coming out of a surmise system (i.e., AND/OR graph). Now, it has been demonstrated how to delineate families of states that are either stable under union or under intersection. There is no formal reason to prefer one kind of stability over the other when working with skill assignments. Only a careful study of the learning situation can give a hint to what should be the most relevant assumptions on the family of states. This versatility of the theory should not be seen as a defect, but rather as a sign of wealth. Even families with no stability property at all can be delineated by skill multiassignments! Once again, decision on the formal approach to follow must be based on experimental considerations.

8.8 References

Barbut, M., & Monjardet, B. (1970). *Ordre et classification: Algèbre et combinatoire.* Paris: Hachette.

Birkhoff, G. (1967). *Lattice theory.* Providence: American Mathematical Society.

Doignon, J.-P., & Falmagne, J.-Cl. (1985). Spaces for the assessment of knowledge. *International Journal of Man-Machine Studies, 23,* 175-196.

Falmagne, J.-Cl., & Doignon, J.-P. (1988a). A class of stochastic procedures for the assessment of knowledge. *British Journal of Mathematical and Statistical Psychology, 41,* 1-23.

Falmagne, J.-Cl., & Doignon, J.-P. (1988b). A Markovian procedure for assessing the state of a system. *Journal of Mathematical Psychology, 32,* 232-258.

Falmagne, J.-Cl., Koppen, M., Villano, M., Doignon, J.-P., & Johannesen, L. (1990). Introduction to knowledge spaces: How to build, test, and search them. *Psychological Review, 97,* 201-224.

Koppen, M., & Doignon, J.-P. (1990). How to build a knowledge space by querying an expert. *Journal of Mathematical Psychology, 34,* 311-331.

Marshall, S.P. (1981). Sequential item selection: Optimal and heuristic policies. *Journal of Mathematical Psychology, 23,* 134-152.

Müller, C.E. (1989). A procedure for facilitating an expert's judgments on a set of rules. In E. Roskam (Ed.), *Mathematical Psychology in Progress* (pp. 157-170). Berlin: Springer-Verlag.

Wille, R. (1987). Bedeutungen von Begriffsverbänden. In B. Ganter, R. Wille, & K.E. Wolff, *Beiträge zur Begriffsanalyse* (pp. 161-211). Zürich: Bibliographisches Institut & Brockhaus AG.

9

Construction of Knowledge Spaces for Problem Solving in Chess

Dietrich Albert[1], Martin Schrepp[1], and Theo Held[1]

ABSTRACT An extension of the theory of knowledge spaces introduced by Doignon and Falmagne (1985) is presented. This extension allows surmise relations and quasi-ordinal knowledge spaces to be constructed with the help of the formal principles 'set inclusion'[2] and 'sequence inclusion' from basic units of knowledge. The basic units for constructing the surmise relation in chess are the tactical elements of the game — the 'motives'. In terms of problem solving, these motives can be regarded as subgoals in the process of problem solving. Two experimental investigations of chess knowledge test the empirical validity of the two principles and show that the theory of knowledge spaces is suitable for testing psychological models.

9.1 Introduction

Chess involves one of the most complex and demanding knowledge domains. Both, the game of chess itself, and also the construction of problems which will serve to assess knowledge concerning the tactical elements of chess require much knowledge and experience and — as we will show here — some special principles.

Both of the experiments reported below are based on an extension of the theory of 'knowledge-spaces' introduced by Doignon and Falmagne (1985). This approach provides the framework for considering the order of a set of chess problems with respect to difficulty and the hypothetical structure of knowledge about the tactical elements of chess.

First we describe some basic elements of the theory of knowledge spaces. One element of this theory is the 'surmise relation'. This is a reflexive and transitive relation defined on a set of problems which are to be given to subjects to solve. If, for example, there is a problem set Q containing problems a and b, the expression $a \preceq b$ signifies:

> If a subject is able to solve problem b, then this subject will also be able to solve problem a.

---

[1]University of Heidelberg, Department of Psychology, Hauptstraße 47–51, D-6900 Heidelberg, Germany
The research reported in this paper was supported by Grant Lu 385/1 of the Deutsche Forschungsgemeinschaft to J. Lukas and D. Albert at the University of Heidelberg. We are grateful to M. Koppen for his important remarks and suggestions, especially with respect to 'Principle 2'. We also thank E. Roskam, J. Lukas, three anonymous referees, and the editors of this volume for their invaluable comments on an earlier draft of this paper, B. Hierholz for constructing the problems of Experiment I and for conducting this experiment, and M. Kadijk for his support in writing the experimental software.

[2]This principle is also reported in Albert and Held (in press).

'$\preceq$' is called a *surmise relation.* The set of problems a subject is capable of solving is called *knowledge state.* From a surmise relation the set of possible knowledge states can be inferred. This set is called a *quasi-ordinal knowledge space.* In the following we will use the term 'knowledge space' as an abbreviation of 'quasi-ordinal knowledge space'.

A crucial problem in the theory of knowledge spaces is how to establish a surmise relation. This may be done in several ways. First, there is the possibility of querying experts. Questioning procedures[3] have been developed by Dowling (1991, in press) and Koppen and Doignon (1988). Another approach is to analyze response patterns as proposed by Airasian and Bart (1973), Bart and Krus (1973), Van Leeuwe (1974), and Villano (1991). A third approach — the one used in our investigations — is to infer the surmise relation from ordering principles which apply to the components of the problems. These ordering principles are described below.

What types of problems might prove suitable for assessing a player's knowledge of chess? Because chess is so very complex we restrict considerations to problems with unique solutions. In addition, we need to be able to specify the particular knowledge of chess, which is needed to solve each problem.

Tactical chess problems fulfill these requirements. Two typical problems of this kind are shown in Figure 9.1. In both positions White moves first, and the problem is to find the best move for White. The solution of problem (a) is:

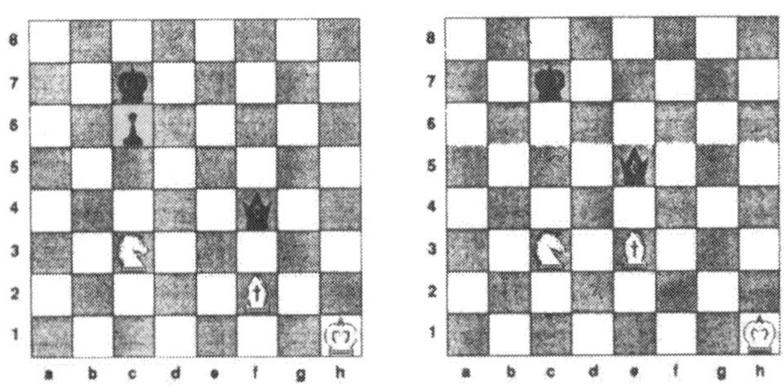

(a) (b)

FIGURE 9.1. Two examples of tactical chess problems

1. Ktd5+ cxd5; 2. Bg3 Qxg3 stalemate. The solution of problem (b) is: 1. Bf4 Qxf4; 2. Ktd5+ K ~ ; 3. Ktxf4 draw. Solving such problems requires knowledge of tactical elements, which are called 'motives' in chess terminology. Definitions of all motives that will be considered are given in Table 9.1. We now give an example of how a chess problem can be coded as a sequence of motives. The goal of White in problem (a) is to achieve a draw by forcing stalemate. To

[3]Note: These procedures are not restricted to the establishment of quasi-ordinal knowledge spaces only.

TABLE 9.1. Description of motives

Motive	Name	Description
F	Fork	One piece simultaneously attacks two opposing pieces of higher value.
P	Pin	A piece prevents an opposing piece from moving.
G	Guidance	An opposing piece is forced to a disadvantageous square.
D	Deflection	An opposing piece is forced to leave an important line or square.
E	Elimination	The elimination of an own piece is forced, e.g., to achieve a stalemate.
C	Clearing	An important line or square is left, e.g., to achieve a stalemate.
T	Promotion	A pawn has to be promoted to the most suitable piece, by moving to the 8th line.
S	Stalemate	A possible stalemate has to be anticipated and has either to be provoked or avoided.

achieve this, White must eliminate the Knight on c3. The move 1. Ktd5+ forces the elimination (cxd5) of the Knight. The move 2. Bg3 forces the black Queen to the disadvantageous square g3 by Qxg3, and reaches a draw by stalemate. Hence, the three motives 'elimination', 'guidance' and 'stalemate' are contained in problem (a) in this sequence. The idea behind problem (b) is (by winning the black Queen) to reach a draw. The move 1. Bf4 forces the black Queen to the disadvantageous square f4 (Qxf4) and the 'fork' 2. Ktd5+ wins the black Queen. Therefore, the motives 'guidance' and 'fork' are contained in problem (b) in this sequence. As can be seen from these examples, the same move Ktd5+ in nearly identical positions is related to different motives. In problem (a) it is based on the motive 'elimination', in problem (b) on the motive 'fork'. Hence, motives are mainly viewed as 'standard ideas' or 'subgoals'. Note that we only use motives which occur as subgoals in the solution plan, e.g., in problem (a) of Figure 9.1, the move 2. Bg3 is not based on the motive 'pin', because 'guidance', and not 'pin' is a subgoal for reaching the terminal goal 'stalemate'. That is because in this case, the Queen is not prevented from moving, but forced to move to g3. From our point of view knowledge or recognition of a motive means that a motive has been learned before and can be detected within a solution process. Recognizing the motives of a problem results in a solution plan. It is our opinion that the solution process for these types of problems is determined mainly by the combination and sequence of motives within a problem. There are probably other factors involved in a chess problem which also have an influence on the solution process. These factors are not the subject of our investigation;

considerations regarding these factors can be found in the general discussion below.

We next introduce the theoretical considerations underlying our investigations. We show how surmise relations may be established on sets of component-based chess problems. These surmise relations are inferred from psychological assumptions which are set out below. These relations, i.e., the corresponding sets of knowledge states, constitute the hypotheses of our empirical investigations. The test of a hypothesis will therefore consist mainly of comparing the set of theoretically inferred knowledge states with the response patterns observed in an experiment.

9.2 Theory

This section sets out two principles for the establishment of a surmise relation on sets of chess problems. Both principles are based on the idea that the difficulty of a chess problem depends mainly on the motives which a subject must recognize in order to find the correct solution of the problem.

Assume a finite set M of motives. Let the *problem space* $P(M)$ be the set of all chess problems which could be characterized by the motives in M.

PRINCIPLE 1 (set inclusion)

For a chess problem p let $F(p)$ be the set of the motives which a subject must recognize in order to find a correct solution for p. Assume that the difficulty of a chess problem p depends only on $F(p)$, i.e., all other factors are taken to be constant for all problems under investigation. To establish a surmise relation on $P(M)$ we can therefore identify a problem $p \in P(M)$ with the set of motives $F(p) \subseteq M$. The relation $\preceq$ is defined for all problems $p, q \in P(M)$ by the following condition,

$$p \preceq q :\Leftrightarrow F(p) \subseteq F(q).$$

This means that if a person is able to solve problem q, then he or she is also able to solve any problem p which can be solved with the knowledge of a subset of $F(q)$. From set theory it is clear that $\preceq$ is a partial order and hence a surmise relation.

Example: Let $M := \{a, b, c\}$. Figure 9.2 shows the resulting surmise relation on $P(M)$.

PRINCIPLE 2 (sequence inclusion)

In Principle 1, the order in which the motives occur in the solution of a chess problem does not affect the resulting surmise relation. Ignoring this order may, however, be problematical for the analysis of (chess) problem solving, e.g., because of forward and backward search strategies in problem solving. In Principle 2, the order in which the motives occur in the solution of a problem plays

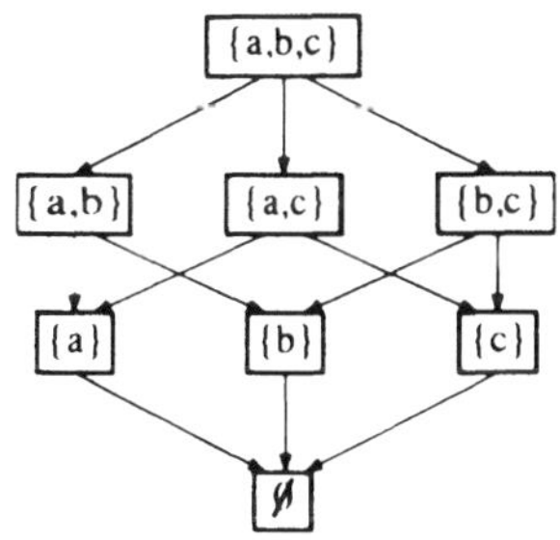

FIGURE 9.2. Surmise relation on $P(M)$, according to 'Principle 1'

a central role in the construction of the surmise relation on $P(M)$. The central idea of Principle 2 is that a problem a is more difficult than a problem b if the sequence of motives which must be recognized by a subject to solve a includes the sequence of motives which must be recognized to solve b. For a chess problem $p \in P(M)$, let $G(p)$ be the ordered tuple of motives which occur in the solution of the problem. $G(p) = (m_1, \ldots, m_k)$ means that the first motive which occurs is m_1, the second is m_2, etc. Note that in contrast to Principle 1, one motive can occur more than once within $G(p)$, i.e., the multiplicity of motives is also covered.

The order in which the motives occur in the solution of a problem[4] is very important for Principle 2. But it will be clear from our ordering principle that the resulting surmise relation will not change if the order $(m_1, \ldots, m_k)$ is, *for all* problems, inverted to $(m_k, \ldots, m_1)$. We define,

$$M_k := \{(m_1, \ldots, m_k) \mid m_1, \ldots, m_k \in M\},$$
$$M_{\mathbb{N}} := \textstyle\bigcup_{k \in \mathbb{N}} M_k.$$

It is clear from chess experience that some motives are easier to recognize than others. This means that the motives themselves can be ordered with respect to their difficulty. Therefore we can assume a quasi-order $\sqsubseteq_M$ on M. For motives $m_i, m_j \in M$ the interpretation of $m_i \sqsubseteq_M m_j$ is "every person who is able to recognize m_i is also able to recognize m_j".

Suppose that under suitable conditions the difficulty of a chess problem $p \in P(M)$ depends only on $G(p)$. To establish a surmise relation $\preceq$ on $P(M)$, we identify a problem p with $G(p)$. It is then sufficient to define a quasi-order $\sqsubseteq$ on $M_{\mathbb{N}}$. For problems $p, q \in P(M)$ define, as in Principle 1,

$$p \preceq q :\Leftrightarrow G(p) \sqsubseteq G(q).$$

[4] The order in which the motives are recognized by a subject is, however, not necessarily identical with the order in the solution of a problem, and in the solution plan, respectively.

We now present a formalization of the assumed quasi-order $\sqsubseteq$ on $M_{\mathbf{N}}$.

Definition 1 *Let* $(m_1, \ldots, m_k), (m'_1, \ldots, m'_l) \in M_{\mathbf{N}}$. *Then* $(m_1, \ldots, m_k) \sqsubseteq (m'_1, \ldots, m'_l)$ *if and only if there exists a function* $f : \{1, \ldots, k\} \rightarrow \{1, \ldots, l\}$ *which satisfies the following conditions:*

1. $\forall i, j \in \{1, \ldots, k\} \quad (i < j \rightarrow f(i) < f(j))$,
2. $\forall j \in \{1, \ldots, k\} \quad m_j \sqsubseteq_M m'_{f(j)}$.

The reflexivity of $\sqsubseteq$ is obvious (choose the identity for f). The transitivity of $\sqsubseteq$ follows because the convolution of two monotonic increasing functions also has this property. Hence, $\sqsubseteq$ is a quasi-order which depends on the relation $\sqsubseteq_M$ on M. The following example illustrates this dependency.

Example: Let $M := \{a, b\}$. Figure 9.3 shows the resulting quasi-order $\sqsubseteq$ on $M_1 \cup M_2 \subseteq P(M)$ for two different assumptions on $\sqsubseteq_M$. In one of them, it is assumed that $a \sqsubseteq_M b$ (all edges); in the other, this assumption is dropped (bold edges).

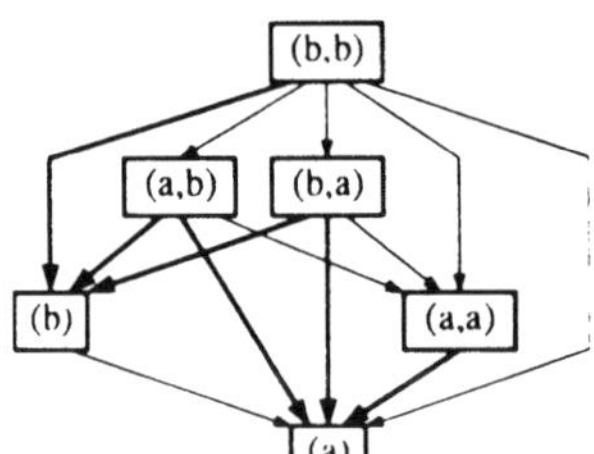

FIGURE 9.3. Bold edges indicate the resulting quasi-order $\sqsubseteq$, if it be assumed that $x \sqsubseteq_M y \Leftrightarrow x = y$, with $x, y \in \{a, b\}$. Thin edges indicate the additional pairs, if it be assumed that $x \sqsubseteq_M y \Leftrightarrow x = y \vee (x = a \wedge y = b)$, with $x, y \in \{a, b\}$.

9.3 Experiment I[5]

For this experiment problem construction and problem ordering were based on the principle of set inclusion (Principle 1) outlined above. The set of motives M contains the four motives F, P, G and D (see Table 9.1). The complete combination of these motives (the power set of M) yields — without the 'empty problem' — 15 different motive sets. The resulting relation on the problem space

[5]Detailed information concerning this experiment as well as Experiment II is available from the authors on request.

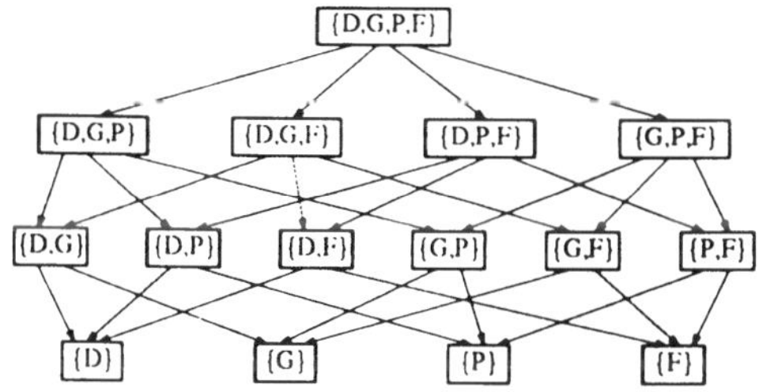

FIGURE 9.4. Problem structure of Experiment I

is depicted in Figure 9.4. The corresponding knowledge space consists of 167 states.

METHOD

For each of the 15 motive sets one possible problem-realization was chosen.

Thirteen male subjects, 19 – 54 years of age, took part in the group investigation. All subjects were members of the Ladenburg chess club in Germany. Their chess playing ability varied widely. The subjects were asked to solve the problems "as accurately and quickly as possible"; however, the solution time was not restricted. The problems were presented in order of decreasing hypothesized difficulty, so that the problem with four motives was first and the problems with one motive were last in succession.

RESULTS

Table 9.2 shows the symmetric distances[6] between each person's response pattern and the closest knowledge state. The response patterns of only four subjects agree with our hypothesis, and two subjects deviated from our hypothesis by only one problem.

DISCUSSION

The results of Experiment I are unsatisfactory since only four of the observed response patterns are elements of the predicted knowledge space, and the distances (see Table 9.2) for most of the other 9 subjects are fairly large. The reasons for this result may be found both in the experimental design and the theoretical approach of 'Principle 1'.

The main problems with the experimental design may be the setting as a group experiment and the absence of any limit on solution time. As a consequence,

[6]The symmetric distance d between two sets A and B is defined as follows: $d(A, B) = | A \Delta B |$, where $A \Delta B = (A \backslash B) \cup (B \backslash A)$.

TABLE 9.2. Frequency of distances between response patterns and states for Experiment I

			Distances				Average
0	1	2	3	4	5	6	Distance
4	2	1	3	2	0	1	2.1

Number of subjects who solved all problems: 3
Number of subjects who solved no problem: 0

some subjects had already completed all of the problems, while others were still working on them. This may have caused motivational difficulties. The order of problem presentation (decreasing difficulty) may also have resulted in difficulties because of declining motivation. Fatigue may have also prevented the correct solution of the simple problems. Another critical point is the inhomogeneity of the problem set which includes endgame studies as well as examples from 'problem chess'.

As already mentioned (see 'Principle 2' above), ignoring the sequence of the motives within the problems seems to be problematic for chess problems.

9.4 Experiment II

The second experiment made use of the ordering method of the principle of sequence inclusion (Principle 2). The motives F, S, E, C, G, and T (see Table 9.1) were used for problem construction. Each problem contains a maximum of four motives. We distinguish between two different types of surmise relations. If it be assumed that none of the motives can be recognized more easily than any other (i.e., $m_i \sqsubseteq_M m_j \Leftrightarrow i = j$), the resulting surmise relation consists of two disjoint digraphs as shown in Figure 9.5. But if it be assumed that Motive F can be recognized more easily than motives S, E, C, G, and T (i.e., $m_i \sqsubseteq_M m_j \Leftrightarrow i = j \vee (m_i = F)$), the surmise relation shown in Figure 9.6 results. This further assumption adds additional pairs of problems to the surmise relation (e.g., $(G, S) \sqsubseteq (G, F)$).

METHOD

The problem set consisted of 17 problems[7]. The number of moves necessary for the solution of the problems ranged from one to four.

The experiment was conducted with 37 male and 9 female subjects. Their ages

[7]Due to an error in presentation one problem had to be excluded from further analysis. So only 16 problems remain.

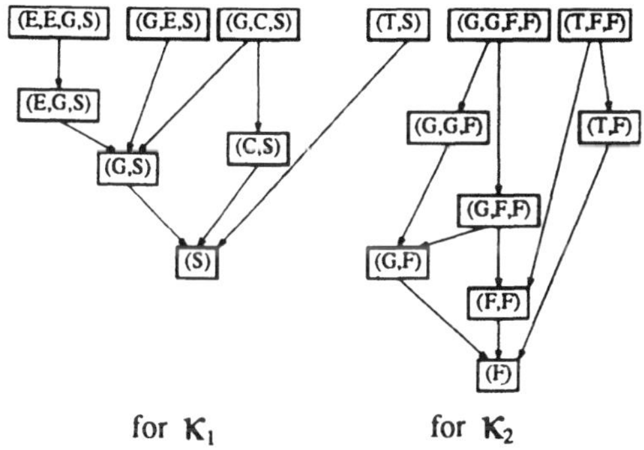

FIGURE 9.5. Surmise relations for Experiment II, assuming that no motive can be recognized more easily than any other

ranged from 14 to 71 years. All of them were familiar with the basic rules of chess. The problems were presented on a computer screen and the subjects were required to make their moves using a mouse. The 17 problems were presented in random order. The subject had 90 seconds to work out a solution plan to a problem. After that, each move had to be made within 30 seconds. The time limitations were chosen to preclude two kinds of strategy on the part of the subjects: (1) to prevent any subject moving too soon — without having thought sufficiently about the solution to the problem, and (2) to prevent the subject continuing to work out a solution while making his or her moves. After each move by the subject, the computer either made a suitable reply (if the subject's move was correct), or the game was terminated (if either the problem was completed or the subject's move was wrong).

RESULTS

First, how well do the data fit our knowledge spaces? The knowledge spaces are labeled $\mathcal{K}_1$, $\mathcal{K}_2$, $\mathcal{K}_{(12)}$, $\mathcal{K}_3$, and correspond to the surmise relations in Figures 9.5 and 9.6. Table 9.3 provides the number of states and 'nonstates'; nonstates are subsets of the problem set which do not belong to the corresponding knowledge space. Table 9.3 also shows the number of subjects with response patterns congruent with a state, response patterns not congruent with any state, as well as the number of *different* congruent, and not congruent response patterns within the data. Table 9.4 presents the symmetric distances between response patterns and states.

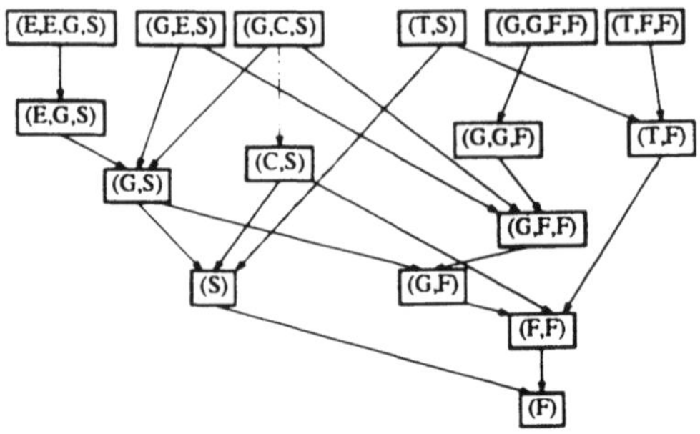

for $\mathcal{K}_3$

FIGURE 9.6. Surmise relation of Experiment II, based on the assumption that motive F can be recognized more easily than all other motives in our motive set

TABLE 9.3. Possible states and response patterns for Experiment II

Knowl. space	Number of states	Number of non-states	Number of congruent patterns	Number of non congruent patterns	Number of different congruent patterns	Number of different non congruent patterns
$\mathcal{K}_1$	41	215	35	11	14	4
$\mathcal{K}_2$	25	231	37	9	13	7
$\mathcal{K}_{(12)}$	1025	64511	28	18	20	16
$\mathcal{K}_3$	368	65168	23	23	15	21

DISCUSSION

On the whole, this experiment can be considered a success. Table 9.4 shows that the symmetric distances between the response patterns and the closest states are (in almost all cases) rather small (≤ 2). Most of the violations may follow from lucky guesses and careless errors (see Falmagne & Doignon, 1988)[8].

In analyzing the results, special attention should be paid to the proportions between the number of states and the number of nonstates, on the one hand, and the proportions between the number of response patterns congruent with states and the number of response patterns not congruent with states, on the other. The sequence inclusion principle seems to be adequate for chess problems.

[8]The probability of lucky guesses and careless errors must be very low. If we assume for instance that the probability of both careless errors and lucky guesses is 0.05, then the probability that the response pattern of the subject is equal to the subject's knowledge state is $0.95^{16} \approx 0.44$.

TABLE 9.4. Frequency of distances between response patterns and closest states for Experiment II

Knowl. space	Distances 0	1	2	3	4	Average Distance
$\mathcal{K}_1$	35	10	1	0	0	0.26
$\mathcal{K}_2$	37	8	0	1	0	0.24
$\mathcal{K}_{(12)}$	28	14	3	1	0	0.50
$\mathcal{K}_3$	23	15	7	0	1	0.72

Number of subjects who solved all problems: 4
Number of subjects who solved no problem: 0

The additional assumption that the motive 'fork' can be recognized more easily than all of the other motives leads to a reduction in the number of states. It therefore comes as no surprise that this assumption leads to a larger number of response patterns which do not agree with the knowledge space. Most of them, however, have a small symmetric distance. In general, the proposed surmise relations following from the assumptions of 'Principle 2' have proved suitable for our set of chess problems.

GENERAL DISCUSSION

The results of Experiment I lead to the assumption that pure set inclusion as introduced in 'Principle 1', seems to be — at least with respect to our experimental conditions — inappropriate for the ordering of component based chess problems. To confirm this assumption, the problems of Experiment II were also ordered by means of 'Principle 1'[9]. The corresponding knowledge space consists of 85 states, the number of nonstates is 65451. The response patterns of only seven subjects (out of 46) are elements of this knowledge space (average distance: 1.43). Comparing this with the result for the knowledge space $\mathcal{K}_{(12)}$ (see Table 9.3 and Table 9.4), Principle 2 is much more successful, with respect to the congruent data patterns, than Principle 1. However, there is an ordering principle which is more specific than set inclusion and less specific than sequence inclusion — the "multiset inclusion"[10]. This principle takes only the multiplicity of motives into account and neglects the motives' sequence. This means that a surmise relation $\trianglelefteq$ can be defined through the condition $x \trianglelefteq y$ if and only if every motive appears at least as often in y as in x, for every pair of tuples of motives (x, y). As can be seen from Figure 9.5, it is necessary to add only the

[9]Because of the previously discussed difficulties concerning the experimental setting and the small number of subjects, a corresponding analysis of Experiment I by means of Principle 2 seems inappropriate.

[10]This hypothesis was put forward by one of the anonymous referees.

pairs $(E,G,S) \trianglelefteq (G,E,S)$, $(G,E,S) \trianglelefteq (E,G,S)$ and $(G,E,S) \trianglelefteq (E,E,G,S)$ to the relation $\sqsubseteq$. The corresponding knowledge space consists of 576 states, the number of nonstates is 64960. The response patterns of 27 subjects are elements of this knowledge space (average distance: 0.57). Compared to the results for $\mathcal{K}_{(12)}$, with respect to the congruent data patterns, the multiset inclusion is almost as successful as the more specific Principle 2. Hence, because of the selected problem set, it is difficult to distinguish a-posteriori between Principle 2 and multiset inclusion in our investigation.

Another point which should be discussed is the suitability of motives for the characterization of chess problems. Motives may only be adequate as 'exclusive' components of simple problems as used in our investigations. However, other features of the problems may influence their difficulty. For example, the difficulty of a problem may also be dependent on the number of pieces which appear on the board. But a surmise relation which is solely based on the number of pieces would lead to an average symmetric distance of 2.74. In Experiment II we tried to minimize the influence of other problem elements by striving for maximal homogeneity of the problem set. With more complex problems, the set of relevant problem components will still contain motives, but other factors which are indispensable in the knowledge of a good chess player may become relatively more important. More information concerning these other factors can be gained from the investigations of De Groot (1965).

Our investigation was based on an extension of the theory of Doignon and Falmagne (1985). We have provided principles for the establishment of surmise relations and knowledge spaces which are primarily dependent on problem components. In addition to Doignon and Falmagne, who do not consider the properties of the problems themselves, we are particularly interested in those components of problems which may be fundamental to the surmise relation and the corresponding knowledge space. In both our theoretical and empirical investigations we have been able to show that the theory of Doignon and Falmagne can be used for testing psychological models.

9.5 References

Airasian, P. W., & Bart, W. M. (1973). Ordering theory: A new and useful measurement model. *Educational Technology, May*, 56–60.

Albert, D., & Held, T. (in press). Establishing knowledge spaces by systematical problem construction. To appear in: D. Albert (Ed.), *Knowledge structures.* Heidelberg, New York: Springer.

Bart, W. M., & Krus, D. J. (1973). An ordering-theoretic method to determine hierarchies among items. *Educational and Psychological Measurement, 33*, 291-300.

Doignon, J.-P., & Falmagne, J.-Cl. (1985). Spaces for the assessment of knowledge. *International Journal of Man-Machine Studies, 23*, 175-196.

Dowling, C. E. (1991). *Constructing knowledge structures from the judgement of experts.* Habilitationsschrift. Braunschweig: Technical University of Braunschweig.

Dowling, C. E. (in press). Applying the basis of a knowledge space for controlling the questioning of an expert. *Journal of Mathematical Psychology.*

Falmagne, J.-Cl., & Doignon, J.-P. (1988). A Markovian procedure for assessing the state of a system. *Journal of Mathematical Psychology, 32 (3),* 232–258.

De Groot, A. D. (1965). *Thought and choice in chess.* The Hague: Mouton & Co.

Koppen, M., & Doignon, J.-P. (1990). How to build a knowledge space by querying an expert. *Journal of Mathematical Psychology, 34,* 311-331.

Van Leeuwe, J. F. J. (1974). Item tree analysis. *Nederlands Tijdschrift voor de Psychologie, 29,* 475-484.

Villano, M. (1991). *Computerized knowledge assessment: Building the knowledge structure and calibrating the assessment routine.* Doctoral dissertation. New York: New York University.

10

The Construction of Knowledge Spaces by Querying Experts

Mathieu Koppen[1]

ABSTRACT In the theory of knowledge spaces the actual construction of a space giving a reasonably valid description of some specific domain of knowledge is a critical problem. A method is presented in which the information needed is obtained from experts in the field who are confronted with a carefully chosen sequence of questions about specific relationships between the problems in the domain. The discussion covers the type of questions which have to be asked, how the responses to those questions permit the construction of the corresponding knowledge space, and how inferences from responses previously obtained can be exploited to make the procedure practicable for a substantial number of items.

10.1 Introduction

Advanced computerized instruction systems or so-called *Intelligent Tutoring Systems* (Sleeman and Brown, 1982) need to include an efficient assessment of knowledge. The design of assessment procedures depends on how a student's knowledge is to be represented and measured. A framework for doing this has been developed by Doignon and Falmagne (1985), and an overview of the ensuing project is presented in Falmagne et al. (1990). The basic assumption of this approach is that any domain can be ultimately decomposed into a collection of *problems* or *items* that can be asked in that domain. These items constitute the basic elements in the representation of knowledge. Assessing the knowledge of a student in this domain means determining the subset of items that the individual has mastered. This subset is called his or her *knowledge state.* The collection of all possible knowledge states is a particular subcollection of subsets of items. This collection captures the cognitive organization of the domain and is called its *knowledge structure.*

Given the knowledge structure of a domain, assessing a student's knowledge amounts to assigning this student to one or another of the knowledge states. Falmagne and Doignon (1989 a,b) have developed procedures to accomplish this when the data are fallible (due to careless errors or lucky guesses on the part of the student). In Kambouri (1991) these routines are compared with

---

[1]University of Nijmegen, NICI, Department of Mathematical Psychology, P.O. Box 9104, 6500 HE Nijmegen, The Netherlands. This paper describes work done in close collaboration with Jean-Paul Doignon, Jean-Claude Falmagne, Maria Kambouri, and Michael Villano. The research was supported by DOD grant MDA903-87-K-0002 to Jean-Claude Falmagne at New York University and by a postdoctoral fellowship of the Royal Netherlands Academy of Arts and Sciences to Mathieu Koppen.

an assessment by human experts. A critical problem now is: How can, for a particular domain, the knowledge structure be constructed that is to be searched by these assessment procedures? A "logical" analysis of the domain can, at best, be only part of the solution. Such an analysis will not contribute much unless the domain is highly formalized. And even when this is so (take some subfield of mathematics, say), there may be many states that are logically feasible but do not occur in practice. (One might say they are "psychologically" infeasible.) For a knowledge structure to be efficient, it is necessary that it represent these *factual* restrictions on the collection of knowledge states. This also shows that the knowledge structure is not just a property of the domain, but of the chosen population of students as well.

These considerations suggest that for information on the knowledge structure one could consult experts in the chosen domain who have much experience with the chosen student population. It is not obvious, however, how to elicit the relevant information from the experts. In this paper a procedure is outlined that was developed for this purpose and has been applied to the questioning of five experts in the field of US high school mathematics. The aim is to present a survey of the results in an easily accessible form. The theoretical and technical background is provided in Koppen and Doignon (1990) and Koppen (1992); an analysis of the application appears in Kambouri et al. (1992). Since it relies formally only on the mathematical properties of *closure systems* (see, e.g., Birkhoff, 1967), the procedure to be described is directly applicable in other contexts, viz. in the determination of any system of sets that is closed under either union or intersection.

10.2 Knowledge spaces and entail relations

As discussed elsewhere (Falmagne et al., 1990; Koppen and Doignon, 1990), it is not realistic to ask an expert directly for a list of possible knowledge states. Instead, recourse must be had to an indirect approach based on an alternative representation of the knowledge structure which is available under some reasonable restriction on the knowledge structures to be considered. This approach is illustrated on a miniature domain of five items from the field of basic arithmetic. Consider the set $I = \{a, b, c, d, e\}$ containing the following problems:

a : Multiply: 378×605 b : $58.7 \times 0.94 =$?

c : $\frac{1}{2} \times \frac{5}{6} =$? d : What is 30% of 34 ?

e : Gwendolyn is $\frac{3}{4}$ as old as Rebecca. Rebecca is $\frac{2}{5}$ as old as Edwin. Edwin is 20 years old. How old is Gwendolyn ?

For the sake of argument, suppose that the knowledge structure for this domain (and for the chosen population of students) is given by the following collection

$\mathbf{K}_0$ of subsets of I:

$$\mathbf{K}_0 = \{\emptyset,\ \{a\}, \{c\}, \{a,b\}, \{a,c\}, \{a,d\}, \{b,c\}, \{c,d\}, \{a,b,c\}, \{a,b,d\},$$
$$\{a,c,d\}, \{b,c,d\}, \{a,b,c,d\}, \{a,b,c,e\}, \{a,b,c,d,c\}\ \}. \tag{10.1}$$

Suppose that an expert in this domain relies implicitly on $\mathbf{K}_0$, but, as indicated at the start of the section, it is considered infeasible to ask him or her to simply say, for every subset of I, whether or not it is a state (i.e., a member of $\mathbf{K}_0$). The problem, then, is to elicit information from the expert sufficient to reconstruct the knowledge structure $\mathbf{K}_0$.

One type of question that comes to mind is this: given two problems x and y within the domain, will a student who solves y correctly also be able to solve x? In view of subsequent developments, the contrapositive form turns out to be more convenient: Will a student who fails on x also fail on y? In order to relate responses directly to the latent knowledge structure, "solving (failing) a problem" is to be read as "(not) mastering a problem". So the expert should disregard the possibility of a careless error or a lucky guess on the student's part. This leads to the following complete version of the question:

[**Q1**] Suppose that a student under examination has just provided an incorrect response to problem x; is it practically certain that this student will then also fail on problem y? Assume that the testing conditions are optimal in the sense that careless errors and lucky guesses do not occur.

In what follows this elaborated question will be abbreviated as: "Does failing on x entail failing on y?".

Clearly, given the knowledge structure, the correct responses to all questions of this form are determined: the answer to the above question is "yes" if and only if there is no state in the knowledge structure containing y and not x. Indeed, any such state implies the (possible) existence of a student failing on x but solving y, and the expert should answer "no". Write $x\mathbf{P}y$ if the expert would respond "yes" to the question "Does failing on x entail failing on y?", $x\mathbf{N}y$ if not. In terms of the underlying knowledge structure,

$$x\mathbf{P}y \quad \text{means} \quad \text{any state containing } y \text{ also contains } x, \tag{10.2}$$
$$x\mathbf{N}y \quad \text{means} \quad \text{there is a state containing } y \text{ and not } x. \tag{10.3}$$

For instance, for $\mathbf{P}$ and $\mathbf{N}$ corresponding to the knowledge structure in (10.1) we have $a\mathbf{N}b$ and $a\mathbf{P}e$. Clearly, $\mathbf{P}$ is a quasi-order (reflexive, transitive relation) on the set of items.

The question at issue, however, requires inference in the contrary direction: can the underlying knowledge structure $\mathbf{K}_0$ be recovered from the knowledge of

TABLE 10.1. The derivation of knowledge structures from questions **[Q1]** and **[Q2]**. A cross signifies that this line's subset of items is excluded as a state by the corresponding positive response on the top line. The starred rows in the columns "$\mathbf{K}_1$" and "$\mathbf{K}_0$" are the states of the knowledge structures obtained from the answers to all questions of the types **[Q1]** and **[Q2]**, respectively.

	aPe	bPe	cPe	K_1	acPb	acPd	K_0
$\emptyset$				*			*
$\{a\}$				*			*
$\{b\}$				*	×		
$\{c\}$				*			*
$\{d\}$				*		×	
$\{e\}$	×	×	×				
$\{a,b\}$				*			*
$\{a,c\}$				*			*
$\{a,d\}$				*			*
$\{a,e\}$		×	×				
$\{b,c\}$				*			*
$\{b,d\}$				*	×	×	
$\{b,e\}$	×		×		×		
$\{c,d\}$				*			*
$\{c,e\}$	×	×					
$\{d,e\}$	×	×	×			×	
$\{a,b,c\}$				*			*
$\{a,b,d\}$				*			*
$\{a,b,e\}$			×				
$\{a,c,d\}$				*			*
$\{a,c,e\}$		×					
$\{a,d,e\}$		×	×				
$\{b,c,d\}$				*			*
$\{b,c,e\}$	×						
$\{b,d,e\}$	×		×		×	×	
$\{c,d,e\}$	×	×					
$\{a,b,c,d\}$				*			*
$\{a,b,c,e\}$				*			*
$\{a,b,d,e\}$			×				
$\{a,c,d,e\}$		×					
$\{b,c,d,e\}$	×						
$\{a,b,c,d,e\}$				*			*

this relation **P**? It is easy to check that, if [**Q1**] be asked for all pairs of items in I, the only positive responses obtained are $a\mathbf{P}e$, $b\mathbf{P}e$, and $c\mathbf{P}e$. Each of these excludes some subsets of I as possible knowledge states according to (10.2). The consequences are set out in Table 10.1. For instance, in the column indexed by $a\mathbf{P}e$ the rows that correspond to subsets containing e but not a are crossed. The knowledge structure derived from the responses to [**Q1**] consists of the subsets having no cross in the first three columns; the corresponding rows are starred in the middle column of Table 10.1. This structure, $\mathbf{K}_1$, is not the same as $\mathbf{K}_0$, but contains some additional states:

$$\mathbf{K}_1 = \mathbf{K}_0 \cup \{\{b\}, \{d\}, \{b, d\}\}. \tag{10.4}$$

Evidently, a unique knowledge structure cannot be inferred from the responses to [**Q1**]: both $\mathbf{K}_0$ and $\mathbf{K}_1$ are compatible with those responses and, in fact, there are other structures as well.

Note that the derived structure $\mathbf{K}_1$ is closed under both union and intersection. (The union and intersection of any two states are again states.) This illustrates Birkhoff's (1937) theorem on the 1-1 correspondence between quasi-orders and systems closed under union and intersection which in general implies that (i) a knowledge structure derived (as in Table 10.1) from the responses to [**Q1**] is necessarily closed under union and intersection, and (ii) among all knowledge structures compatible with the responses to [**Q1**], this is the only one having these closure properties.

If there is reason to suppose that, in practice, any knowledge structure will be closed under union and intersection, then implication (ii) shows that the information provided by [**Q1**] is, in fact, sufficient. Restricted to such knowledge structures, $\mathbf{K}_1$ in (10.4) is the unique solution consistent with the data $a\mathbf{P}e$, $b\mathbf{P}e$, $c\mathbf{P}e$, and $x\mathbf{N}y$ for all other pairs (x, y). As has been argued in detail elsewhere (Doignon and Falmagne, 1985; Falmagne et al., 1990), the assumption of closure under union is acceptable but that of closure under intersection is not, at least, not in general. A knowledge structure closed under union is called a *(knowledge) space*.

It can be checked that $\mathbf{K}_0$ is a space, but it is not closed under intersection (e.g., $\{a, d\} \cap \{b, c, d\} \notin \mathbf{K}_0$). For a knowledge space, the responses to [**Q1**] do *not* give complete information. For instance, although both $a\mathbf{N}b$ and $c\mathbf{N}b$, the expert relying on $\mathbf{K}_0$ would answer "yes" to the question

> "Does failing both on a and c entail failing on b?",

since any state in $\mathbf{K}_0$ containing b contains either a or c (or both). This suggests a more general type of question to ask the expert:

[**Q2**] Suppose that a student under examination has just provided an incorrect response to all problems in a subset A of I; is it practically certain that this student will then also fail on problem y? Assume that the testing conditions are optimal in the sense that careless errors and lucky guesses do not occur.

When, given a knowledge structure, the answer to this question is "yes", write $A\mathbf{P}y$, otherwise $A\mathbf{N}y$, thereby extending the definition of $\mathbf{P}$ to a relation between subsets of items and single items:

$$A\mathbf{P}y \quad \text{means that} \quad \text{any state containing } y \text{ contains an element of } A, \tag{10.5}$$

$$A\mathbf{N}y \quad \text{means that} \quad \text{there is a state containing } y \text{ and no element of } A. \tag{10.6}$$

The relation $\mathbf{P}$ as defined in (10.5) is called the *entail relation* (Koppen and Doignon, 1990).

By asking all questions [**Q2**], $\mathbf{K}_1$ is, indeed, further reduced to $\mathbf{K}_0$. Among others, the positive responses $\{a, c\}\mathbf{P}b$ and $\{a, c\}\mathbf{P}d$ will be obtained. (Any state in $\mathbf{K}_0$ containing b or d also contains either a or c.) The consequences of these additional questions are also displayed in Table 10.1. The first response excludes $\{b\}$ as a state, the second excludes $\{d\}$, and $\{b, d\}$ is incompatible with both responses. From (10.4) it is clear then that $\mathbf{K}_0$ eventuates. There are more positive responses to the questions [**Q2**], but these do not delete any more states.

Koppen and Doignon (1990) proved, in a generalization of Birkhoff's (1937) result, that (i) any knowledge structure derived from the responses to [**Q2**] is a space, and (ii) among all knowledge structures that are compatible with the responses to [**Q2**], this is the only one that is a space. (Müller, 1989, and Burigana, 1991, both independently obtained an equivalent result, the latter, interestingly enough, in a completely unrelated context.) Consequently, if the knowledge structure to be discovered is a space, it can, indeed, be reconstructed from the entail relation $\mathbf{P}$ coding the responses to [**Q2**].

10.3 Outline of the algorithm

The preceding discussion suggests the following straightforward procedure to obtain the knowledge space from responses to [**Q2**].

Step 1: Draw up the list of all subsets of I; this constitutes the initial knowledge space.

Step 2: Ask all questions of the form [**Q2**], and whenever a positive response $A\mathbf{P}y$ is obtained, remove from the current list of states all subsets that contain y but none of the items in A.

This procedure is correct, but not practicable. With n items there are $n2^{n-1}$ questions of the form [**Q2**], excluding the trivial ones where $y \in A$. Even *Step 1* becomes quickly infeasible: for 50 items, say, the list would contain over 10^{15} entries. Fortunately, many queries can be eliminated by drawing appropriate inferences from the already obtained responses. The potential for inference is so

TABLE 10.2. The four inference rules used to avoid redundant questions to the expert

	From	one can infer	whenever it has been established that
IR1:	$A\mathbf{P}x$	$B\mathbf{P}y$	$(A \cup \{x\})\mathbf{P}y$ and $B\mathbf{P}A$
IR2:	$A\mathbf{P}x$	$B\mathbf{N}y$	$A\mathbf{N}y$ and $(A \cup \{x\})\mathbf{P}B$
IR3:	$A\mathbf{P}x$	$B\mathbf{N}y$	$B\mathbf{N}x$ and $(B \cup \{y\})\mathbf{P}A$
IR4:	$A\mathbf{N}x$	$B\mathbf{N}y$	$(B \cup \{y\})\mathbf{P}x$ and $A\mathbf{P}B$

great that, in many cases, the number of subsets that need to be generated can be dramatically reduced. Instead of starting with a complete list of subsets, a refined version of the algorithm produces subsets step by step. The rest of this section summarizes how this refined algorithm works. Full details can be found in Koppen (1992).

One basic inference, obvious from the definition of $\mathbf{P}$, is that an observation $A\mathbf{P}b$ implies $A'\mathbf{P}b$ for any $A' \supseteq A$. Thus, any "new" observation of the form $A'\mathbf{P}b$ would not reveal anything not already known. Since, restricted to single items, $\mathbf{P}$ is transitive, one can also infer $a\mathbf{P}c$ from the observations $a\mathbf{P}b$ and $b\mathbf{P}c$. These basic examples can also be formulated negatively (e.g., given $a\mathbf{P}b$, a new observation $a\mathbf{N}c$ implies $b\mathbf{N}c$), and both positive and negative versions generalize to more complex situations. In the refined version of the algorithm there are four inference rules to be invoked after each response (positive or negative) from the expert. These rules specify under what conditions an inference $B\mathbf{P}y$ or $B\mathbf{N}y$ is possible after obtaining the response $A\mathbf{P}x$ or $A\mathbf{N}x$ from the expert. They are set out in Table 10.2, where the notation "$A\mathbf{P}B$" is used as an abbreviation for "$A\mathbf{P}y$ for all $y \in B$". The source of these rules may be found in Koppen (1992).

To apply these inference rules it is necessary to maintain a subsets-by-items table in which it is recorded, for each subset A and item y, whether $A\mathbf{P}y$ or $A\mathbf{N}y$ is known to hold, or whether the question is still open. At first sight this appears to involve a substantial amount of extra bookkeeping, but it turns out to be labor well spent. First, with the table of inferences available, it is no longer necessary to keep track of a list of states (cf. *Step 2*). The final list is easily obtained from the completed table: K is a state if, and only if, for some subset of items A, $K = \{x : A\mathbf{N}x\}$. Second, all the information in the table of inferences is generally already contained in a subtable of dramatically smaller size and the construction of this smaller subtable is sufficient, both for applying the inference rules and for extracting the knowledge states at the end.

The method chosen to build a minimal subtable of inferences is to proceed by *blocks*, which are defined by the number of items in the subset A in **[Q2]**. First the expert is asked, for any single item a and any b, whether failing on a entails failing on b. These are in fact the questions **[Q1]** and they constitute Block 1. Next comes Block 2 where the generic question is whether failing on two items

TABLE 10.3. The subsets-by-items table, consisting of successive "blocks"

	a	b	c	...	y	z	
$\{a\}$	P						
$\{b\}$		P					
$\vdots$							Block 1
$\{z\}$						P	
$\{a,b\}$	P	P					
$\{a,c\}$	P		P				
$\vdots$							Block 2
$\vdots$							
$\{y,z\}$					P	P	
$\{a,b,c\}$	P	P	P				
$\{a,b,d\}$	P	P					
$\vdots$							
$\vdots$							Block 3
$\vdots$							
$\{x,y,z\}$					P	P	
$\{a,b,c,d\}$	P	P	P				
$\vdots$							Block 4

a_1 and a_2 entails failing on item b. Block 3 deals with triplets a_1, a_2, a_3, and so on. Table 10.3 shows this partitioning of the table into blocks. The reasons for this approach have been explained by Koppen (1992).

In fact, Block k of the table is generated only after Blocks 1 to $k-1$ have been completed and will, generally, consist of only a fraction of the total number of subsets of size k. Which subsets need to be included is determined on the basis of information collected from the previous blocks. In this design, these inference rules of Table 10.2 can only be applied to the subsets appearing in the current block; so, before questions in the newly constructed Block k are asked, those inferences for this block are implemented that can be drawn from the reponses obtained in the preceding blocks. Then the remaining open questions in Block k are asked, with inferences being drawn simultaneously with respect to other subsets in the block. This process of constructing the table block by block terminates when the new block does not contain any subset at all. This state of affairs indicates that the subtable constructed so far contains all information about the complete table and, in particular, the knowledge space to be discovered is given by the collection of all sets $\{x : A\mathrm{N}x\}$, where A runs through the rows of this subtable.

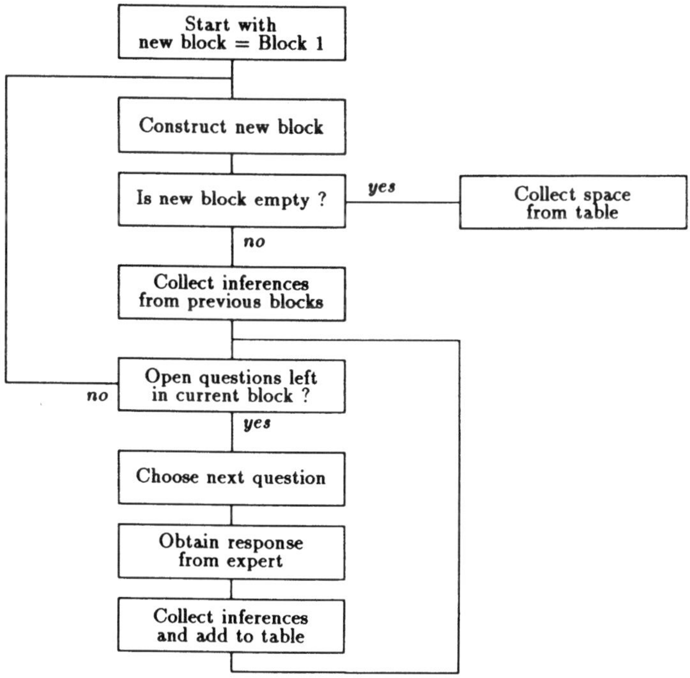

FIGURE 10.1. The overall design of the algorithm. For explanation, see text.

Details on how the information in the previous blocks determines which subsets will appear in the new block, or how to find the postponed inferences for the new block, are given in Koppen (1992). Figure 10.1 shows (in the form of a flow-chart) the overall design of the algorithm, as it has been outlined above.

Finally, a few words on the selection of the next open question to be presented to the expert. The objective is to minimize the number of queries, and various approaches are conceivable. In the current version of the algorithm a "maximin" criterion is used, trying specifically to minimize the possible costs of the more unfavorable (in terms of efficiency) response to the question. For each question, the number of inferences to be drawn from a positive response, and that from a negative response, are computed, and first the questions are selected for which the smaller of these two numbers is maximal over all questions. From among these questions, one is picked for which the other number is highest. When it is not feasible to apply this selection process to all open questions, the choice is based on a pseudorandom sample of some predetermined maximum size.

10.4 An application

This algorithm has been tested on a real-life domain of 50 items in US high school mathematics. (The items I of Section 10.2 are a sample from this domain.) Five experts (experienced tutors) underwent the querying procedure. The results are described in Kambouri et al. (1992); here the discussion will be restricted to a few data which give indications of the feasibility of this whole approach.

Four out of the five experts finished the procedure after completing only 4 or 5 blocks (in principle 49 blocks were possible) and by answering fewer than 2500 questions of the type [**Q2**] (the theoretical maximum was $50 \cdot 2^{49}$, a 17-digit number). They ended up with knowledge spaces containing between 1000 and 3000 states (out of a theoretical maximum of 2^{50}). The whole procedure took them between 10 and 20 hours of work (excluding interruptions of arbitrary length and taken at will). One expert could not finish within a reasonable time limit. After 20 hours of work she had completed 3 blocks by answering about 3000 questions and still had a large Block 4 of questions to come. The unfinished space computed from her data up to Block 3 contained almost 8000 states. This space contains the one she would have produced, had she been able to finish the procedure, but the extent of this further reduction is of course unknown.

10.5 References

Birkhoff, G. (1937). Rings of sets. *Duke Mathematical Journal, 3,* 443-454.

Birkhoff, G. (1967). *Lattice theory.* American Mathematical Society Colloquium Publications, Vol. 25. Providence, RI: Amererican Mathematical Society.

Burigana, L. (1991). Organization by rules in finite sequences. *Journal of Mathematical Psychology, 35,* 345-370.

Doignon, J.-P., & Falmagne, J.-Cl. (1985). Spaces for the assessment of knowledge. *International Journal of Man-Machine Studies, 23,* 175-196.

Falmagne, J.-Cl., & Doignon, J.-P. (1988a). A class of stochastic procedures for the assessment of knowledge. *British Journal of Mathematical and Statistical Psychology, 41,* 1-23.

Falmagne, J.-Cl., & Doignon, J.-P. (1988b). A markovian procedure for assessing the state of a system. *Journal of Mathematical Psychology, 32,* 232-258.

Falmagne, J.-Cl. (1989). A latent trait theory via a stochastic learning theory for a knowledge space. *Psychometrika, 53,* 283-303.

Falmagne, J.-Cl., Koppen, M., Villano, M., Doignon, J.-P., & Johannesen, L. (1990). Introduction to knowledge spaces: How to build, test and search them. *Psychological Review, 97,* 201-224.

Kambouri, M. (1991). *Knowledge assessment: A comparison of human experts and computerized procedures.* Doctoral dissertation. New York: New York University.

Kambouri, M., Koppen, M., Villano, M., & Falmagne, J.-Cl. (1992). Knowledge assessment: Tapping human expertise by the QUERY routine. (Submitted.)

Koppen, M. (1992). Extracting human expertise for constructing knowledge spaces: An algorithm. *Journal of Mathematical Psychology.* (In press.)

Koppen, M., & Doignon, J.-P. (1990). How to build a knowledge space by querying an expert. *Journal of Mathematical Psychology, 34,* 311-331.

Müller, C.E. (1989). A procedure for facilitating an expert's judgement on a set of rules. In E. E. Roskam (Ed.), *Mathematical psychology in progress* (pp. 157-170). Berlin: Springer-Verlag.

Sleeman, D., & Brown, J.S. (1982). *Intelligent tutoring systems.* London: Academic Press.

11

Integrating Different Knowledge Spaces

Cornelia E. Dowling [1]

ABSTRACT Knowledge spaces on a set of items from a given branch of knowledge are structures for the assessment of students' or pupils' knowledge. In recent studies, knowledge spaces have been derived from the judgments of teachers. In analyzing the data, the problem arises that different experts have different opinions and may therefore yield different knowledge spaces on the same set of items. The problem of integrating knowledge spaces from different experts into a common knowledge space is addressed. This common space represents opinions on which the experts agree.

11.1 Introduction

Procedures for the assessment of students' knowledge on a given subject have been suggested (Doignon and Falmagne 1985; Degreef, Doignon, Ducamp, and Falmagne, 1986; Falmagne, 1988; Falmagne and Doignon 1988a and 1988b). The assessment algorithms are constructed so that it is possible to determine quickly and correctly what a pupil knows and where there are gaps in his or her knowledge. These algorithms use a family of *knowledge states* for the assessment. Here, a knowledge state is defined to be a subset of items from the branch of knowledge under consideration that could be answered correctly by some student. The assessment algorithms assign such a knowledge state to each student examined. It is important to note that, in practice, the number of knowledge states is often far smaller than the number of subsets of the item set. The following example illustrates items and knowledge states.

Example 1 In a continuing project, a particularly well trained and experienced teacher selected 48 items from a set of skills considered to be important for the diagnosis and training of elementary reading and writing abilities. For clarity, only a small subset $I = \{a, b, c, d, e, f, g\}$ comprising the following seven abilities[2] is used as an example. In this subset

- a is being able to build meaningful words from letters on the blackboard for which the child has learned the corresponding phonemes (e.g., synthesizing "Cat" or "Cup" from A, P, U, C, T),

---

[1] Technische Universität Braunschweig, Institut für Psychologie, Postfach 3329, 3300 Braunschweig, Germany. E-MAIL: i3160501 at dbstu1.rz.tu-bs.de

[2] The descriptions of the abilities are translated from German.

b is being able to pick out a printed character from a collection of visually similar characters (e.g., marking all occurrences of the letter M in the sequence "M M N N M N N N M M"),

c is being able to pick out a specific printed character from a collection of visually different characters (e.g., marking all occurrences of the letter A in the sequence "I E A O U A E I A U E"),

d is being able to identify phonemes in initial, middle, and final sounds of a short spoken word (e.g., C-A-T, D-O-G).

e is being able to continue a sequence of printed characters from left to right (e.g., continue writing "ALIALI..."),

f is being able to continue placing geometric symbols in a given sequence from left to right (e.g. continue placing the plastic shapes "circle", "triangle", "square" in a sequence from left to right),

g is being able to imitate placing geometric symbols in a given sequence from left to right (e.g. imitate placing the plastic shapes "circle", "triangle", "square" in a sequence from left to right).

For this set Q of items, those of its subsets which are knowledge states were determined according to the judgments of another experienced teacher. The following 16 subsets of Q were selected as knowledge states from the family of the $2^7 = 128$ possible subsets of the set Q:

$$\begin{gathered}\{a,b,c,d,e,f,g\},\\ \{a,b,c,e,f,g\},\{b,c,d,e,f,g\},\\ \{b,c,e,f,g\},\{c,d,e,f,g\},\\ \{b,c,f,g\},\{c,e,f,g\},\\ \{b,c,g\},\{c,f,g\},\{e,f,g\},\\ \{b,c\},\{c,g\},\{f,g\},\\ \{c\},\{g\},\\ \emptyset.\end{gathered}$$

To illustrate a principle used by assessment algorithms, assume that only the states above are valid knowledge states, and that a certain pupil has just mastered ability a. Since there is nò knowledge state which contains a and does not also contain the abilities b, c, e, f, and g, mastering a implies the mastering also of b, c, e, f, and g, or, alternatively, abilities b, c, e, f, and g are prerequisites for ability a. If, in this example, a pupil who has mastered a has not also mastered ability d, then the knowledge state $\{a,b,c,e,f,g\}$ is assigned to this student. If, on the other hand, the student has mastered d as well as a, then the student's knowledge state is the total set $\{a,b,c,d,e,f,g\}$. This conclusion can be drawn even without testing the abilities b, c, e, f, g under the validity assumption specified at the beginning of this paragraph.

There are two methods for constructing a set of knowledge states for a given set of items. The first is to question experienced teachers about prerequisite relationships between the items, and to derive the knowledge states from their judgements (Dowling, in press a; Koppen, in press). The second is to estimate which knowledge states are likely to occur in a large sample of students from a specified population (Falmagne, 1989). In practice, both methods together constitute a useful approach. One derives a set of knowledge states from teachers' judgements, and validates it with pupils' data (Falmagne et al, 1990).

There is, however, the problem that different teachers have different opinions, and therefore yield different sets of knowledge states for the same set of items. A procedure for integrating the knowledge states from different teachers into a common knowledge structure representing opinions on which the teachers agree is the subject of this paper.

11.2 Basic Concepts

For the assessment of knowledge, Doignon and Falmagne (1985) have suggested a family of knowledge states with the additional properties given in Definition 1 below. These properties are particularly useful for the representation of prerequisite relationships between the items (Koppen and Doignon, 1990, Müller, 1989).

Definition 1 *If $\mathcal{K}$ is a family of subsets of the finite set Q with the two properties (i) and (ii) below, then the family $\mathcal{K}$ is called a knowledge space on Q.*

(i) The empty set and Q are members of the family $\mathcal{K}$.

(ii) The family $\mathcal{K}$ is closed under union.

Example 2 The knowledge states from Example 1 constitute the following knowledge space $\mathcal{L}$ on the set $I = \{a, b, c, d, e, f, g\}$ of items.

$$\mathcal{L} = \{I, \{a,b,c,e,f,g\}, \{b,c,d,e,f,g\}, \{b,c,e,f,g\}, \{c,d,e,f,g\}, \{b,c,f,g\}, \{c,e,f,g\}, \{b,c,g\}, \{c,f,g\}, \{e,f,g\}, \{b,c\}, \{c,g\}, \{f,g\}, \{c\}, \{g\}, \emptyset\}$$

Examining the last example more closely, there are two types of knowledge states in a knowledge space, the states which can be obtained as the union of other, different, members of the knowledge space, and those which cannot. This observation leads to the Definitions 2 and 3 below.

Definition 2 *Those elements of a family $\mathcal{F}$ of subsets of a finite set Q which cannot be obtained as the union of other, different, members of the family $\mathcal{F}$, constitute the basis $\mathcal{B}$ of the family $\mathcal{F}$.*

Example 3 The family $\mathcal{C}$ below is the basis of the knowledge space $\mathcal{L}$ from Example 2:

$$\mathcal{C} = \{\{a,b,c,e,f,g\},\{c,d,e,f,g\},\{e,f,g\},\{b,c\},\{f,g\},\{c\},\{g\}\}.$$

The basis of a knowledge space is the smallest subset of the space from which the knowledge space itself can be reconstructed by computing the closure of the basis under union. The concept of the basis of a knowledge space was introduced by Doignon and Falmagne (1985).

Definition 3 *A minimal element of a knowledge space $\mathcal{K}$ on Q containing some given item $q \in Q$ is called a minimal state for q in $\mathcal{K}$.*

If $\mathcal{B}$ is the basis of the knowledge space $\mathcal{K}$, and if A is a minimal state for q in $\mathcal{K}$, then it is easy to see that A is also a minimal state for q in $\mathcal{B}$.

Proposition 1 *(Doignon & Falmagne, 1985) The members of a knowledge space $\mathcal{K}$ on Q which are minimal for some $q \in Q$ in $\mathcal{K}$ constitute the basis $\mathcal{B}$ of this knowledge space.*

This proposition leads to an interesting interpretation of the knowledge space and, in particular, of its basis. The basis of a knowledge space can be considered as a structure for the representation of *prerequisite relationships*. Whenever a subset A of the set Q of items is minimal for q in $\mathcal{K}$, then the members of A can be considered to be *prerequisites* for q. Since every state that is minimal for some q contains q itself, each item q is considered a prerequisite for itself.

Example 4 In Example 3 of the basis $\mathcal{C}$ of the knowledge space $\mathcal{L}$ on the set $I = \{a,b,c,d,e,f,g\}$ from Example 2, exactly one knowledge state is minimal for each $q \in Q$. As always, each element $q \in Q$ is a prerequisite for itself. There are no other prerequisites for c and g since the minimal states for these items are $\{c\}$ and $\{g\}$, respectively. The minimal state for f is $\{f,g\}$, i.e., item g is a prerequisite for item f. Similarly, the prerequisites for b, e, d, and a are item c, items f and g, items c, e, f, and g, and items b, c, e, f, and g, respectively.

Using the basic concepts introduced in this chapter, the subject of this paper can be formulated more precisely. Interactive procedures for constructing a knowledge space (Müller, 1989; Koppen and Doignon, 1990; Koppen, in press) or the basis of a knowledge space (Dowling, in press a) from a teacher's judgements have been suggested. These procedures have been applied to questioning of teachers on items from different branches of knowledge such, as for example, high school mathematics (Falmagne et al, 1990), Computer Aided Design (Dowling and Malinowski , in preparation), and elementary reading and writing (Example 1). Since different teachers yield different knowledge spaces on the same set of items, a procedure is needed which integrates knowledge spaces from different experts into a common knowledge space representing prerequisite relationships on which the experts agree.

11.3 Determining Common Knowledge States

Suppose the knowledge space $\mathcal{K}_1$ has been derived from the judgments of one expert, and the knowledge space $\mathcal{K}_2$ from the judgments of another expert. The two spaces may be compared by determining the knowledge states which are in common to both spaces, i. e., by computing the intersection $\mathcal{K}_1 \cap \mathcal{K}_2$. Note that the intersection of knowledge spaces on the same set of items is itself a knowledge space.

The point to be made in this section is that this intersection of two knowledge spaces does *not* represent those prerequisite relationships which are represented by the knowledge space $\mathcal{K}_1$ as well as by the knowledge space $\mathcal{K}_2$. The intersection does, however, capture the prerequisite relationships represented by at least one of the two spaces, i.e., by $\mathcal{K}_1$ or $\mathcal{K}_2$. These facts are best illustrated using another example.

Example 5 Let $J = \{u, v, w, x, y\}$ be a set of items, and let prerequisite relationships between these items be represented by the following examples of three statements:

(1) If a pupil masters item u, then this pupil also masters the items v and w, or the items x and y.

(2) If a pupil masters item v, then this pupil also masters item w.

(3) If a pupil masters item x, then this pupil also masters item y.

Suppose, two experts have been asked about prerequisite relationships between these five items. Expert 1 judges that the statements (1) and (2) are correct, whereas statement (3) is incorrect. Expert 2, however, believes the statements (1) and (3) to be true, and statement (2) to be false. Statements involving the connectives "if, then", "and", and "or" can be displayed as an AND/OR GRAPH, where "and" is represented by $\wedge$, and "or" by $\vee$. Graph 1 below represents the statements (1) and (2) considered to be true by Expert 1, graph 2 shows the corresponding statements (1) and (3) from Expert 2.

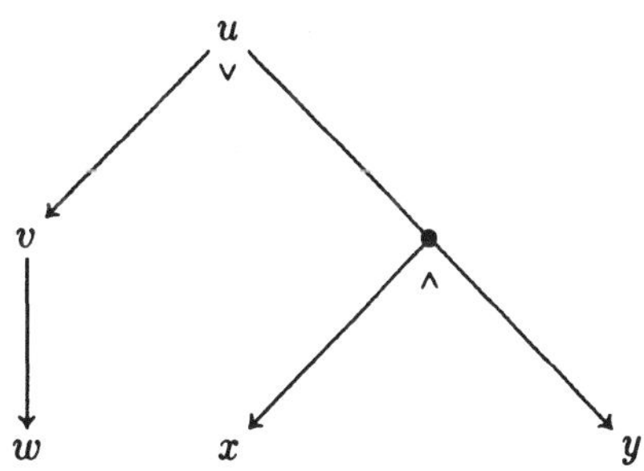

Graph 1

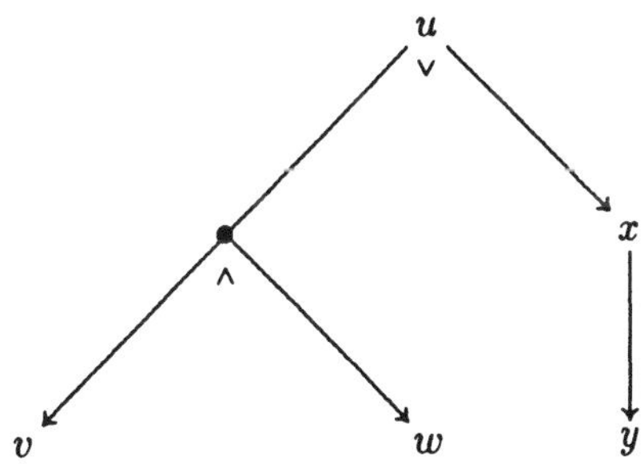

Graph 2

The knowledge space $\mathcal{M}_1$ of Expert 1 is comprised of all knowledge states, which are compatible with the statements (1) and (2),

$$\begin{aligned}\mathcal{M}_1 = \ & \{\emptyset, \{w\}, \{x\}, \{y\}, \{v,w\}, \{w,x\}, \{w,y\}, \{x,y\}, \\ & \{u,v,w\}, \{v,w,x\}, \{v,w,y\}, \{w,x,y\}, \{u,x,y\}, \\ & \{u,v,w,x\}, \{u,v,w,y\}, \{v,w,x,y\}, \{u,w,x,y\}, J\}.\end{aligned}$$

The set $\{v,x,y\}$, for example, is not a knowledge state, since it "contradicts" statement (2) as follows: Item v is an element of $\{v,x,y\}$, but its prerequisite w is not an element of this set. The set $\{u,w,y\}$ is not compatible with statement (1), since item u is an element of this set, but neither the set $\{v,w\}$ nor the set $\{x,y\}$ of its prerequisites is a subset of $\{u,w,y\}$. In a similar manner the knowledge space $\mathcal{M}_2$ of Expert 2 is constructed; this has only states compatible with the statements (1) and (3) as its elements.

$$\begin{aligned}\mathcal{M}_2 = \ & \{\emptyset, \{v\}, \{w\}, \{y\}, \{v,w\}, \{v,y\}, \{w,y\}, \{x,y\}, \\ & \{u,v,w\}, \{v,w,y\}, \{v,x,y\}, \{w,x,y\}, \{u,x,y\}, \\ & \{u,v,w,y\}, \{v,w,x,y\}, \{u,v,x,y\}, \{u,w,x,y\}, J\}.\end{aligned}$$

The intersection of the two knowledge spaces $\mathcal{M}_1$ and $\mathcal{M}_2$ is the knowledge space

$$\begin{aligned}&\mathcal{M}_1 \cap \mathcal{M}_2 = \{\emptyset, \{w\}, \{y\}, \{v,w\}, \{w,y\}, \{x,y\}, \{u,v,w\}, \\ &\{v,w,y\}, \{w,x,y\}, \{u,x,y\}, \{u,v,w,y\}, \{v,w,x,y\}, \{u,w,x,y\}, J\}.\end{aligned}$$

The basis of the knowledge space $\mathcal{M}_1 \cap \mathcal{M}_2$ is the family

$$\mathcal{D}_{1,2} = \{\{w\}, \{y\}, \{v,w\}, \{x,y\}, \{u,v,w\}, \{u,x,y\}.$$

Observe that this basis represents the same prerequisite relationships as the statements (1), (2), and (3) together. These statements are those considered to be true by Expert 1 or Expert 2. In contrast, consider the family

$$\begin{aligned}\mathcal{M}_1 \cup \mathcal{M}_2 = \ & \{\emptyset, \{v\}, \{w\}, \{x\}, \{y\}, \{v,w\}, \{v,y\}, \{w,x\}, \{w,y\}, \{x,y\}, \\ & \{u,v,w\}, \{v,w,x\}, \{v,x,y\}, \{w,x,y\}, \{u,x,y\}, \\ & \{u,v,w,x\}, \{u,v,w,y\}, \{v,w,x,y\}, \{u,v,x,y\}, \{u,w,x,y\}, J\},\end{aligned}$$

and the basis $\mathcal{E}_{1,2}$ of the family $\mathcal{M}_1 \cup \mathcal{M}_2$,

$$\mathcal{E}_{1,2} = \{\{v\}, \{w\}, \{x\}, \{y\}, \{u,v,w\}, \{u,x,y\}.$$

The basis $\mathcal{E}_{1,2}$ represents only the prerequisite relationships summarized by statement (1) alone, which is the only statement on which the two experts agree.

The previous example suggests that the intersection of two knowledge spaces is not suitable for representing only those prerequisite relationships on which the experts agree, whereas the union of two knowledge spaces does represent exactly those prerequisite relationships.

The basis of a knowledge space is one structure for representing prerequisite relationships; another such structure is the AND/OR GRAPH (Doignon and Falmagne, 1985; Heines and O'Shea, 1985) briefly introduced in Example 5. Further structures for representing prerequisite relationships are entail and implication relations. Their relationship to knowledge spaces and their bases are established in papers by Müller (1989), Koppen and Doignon (1990), Koppen (in press), and Dowling (in press a).

11.4 Integrating Knowledge Spaces and their Bases

In this section a procedure is suggested for integrating knowledge spaces constructed from the judgments of different experts into a common knowledge space. For simplicity, the integration of only two different knowledge spaces is considered, but the method suggested can be generalized to integrate a larger number of spaces. The following definition was used by Dowling (in press b).

Definition 4 *Let $\mathcal{F}$ be a family of subsets of a finite set Q, let $\mathcal{S}$ be a subfamily of the family $\mathcal{F}$ so that $\mathcal{S} \in 2^{\mathcal{F}}$, and set*

$$\bigcup \mathcal{S} = \bigcup_{X \in \mathcal{S}} X.$$

Then the family $\mathcal{F}^u = \{\bigcup \mathcal{S} \mid \mathcal{S} \in 2^{\mathcal{F}}\}$ is called the closure of $\mathcal{F}$ under union.

In particular, if $\bigcup \mathcal{F} = Q$, then $\mathcal{F}^u$ is a knowledge space. Suppose $\mathcal{K}_1$ and $\mathcal{K}_2$ are two knowledge spaces representing prerequisite relationships determined by two different experts. Then the knowledge space

$$\mathcal{K}_{1,2} = (\mathcal{K}_1 \cup \mathcal{K}_2)^u$$

can be interpreted as a structure representing the prerequisite relationships on which the different experts agree. These common prerequisite relationships are, in particular, represented by the basis $\mathcal{B}_{1,2}$ of the knowledge space $\mathcal{K}_{1,2}$. This follows from the fact that there exists a combinatorial Galois connection between a structure representing statements on prerequisite relationships, and knowledge spaces on a set of items (Koppen and Doignon, 1990; Müller, 1989). It is illustrated by the following example.

Example 6 Return to the introductory Example 1 and its set $I = \{a, b, c, d, e, f, g\}$ of seven reading and writing abilities. The knowledge space $\mathcal{L}_1$ has been derived from the judgments of one expert and the knowledge space $\mathcal{L}_2$ constructed of the judgments of another expert. This knowledge space $\mathcal{L}_2$ is identical with the knowledge space $\mathcal{L}$ from Example 2.

$$\begin{aligned}\mathcal{L}_1 = \; & \{I, \{a,b,c,d,f,g\}, \{b,c,d,e,f,g\}, \{b,c,e,f,g\}, \{b,c,d,f,g\}, \{b,c,f,g\}, \\ & \{c,d,f,g\}, \{d,f,g\}, \{c,f,g\}, \{f,g\}, \emptyset\}.\end{aligned}$$

$$\mathcal{L}_2 = \{I, \{a,b,c,e,f,g\}, \{b,c,d,e,f,g\}, \{b,c,e,f,g\}, \{c,d,e,f,g\}, \{b,c,f,g\} \{c,e,f,g\}, \{b,c,g\}, \{c,f,g\}, \{e,f,g\}, \{b,c\}, \{c,g\}, \{f,g\}, \{c\}, \{g\}, \emptyset\}.$$

The union of the two knowledge spaces is,

$$\mathcal{L}_1 \cup \mathcal{L}_2 = \{I, \{a,b,c,d,f,g\}, \{a,b,c,e,f,g\}, \{b,c,d,e,f,g\}, \{b,c,d,f,g\}, \{b,c,e,f,g\}, \{c,d,e,f,g\}, \{b,c,f,g\}, \{c,d,f,g\}, \{c,e,f,g\}, \{b,c,g\}, \{c,f,g\}, \{d,f,g\}, \{e,f,g\}, \{b,c\}, \{c,g\}, \{f,g\}, \{c\}, \{g\}, \emptyset\},$$

and the closure under union of the family $\mathcal{L}_1 \cup \mathcal{L}_2$,

$$(\mathcal{L}_1 \cup \mathcal{L}_2)^u = \mathcal{L}_1 \cup \mathcal{L}_2 \cup \{\{d,e,f,g\}\}.$$

The prerequisite relationships represented by the three knowledge spaces $\mathcal{L}_1$, $\mathcal{L}_2$, and $\mathcal{L}_{1,2}$ can be recognized best by looking at the corresponding bases $\mathcal{C}_1$, $\mathcal{C}_2$, and $\mathcal{C}_{1,2}$, respectively.

$$\begin{aligned}
\mathcal{C}_1 &= \{\{a,b,c,d,f,g\}, \{b,c,e,f,g\}, \{b,c,f,g\}, \{c,f,g\}, \{d,f,g\}, \{f,g\}\}.\\
\mathcal{C}_2 &= \{\{a,b,c,e,f,g\}, \{c,d,e,f,g\}, \{e,f,g\}, \{b,c\}, \{f,g\}, \{c\}, \{g\}\}.\\
\mathcal{C}_{1,2} &= \{\{a,b,c,d,f,g\}, \{a,b,c,e,f,g\}, \{d,f,g\}, \{e,f,g\},\\
&\quad \{b,c\}, \{f,g\}, \{c\}, \{g\}\}.
\end{aligned}$$

Comparing these bases, it can be seen, that the basis $\mathcal{C}_{1,2}$ represents exactly those prerequisite relationships on which the two experts agree. For example, the minimal state for item e is $\{e,f,g\}$ in the basis $\mathcal{C}_2$, i.e., Expert 2 considers only the items f and g as nontrivial prerequisites for item e. In the basis $\mathcal{C}_1$, however, the minimal state for item e is $\{b,c,e,f,g\}$; i.e., Expert 1 also regards the items b and c as prerequisites for item e. The minimal state for item e in the common basis $\mathcal{C}_{1,2}$ is again $\{e,f,g\}$; only the items f and g, on which both experts agree, are prerequisites for item e. The prerequisites for item a constitute another interesting example. Both experts agree that the items b, c, f, and g are prerequisites for item a, but Expert 1 selects item d as an additional prerequisite for item a, whereas Expert 2 selects item e as additional prerequisite for item a. The prerequisite relationships of item a on which the two experts agree are therefore: "the items a, b, c, f, g, and d, OR the items a, b, c, f, g, and e are prerequisites for item a." These prerequisite relationships are represented by the fact that there are two minimal states for a in the common basis $\mathcal{C}_{1,2}$, the set $\{a,b,c,d,f,g,\}$ and the set $\{a,b,c,e,f,g\}$.

If $\mathcal{B}_1$ is the basis of the knowledge space $\mathcal{K}_1$, if $\mathcal{B}_2$, is the basis of space $\mathcal{K}_2$, and if $\mathcal{B}_{1,2}$ is that of the common space $\mathcal{K}_{1,2}$, then

$$\mathcal{B}_{1,2}^u = (\mathcal{B}_1 \cup \mathcal{B}_2)^u = (\mathcal{K}_1 \cup \mathcal{K}_2)^u.$$

The nested families below all represent the prerequisite relationships which are common to the bases $\mathcal{B}_1$ and $\mathcal{B}_2$:

$$\mathcal{B}_{1,2} \subseteq (\mathcal{B}_1 \cup \mathcal{B}_2) \subseteq (\mathcal{K}_1 \cup \mathcal{K}_2) \subseteq \mathcal{K}_{1,2}.$$

In practical applications, an algorithm is often needed for computing the basis $\mathcal{B}_{1,2}$ from the family $\mathcal{B}_1 \cup \mathcal{B}_2$, the family $\mathcal{K}_1 \cup \mathcal{K}_2$, or the family $\mathcal{K}_{1,2}$. One such algorithm has been suggested by Dowling (in press b).

The method for integrating the knowledge spaces of two experts can, of course, be extended to a number N of experts. The knowledge space representing only prerequisite relationships on which N experts agree is

$$(\bigcup_{i=1}^{N} \mathcal{K}_i)^u.$$

The computation of the closure under union of a family $\mathcal{F}$ of sets can be extremely costly computationally, so that algorithmic efficiency is paramount. An algorithm which obtains optimal efficiency by eliminating redundant computation has been suggested by Dowling (in press b).

11.5 References

Birkhoff, G. (1979). *Lattice theory*. Providence: American Mathematical Society.

Degreef, E., Doignon, J.-P., Ducamp, A., & Falmagne, J.-Cl. (1986). Languages for the assessment of knowledge. *Journal of Mathematical Psychology, 30,* 234-256.

Doignon, J.-P., & Falmagne, J.-Cl. (1985). Spaces for the assessment of knowledge. *International Journal of Man-Machine Studies, 23,* 175–196.

Dowling, C. E. (in press a). Applying the basis of a knowledge space for controlling the questioning of an expert. *Journal of Mathematical Psychology.*

Dowling, C. E. (in press b). On the irredundant construction of knowledge spaces. *Journal of Mathematical Psychology.*

Falmagne, J.-Cl. (1988). A latent trait theory via a stochastic learning theory for a knowledge space. *Psychometrika, 53,* 283-303.

Dowling, C.E., & Malinowski, U. (in preparation). Determining knowledge structures for a CAD tutorial.

Falmagne, J.-Cl., & Doignon, J.-P. (1988a). A class of stochastic procedures for the assessment of knowledge. *British Journal of Statistical and Mathematical Psychology, 41,* 1–23.

Falmagne, J.-Cl., & Doignon, J.-P. (1988b). A Markovian procedure for assessing the state of a system. *Journal of Mathematical Psychology, 32,* 232–258.

Falmagne, J.-Cl. (1989). A latent trait theory via a stochastic learning theory for a knowledge space. *Psychometrika, 54,* 283-303.

Falmagne, J.-Cl., Koppen, M., Villano, M., Doignon, J.-P., & Johannesen, L. (1990). Introduction to knowledge spaces: How to build, test, and search them. *Psychological Review, 97,* 201-224.

Heines, J.M., & O'Shea, T. (1985). The design of a rule-based CAI tutorial. *International Journal of Man-Machine Studies*, *23,* 1-25.

Koppen, M. (in press). Extracting human expertise for constructing knowledge spaces: An algorithm. *Journal of Mathematical Psychology*.

Koppen, M., & Doignon, J.-P. (1989). How to build a knowledge space by querying an expert. *Journal of Mathematical Psychology*, *34,* 311-331.

Müller, C. E. (1989). A procedure for facilitating an expert's judgments on a set of rules. In E.E. Roskam, (Ed.), *Mathematical Psychology in Progress* (pp. 157-170), Berlin / New York: Springer.

12

Abstract Decomposition of Spelling Errors

Anton J. Smolenaars[1]

ABSTRACT
There has, hitherto, been no adequate method to found a taxonomy of spelling errors based on any substantial cognitive theory of spelling behavior, because there has been no theory adequate for the purpose. But such a taxonomy is needed. This paper begins, as it were, at the other end, by creating an 'abstract' taxonomy that might lead to an adequate theory. Suppose the spelling of a word to be determined by a collection of abstract *conditions*. If a subject satisfies all the relevant conditions, the word is spelled correctly. If two subjects satisfy an identical pattern of conditions they spell the word identically (though the converse need not be true).
Formal development leads to an algorithm for constructing a minimal set of conditions to represent formally a subjects × words arrangement of spellings. The algorithm is applied to the spellings of 109 words from two groups ('dyslectics' and 'normals') of 26 schoolchildren. Assumptions are sufficiently elementary that they do not invoke any semantic link to spelling errors as such. Consequently, the reasoning and the algorithm ought to have wide applicability.

12.1 Introduction

Confronted with the task of creating order within a corpus of spelling errors, it rather immediately becomes apparent that existing taxonomies of spelling errors are unsatisfactory: they lack theoretical foundation and are empirically ambiguous. Indeed, spelling research in general needs some procedure to give structure to the abundant data on spelling errors. Since it proves impossible to base such a procedure on some substantial cognitive theory of spelling behavior – there is none fitted to the purpose – it is better to start at the other end by creating something like an 'abstract' taxonomy that might eventually lead to an adequate theory. It is only necessary, then, to avoid a theoretical *parti pris* as far as possible and to take 'abstraction' to imply an absolute minimum of substantive hypotheses. In this vein, one might think of the orthography of a word as a *set*[2] of atomic spelling rules or spelling abilities of unspecified nature. After some superficial assumptions, which are not specific to spelling behavior at all, it becomes possible to describe relations among orthographies in set-relational terms like *in-*

[1]Psychology Laboratory, University of Amsterdam, Amsterdam, The Netherlands.
This research was conducted under the 'VF - project' "Knowledge Acquisition in Formal Domains" at the University of Amsterdam. Requests for reprints should be directed to Dr A. J. Smolenaars, Psychology Laboratory, University of Amsterdam, Dept. of Psychonomics, Roetersstraat 15, 1018 WB Amsterdam; e-mail : smol@uvapsy.psy.uva.nl.

[2]There is no order, or iteration among the elements. Consequently, any 'meta-rule' that prescribes structure over the application of some rules, is taken to be an 'atom' in itself.

clusion, overlap, exclusion and *identity* (Smolenaars, 1989). These results, which show some homomorphism with packages of spelling rules that were developed independently for the purposes of spelling software (Smolenaars), are, however, too restricted to provide a foundation for a taxonomy (Smolenaars, 1990). In the next section a model is developed that is an elaboration of the notion above. As the previously mentioned set relations were derived independently of this model, they will be treated, together with the actual (mis)spellings, as empirical data.

12.2 The Model

Keeping to the idea of the orthography of a word as a set of *conditions* that needs to be fulfilled in order for the word to be spelled correctly, the idealization of orthographic competence is an incidence matrix, C, with the rows corresponding to m subjects and the colums representing an exhaustive enumeration of n conditions, such that

$$c_{ij} = \begin{cases} 1 \text{ iff subject } i \text{ 'satisfies' condition } j\text{, that is, either} \\ \text{has the knowledge or the ability that is} \\ \text{represented by condition } j, \\ \\ 0 \text{ otherwise.} \end{cases}$$

The orthography of a specific word consists of a subset of the n conditions. It is convenient to have this subset represented by an $m \times n_k$ incidence matrix W for a particular word k; that is, the m rows of W again describe the subjects, and the n_k columns the conditions that are relevant to word k. So, the columns of W are a nonempty subset S_k of the columns of C with $0 < n_k = \mid S_k \mid \leq n$.

Clearly, there have to be constraints on this idealization in order to produce a model of some data. There are four constraints of which the first is no more than common sense: it protects the idealization against triviality. The other three constraints establish a *morphism* between C and the empirical data, viz. the actual corpus of spellings and the aforementioned set-relations among orthographies.

If

$\sigma_i(k)$ denotes the spelling of word k by subject i,
$\sigma_i^+(k)$ indicates that this spelling is correct, and
$\sigma_i(k) = \sigma_j(k)$ denotes that two spellings are identical 'phenotypically',

then the four constraints are

1. n and the n_k are minimal;
2. if the orthography of word a is a proper *subset* of the orthography of word b, then $S_a \subset S_b$ (and likewise for *overlap, exclusion,* and *identity*);

TABLE 12.1. Condition and spellings of 'weer'

	q	
a	1	weer
b	1	weer
c	1	weer
d	1	weer
e	0	vir
f	1	weer

3. $\forall_{j \in S_k} w_{ij} = 1 \Leftrightarrow \sigma_i^+(k)$;
(When a row of W contains only 1's, this is equivalent to saying that the corresponding subject spells the word, represented by W, correctly);
4. $\forall_{j \in S_k} w_{pj} = w_{qj} \Rightarrow \sigma_p(k) = \sigma_q(k)$.
(Whenever two rows in W are identical, the spellings of the word are identical for the two corresponding subjects. Note that, unlike (3), there is only implication, and not equivalence, here: it does not follow that, because two spellings are identical, they stem from an identical sets of condition-values, except, of course, for correct spellings. Different spelling processes can lead to phenotypically identical errors.)

I conjecture that these four constraints will prove sufficient to generate a limited number of incidence matrices of practical usefulness. This set of constraints seems to be the minimum to confine attention to some meaningful idealization; and the constraints themselves do not compromise the outcome since they prescribe nothing that is specific to spelling in them: they could, in principle, be used to idealize any 'discrete' cognitive domain.

12.3 An Example

Since the model, by virtue of its very abstractness, can be understood without details of the orthography of the language to which it is applied, it can be illustrated here with original misspellings of Dutch words and their representation in the terms of the model as an extended example.

The data come from six subjects, a to f. Begin with the word *'weer'* ("again"); the orthography of this word is represented by one condition labeled q.

This word provides a useful starting point, since it illustrates only Constraint 3. Subject e spells *'weer'* incorrectly, all other subjects spell it correctly, so condition q has a '0' in row e and a '1' in all other rows.

TABLE 12.2. Conditions and spellings of 'herfststorm'

	p	q	r	s	t	
a	0	1	1	1	1	herfst*storm
b	0	1	0	1	1	herft*stoorm
c	1	1	1	1	1	herfststorm
d	1	1	1	0	1	herfst*storm
e	1	0	1	1	1	herftstoorm
f	0	1	1	1	1	herfst*storm

Next the noun *'herfststorm'* ("autumnal storm" – one word in Dutch) is more complicated orthographically: it is represented by five conditions, p to t. (A'*' indicates a blank *within* a word which, in Dutch, is always an error).

Constraint 3 is satisfied: subject c is the only subject to spell this word correctly and, consequently, row c is the only row consisting solely of '1's. Row a and row f illustrate Constraint 4: they are identical, as also are the spellings of subjects a and f .

Note that subject d exhibits the same spelling as a and f, yet is represented by a different set of conditions. This is not, however, a violation of Constraint 3, which is an implication, not an equivalence. Different incompetences can lead to the same misspelling. Suppose that condition p embodies some orthographic *knowledge* about multi-morphemic words and how they are spelled; because neither subject a nor f has mastered this specific knowledge, both make the same error. Condition s, on the other hand, might embody some psychological 'Gestalt'-like ability; because subject d, who has acquired all the relevant knowledge, nevertheless has a deficiency here, he makes an error that is 'phenotypically' the same as that of subjects a and f.

Component t appears redundant in Table 12.2, or in any other representation for that matter, in the sense that it contributes nothing to either the satisfaction or violation of the model. Component t will be examined in Section 12.5.

The past participle *'verzameld'* ("collected"; Table 12.3) is represented by five conditions, p, q, r, s, u. Constraint 3 again differentiates between subject c and the others and, again, subject d makes the same error as subject a but, on different grounds. The complete incidence matrix for these three words and these six subjects now appears in Table 12.4.

It is known from earlier work (Smolenaars, 1989) that the orthography of *'weer'* is a subset of that of *'verzameld'*. So Constraint 2 requires that all the conditions of *'weer'* should reappear as conditions of *'verzameld'*. This Constraint is satisfied in Table 12.4; condition q is sufficient by itself for *'weer'* and is one of the conditions of *'verzameld'*.

It is also known from the same previous analysis that the orthographies of

TABLE 12.3. Conditions and spellings of *'verzameld'*

	p	q	r	s	u	
a	0	1	1	1	0	verzamelt
b	0	1	0	1	1	ver*zaamelt
c	1	1	1	1	1	verzameld
d	1	1	1	0	1	verzamelt
e	1	0	1	1	1	versamlt
f	0	1	1	1	1	versamelt

TABLE 12.4. Conditions for all three words

	p	q	r	s	t	u
a	0	1	1	1	1	0
b	0	1	0	1	1	1
c	1	1	1	1	1	1
d	1	1	1	0	1	1
e	1	0	1	1	1	1
f	0	1	1	1	1	1

'herfststorm' and *'verzameld'* overlap so that again according to Constraint 2, *'herfststorm'* and *'verzameld'* should have some condition(s) in common. Those conditions are p, q, r and s.

Following this illustration of the model the discussion now leads to the construction of an incidence matrix from given data.

12.4 The Algorithm

So far, nothing has been said about the *uniqueness* of the matrix C as a representation of orthographic competence. There is no *a priori* reason why there should not be several different matrices, each consistent with the four constraints of the model, representing the same set of misspellings. Of course, if the number of different matrices proved excessive, something is wrong and the constraints do not constrain sufficiently. But, for the moment, suppose that only a small number of matrices comply with the model constraints for certain given data. These different matrices might be regarded as discrete-logical 'rotations', one of another, by analogy with factor analysis, and it might be possible to choose a 'best' solution using criteria like 'Procrustes' or 'simple structure'. There is a clear need of an algorithm that will generate the full set of possible matrices C for given data. The data, as has been remarked before, consist of two given components: (1) an array of spellings of words by subjects, and (2) for some pairs of words a set-theoretical characterization of the relation between their orthographies.

For the sake of exposition, one starts with a fiction that, peculiar as it may appear at first sight, will turn out to be completely empty. One assumes a solution C_0 to our data, an incidence matrix consisting of m different rows (subjects) and n different columns (conditions), satisfying all four constraints. The task now is the generation of all *other* solutions, i. e., matrices $C_1, \ldots, C_T$, complying with the constraints of the model. Interpret this operation as the systematic *transformation* of C_0 into the $C_1, \ldots, C_T$. This interpretation allows the algorithm to be described in terms of Boolean functions. A Boolean function in n variables, of the format

$$f : \mathcal{B}^n \to \mathcal{B}, \qquad \mathcal{B} = \{0, 1\},$$

i. e., a mapping of a string of n '0's and '1's into a single '0' or '1', turns each row of C_0 into a '0' or '1'; as a consequence it will turn the complete C_0 into one column of '0's and '1's, i. e., one *new condition*. Thus the transformation into C_1 can be equated to a collection of Boolean functions in n variables, as can the transformation into C_2, C_3 etc. Note that, as the *order* of columns in a C has no significance, one can speak of a *collection* of Boolean functions. Because of the unambiguous specification of a Boolean function, the order of columns in C_0 must be fixed. Note furthermore that a solution implies that the relevant conditions to each word are known, and that the transformation should also 'transform' this information . The algorithm by now can be required to generate sets of Boolean functions.

12.4.1 Implementation of Constraints 3 and 4

What are the characteristics of the Boolean functions that comply with constraints 3 and 4 of our morphism, i. e., those handling correctness and differences in spelling? First regard the individual words separately. Suppose that at word i the (wrong) spelling of subject a differs from the incorrect spelling of subject b, and that these two differ from the spelling of subject c, who spells the word correctly.

Let F_i be the (sub)set of Boolean functions handling word i; then, the Constraints 3 and 4 lead to the following condition on F_i:

$$\underset{f \in F_i}{\exists} f(a) \neq f(b) \wedge \underset{g \in F_i}{\exists} g(a) \neq g(c) \wedge \underset{h \in F_i}{\exists} h(b) \neq h(c) \wedge \underset{l \in F_i}{\forall} l(c) = 1, \qquad (12.1)$$

that is, there is a function f within F_i that transforms row a and b of C into something *different* (Constraint 4), *and* there are functions g and h which work analogously; moreover, *all* functions that handle word i have a '1' in row c (Constraint 3).

At this point, there is no occupation whatsoever with the cardinality of F_i, that is, it is completely irrelevant how many Boolean functions are needed in order to satisfy (12.1).

The set of all Boolean functions in n variables is a Boolean algebra[3], in which the *atoms* are the minterms (Grimaldi, 1989, p.596, p.620). Adopting a common symbolism of Boolean algebras, a *minterm* can be defined as a Boolean product of the form $y_1y_2 \ldots y_m$, in which y_i is either x_i or $\bar{x}_i$, and the x_i are the original variables. The image of a minterm is '1' for one specific set of values of the x_i, and '0' for all other sets. This definition nicely fits to our design, and, consequently, $minterm_k$ is defined as the Boolean function that transforms row k of C into '1' and all other rows of C into '0' (this includes the 'virtual' rows, i. e., rows that are not represented by some subject; remember that all rows of C_0 are assumed different).

As each element $\neq$ **0** (which denotes the Boolean function whose value is always '0') of a finite[4] Boolean algebra can be written as a sum of atoms, uniquely up to order (Grimaldi p.620), each of these Boolean functions can be written as a sum of minterms, uniquely up to order.

There are now two Boolean levels, so to speak: the original Boolean functions and the set of all these, constituting a Boolean algebra with the minterms as atoms. These *atoms* will help to formulate the requirements that the Boolean functions must fulfill. The Boolean expressions that formulate these requirements *about* the Boolean functions form another meta-description that may be termed the third Boolean level. These abstractions appear to be prohibitive to understanding, yet they will soon turn out to effect a simple and elegant tool to design the four constraints of the model in the context of empirical data.

In order to simplify things rewrite

$$\underset{f \in F_i}{\exists} f(a) \neq f(b) \quad \text{into} \quad [a\bar{b} + \bar{a}b],$$

which becomes transparant if one takes the righthand term to mean: "there is a Boolean function which either contains atom 'a' but not atom 'b'[5], or atom 'b' but not atom 'a' ".

Translating this back to the original data, one gets: "if subject a spells word i differently from subject b, there is need for a Boolean function (a column in C) with either '1' at row a and '0' at row b, or *vice versa.*"

The square brackets '[', ']' denote a specific notion: they delineate the *scope*, *viz.* that of one and the same Boolean function. These brackets entail an 'elementary Boolean condition', (EBC), to a Boolean function. Note that the concept

[3] $\langle B, +, \cdot, - \rangle$ in which B is a nonempty set that contains two special elements 0, 1, and which is closed with respect to the binary operations '+' and '$\cdot$' and the unary operation '$-$', is a Boolean algebra iff certain axioms hold (after Grimaldi, 1989,p. 615). '$a \cdot b$' is often written as 'ab'. Any element $\neq$ **0** of a Boolean algebra can be written as a sum of *atoms*, uniquely up to order (Grimaldi, p.620).

[4] As there is a finite number of possible Boolean functions for our C_0, there is a finite Boolean algebra.

[5] which is different from "contains $\bar{b}$"

of scope implies that the EBC's must not be split up, i. e., $[a\bar{b}+\bar{a}b]$ denotes something quite different from either $[a\bar{b}]+[\bar{a}b]$ or $[a][\bar{b}]+[\bar{a}][b]$: the former allows *two* Boolean functions, which is not concise enough in this case, whereas the latter completely misses the point.

Alternatively, combinations of EBC's are not forbidden, but are rather desirable, in the light of Constraint 1. One will return to this point later, but for the moment will note that

$$\{[x][y]\}$$

means: "there should be a Boolean function that satisfies x, and there should be one that satisfies y, whether these are in fact one and the same or two separate ones, is not at issue at the moment". Note that the concatenation of the two EBC's does not deviate from the standard Boolean algebra terminology.

Following these conventions (12.1) can be rewritten as

$$[(a\bar{b}+\bar{a}b)c][(a\bar{c}+\bar{a}c)c][(b\bar{c}+\bar{b}c)c]. \tag{12.2}$$

The universal quantifier in (12.1) requires the containing of atom 'c' into all EBC's. Clearly, (12.2) can be simplified to

$$[a\bar{b}c+\bar{a}bc][\bar{a}c][\bar{b}c], \tag{12.3}$$

(but not further, as combinations of EBC's are not at issue yet), and it is obvious that for every word one can rewrite the implications of Constraints 3 and 4 together in some 'EBC-form' like (12.3) that represents F_i . The next step consists in the merging of the sets F_i , i. e., joining the demands of all words together. As $F=\cup_{i=1}^{n}F_i$ this is elementary as it consists of the Boolean multiplication of the 'EBC-forms', so that, finally, there is a – presumably rather long – product of EBC's.

12.4.2 Implementation of Constraint 2

The formalism shows its usefulness in its feasibility to incorporate the implications of Constraint 2, thus bringing the complete morphism into one logical format.

Suppose words w_a and w_b have led to

$$\left.\begin{array}{ll} & F_a=\{[x][y]\} \\ \text{and} & \\ & F_b=\{[p][q]\}\end{array}\right\}, \tag{12.4}$$

in which x, y, p, q are Boolean expressions like in (12.3), or to any degree of complication.

Now suppose that the conditions on word a form a *proper subset* of those of word b, thus $F_a\subset F_b$, then (12.4) can be *replaced* by

$$\{[x(p+q)][y(p+q)][(p+q)(\bar{x}\bar{y})][p][q]\}. \tag{12.5}$$

The first two EBC's express the fact that here *every* condition pertaining to w_a should do so to w_b: those EBC's belong to F_a as well as to F_b; the third factor represents the fact that there is a *proper* subset relation, it belongs exclusively to F_b, as do the last two EBC's, which guarantee the presence of both p and q.

In case of overlap of w_a and w_b, i. e., $F_a \cap F_b \neq \mathbf{0}$, the implementation is simpler: add to (12.4)

$$[xp + yp + xq + yq], \tag{12.6}$$

which belongs both to F_a and F_b.

The set relation of *exclusion*, i. e., $F_a \cap F_b = \mathbf{0}$, of w_a and w_b, finally, should replace (12.4) by

$$[xpq][ypq][p\bar{x}\bar{y}][q\bar{x}\bar{y}], \tag{12.7}$$

in which the first two EBC's belong to F_a, and the last two to F_b.

Thus, the implementation of Constraint 2 results in a somewhat more complicated product of EBC's.

12.4.3 Implementation of Constraint 1

So far, a (presumably long) sequence of EBC's has been generated. The implementation of Constraint 1 now requires the reduction of this sequence to a sequence of EBC's that has minimal length. 'Reduction', of course, consists of the 'merging' of several EBC's into one, i. e., of materializing the fact that certain elementary Boolean conditions to a Boolean function can be satisfied by one and the same Boolean function; an issue that up to this point was deliberately and explicitly neglected.

Call a number of EBC's *compatible* iff they can be satisfied by one and the same Boolean function. Compatibility is achieved whenever the Boolean multiplication of the EBC's, being both commutative and associative, results in an EBC which contains a Boolean expression that is unequal to *false*. For instance, $[a\bar{b} + \bar{a}b]$ and $[ac]$ are compatible and unite into $[a\bar{b}c]$, while $[a\bar{b} + \bar{a}b]$ and $[\bar{a}\bar{b}]$ are incompatible.

As the way in which the EBC's can be merged is far from unique, there is a genuine minimalization problem here that as such is rather opaque. Thus, one resorts to a *greedy* algorithm, viz., a procedure in which a *local* minimalization is achieved; fortunately, these algorithms have quite a reputation in approximating *global* minimalizations (Azmoodeh, 1988, p. 334). In this case, the greediness consists in merging as many EBC's into one as possible.

To this aim the (undirected) *compatibility graph* G, in which the EBC's are the vertices, and the compatibility relations correspond to the edges must be defined. An outline of this sub-algorithm can now be given.

1. Detect the isolated vertices in G and set them apart as results.
2. If there are vertices left, find a *maximal fully compatible subgraph* G' *within* G.

3. Transform G' into one EBC that is then added to the results. Delete G' from G.
4. If there are any vertices left, go to 1.

Step 2 is the heart of the algorithm; its realization is, in itself, far from simple, as compatibility is not transitive, witness the triple

$$[abc][def][\bar{a}\bar{b}\bar{c}],$$

but *in*compatibility is neither,

$$[abc][\bar{c}][cef].$$

Taking the *clique number* of G (Grimaldi, 1989, p. 472) will not be sufficient, as, for instance, the clique

$$[abc + xyz][\bar{a}\bar{b}\bar{c} + xyz][\bar{x}\bar{y}\bar{z}]$$

is not *fully* compatible in that it cannot be reduced to *one* EBC. There seems to be no alternative to a sophisticated systematic search for G', departing from a maximal clique from G. To this end, a 'branch-and-bound' algorithm (Azmoodeh, 1988, p.331) is used, but further details are omitted here.

There are altogether three sources of *non-uniqueness* within the algorithm as sketched above: (1) Step 2 of the algorithm could find more than one G', leading to different 'routes' in the algorithm that can end up in different equivalent solutions . (2) Solutions can be of the type $[abc+def]$, implying that a Boolean function with the atoms '*abc*' is an equivalent to one with the atoms '*def*'. (3) The final EBC's presumably do not specify the presence or absence of *all* atoms, implying that there are 'open spots' in the pertaining condition, leading to free choices, or, alternative versions.

By now, the fictitious character of the assumption of the existence of C_0 at the beginning of Section 12.4 will be clear. It gave the opportunity to put the whole reasoning in terms of Boolean functions, Boolean algebra, etc., which have properties that make up a substantial expository utility. In reality, data will lead directly to the creation of EBC's or to transformations on them, and the final EBC's can immediately be translated into the incidence vectors that constitute the solution.

12.5 An Application and Some Conclusions

An analysis was performed along the lines of the preceding sections on the spellings of 109 words by two groups of 26 subjects each. The subjects were schoolchildren aged about 10 years, including a group of 'dyslectics', and a matched group of 'normals'[6]. The first group, generating 1147 errors, effected

[6]Details and an elaboration on the results will be the subject of a separate paper.

45 conditions; the second group, misspelling 380 times, could be described by 28 conditions. In fact, the example in Section 12.3 is not fictional, but a slice of the empirical results for six 'dyslectic' subjects. This explains the – illusory – redundancy of component t in the example: it does not have 1's for all twenty-six subjects.

The general – rather abstract – approach that is formulated in the above sections can be put to work, so it seems. At the very least there is a crisp and concise representation of the spelling behavior of some subjects. Perhaps the algorithm itself is the limit to a formal taxonomy for spelling errors, in the sense that, complying with Constraint 4, it is impossible, by definition, to regard two different spellings as effected by the same syndrome of conditions, and, as such, belonging to the same 'class'; worse: even two identically incorrect spellings cannot be supposed to be determined in an identical way. One should realize, however, that, as far as the abstract conditions that result from the algorithm have a psychological/orthographical reality, a specific misspelling leads to a 'differential diagnosis' that makes a taxonomy as such superfluous.

One of the first steps of further analysis should, therefore, be the 'embodiment' of the abstract conditions. Fortunately, a wide variety of test data is available from the same subjects, so such a project seems feasible.

As a final remark, emphasis must be placed on the fact that the model described has minimal semantic links to spelling behavior as such. Consequently, it could be applied to any (cognitive) domain in which correct, i. e., rule-obeying, responses could be differentiated from incorrect ones, and in which the latter could become unambiguously partitioned into 'same' or 'different'.

12.6 References

Azmoodeh, M. (1988). *Abstract data types and algorithms.* Houndmills & London: McMillan.

Grimaldi, G. (1989). *Discrete and combinatorial mathematics* (2nd ed.). Reading etc.: Addison Wesley .

Smolenaars, A. J. (1989). An elementary formal categorization of a corpus of spelling errors. In E.E. Roskam (Ed.), *Mathematical Psychology in Progress* (pp. 197 - 213). Berlin, etc.: Springer Verlag.

Smolenaars, A. J. (1990). *A sequel to an elementary formal categorization of a corpus of spelling errors.* Unpublished manuscript. Report nr. 114 of the project 'Knowledge Acquisition in Formal Domains.' Amsterdam: Department of Psychology, University of Amsterdam.

Part II

Choice and Reaction Time

13

From Binary Choices to Random Utility Representations

Reinhard Suck[1]

ABSTRACT In this paper a procedure is outlined by which a paired comparison matrix can be represented as a convex combination of order matrices that can be transformed into a random utility representation. Furthermore, the set of all distributions over the permutations of the alternatives which yield the data matrix is characterized.

13.1 Introduction

13.1.1 Random utility representations

A binary choice system is a pair (A, p) consisting of a set A of n alternatives (for simplicity, $A = \{1, \ldots, n\}$) and a mapping $p : A \times A \longrightarrow [0, 1]$ such that $p(i, j) + p(j, i) = 1$ for $i \neq j$. This situation is often encountered in choice theory; in books on data theory and scaling p is called a *paired comparison matrix.*

A classical problem of choice theory is the search for conditions under which (A, p) is representable as a *random utility model.*

Definition 1 *A random utility representation of a binary choice system (A, p) is a random vector $(U_1, \ldots, U_n)$ such that, for all $i, j = 1, \ldots, n$ with $i \neq j$,*

$$p(i, j) = P(U_i > U_j). \tag{13.1}$$

Recently the question of the existence of random utility representations for paired comparison data has attracted much interest. A testable set of necessary and sufficient conditions is still lacking, although there has been much progress. (The history of this problem has been summerized by Marley, 1990; Fishburn, 1992, and Suck, 1992, present results and techniques not covered in Marley's overview).

In contrast to scaling methods, such as Thurstone's law of Comparative Judgement which mainly consists of a procedure for calculating the distributions of thc random variables, a solution of the existential question can be used to distinguish those situations in which the observed data matrix $p = (p(i, j))$ could be generated by realizations of random processes from those which are incompatible with such a model.

In the present paper the focus is on the actual determination of the random utility representation. Both questions, the existence and the determination, can

[1]University of Osnabrück, Postfach 4469, D-4500 Osnabrück, FRG

be reduced to a problem of convex analysis. To explain this, we observe that an ordering of $\{1,\ldots,n\}$ can be regarded as an extreme case of a random utility representation. Let σ be a permutation of $1,\ldots,n$; then define

$$\pi_\sigma(i,j) = \begin{cases} 1 & \text{if } \sigma^{-1}(i) \leq \sigma^{-1}(j), \\ 0 & \text{otherwise.} \end{cases} \tag{13.2}$$

The matrix $\pi_\sigma = (\pi_\sigma(i,j))$ is a point in $\mathbf{R}^{n^2}$ (where $\mathbf{R}$ is the set of real numbers). The convex hull of these points is a polytope called *the linear ordering polytope*, denoted by P^n_{LO}.

If $p \in P^n_{LO}$, then there exist, by definition of the convex hull, real numbers λ_σ such that

$$p = \sum_{\sigma \in S_n} \lambda_\sigma \pi_\sigma \quad \text{with} \quad \sum_{\sigma \in S_n} \lambda_\sigma = 1, \quad \lambda_\sigma \geq 0, \tag{13.3}$$

where S_n denotes the permutation group, and the vector $(\lambda_\sigma; \sigma \in S_n)$ can be thought of as a probability distribution over the orderings of $\{1,\ldots,n\}$.

It is well known (cf. Luce & Suppes, 1965, or Colonius, 1984, Ch. 7, and for a more general framework, Heyer & Niederée, 1992) that (13.3) is necessary and sufficient for the existence of a random utility representation.

Thus, our problem of actually finding a random utility representation of p is solved if we can determine $(\lambda_\sigma; \sigma \in S_n)$ satisfying (13.3). Since, in most cases, the solution is not unique, there is a second question: What is the complete set of solutions? Both problems are addressed in the sequel.

Clearly, the determination of the face structure of the polytope P^n_{LO}, in particular, the faces of maximal dimension (called *facets*), plays a crucial role in solving the existence and determination problem (cf. Suck 1992, Section 1, for a more detailed explanation of the correspondence between facets and conditions).

In questions of integer optimization P^n_{LO} has been extensively investigated (cf. Grötschel, Jünger & Reinelt, 1985), and recent results are derived in McLennan (1990), Leung & Lee (1990), Reinelt (1991), and Suck (1992). A summary is given in Fishburn (1992).

If the facets of P^n_{LO} were known, then (13.3) could be decided by checking whether p satisfies a system of linear inequalities. However, optimization problems concerning P^n_{LO} can be solved without complete knowledge of the facets. The idea behind this is that a relaxation of the optimization problem (in which the same objective function is maximized over a polytope T_n which covers P^n_{LO} and whose integer vertices coincide with the vertices of P^n_{LO}) is solved. If the solution of the relaxation entails integer valued points, then these belong also to P^n_{LO}, and the convex hull of all these solutions comprises the complete solution of the original problem (cf. Theorem 5).

Grötschel et al. (1985) developed the idea sketched above into a numerical procedure and were able to solve "real world" optimization problems over P^n_{LO} for n up to 80.

In Section 13.2 I investigate how a convex representation (13.3) of a given paired comparison matrix can be found using a similar idea, working with the vertices of P^n_{LO} and the facets of T_n (cf. Definition 2d for T_n).

Section 13.3 describes the set of all convex representations of $p \in P^n_{LO}$ denoted by K_p (cf. Definition 3). This is a kind of uniqueness problem. It should, however, be noted that, for a given vector $(\lambda_\sigma, \sigma \in S_n)$ satisfying (13.3), there are numerous random utility representations. There is, therefore, a twofold nonuniqueness: The set K_p and, for a given element from K_p, the set of random variables $U = (U_1, \ldots, U_n)$ satisfying (13.3). While Theorems 6 and 7 deal with the first question (characterization of K_p), the second remains an interesting open problem.

13.1.2 Notations and preliminaries

This subsection sets out the notation used in the sequel and discusses a few basic results of the theory of convexity. A brief introduction is given in Suck (1992); for a comprehensive exposition Grünbaum (1967) or Brøndsted (1983) is recommended.

Let X be a subset of $\mathbb{R}^d$. The *convex hull* of X, i.e., the smallest convex set containing X, will be denoted conv$X$. The dimension of X, i.e., $\dim X$, is the dimension of aff$X$, i.e., of the affine hull of X. The following celebrated characterization of conv$X$ is all important and will be used in Section 13.3.

Theorem 1 *(Carathéodory) The convex hull of a set $X \subseteq \mathbb{R}^d$ is the set of all convex combinations of subsets of X consisting of at most $d+1$ points, i.e.,*

$$\text{conv}X = \left\{x \in \mathbb{R}^d; \exists x_0, \ldots, x_d \in X, \lambda_i \geq 0, \sum_{i=0}^{d} \lambda_i = 1, x = \sum_{i=0}^{d} \lambda_i x_i\right\}.$$

If $\dim X = d$, then the number $d+1$ cannot be reduced.

For a proof see Grünbaum (1967, p. 15). Other proofs using linear programming results can be found in Schrijver (1986, p. 94) or Chvátal (1983, p.265). The latter proof, in particular, stimulated the investigation in Section 13.3 of the present paper.

A closed convex set K is equal to the intersection of the closed halfspaces in which it is contained. Hyperplanes which intersect K but not its relative interior are *supporting hyperplanes.* The intersection of a supporting hyperplane and K is a *face* of K; a face of dimension $\dim K - 1$ is a *facet.*

If K is the intersection of a finite number of closed halfspaces, then it is called a *polyhedral set.* Thus, a polyhedral set is characterized by a finite number n, say, of linear inequalities $\langle b_i, x\rangle \leq c_i$, for $i = 1, \ldots, n$, for some vectors b_i and real numbers c_i. If, for all $x \in K$, the inequality $\langle b, x\rangle \leq c$ is satisfied, then it is a *valid inequality;* if those $x \in K$ for which $\langle b, x\rangle = c$ form a facet of K, then the inequality is called *facet–defining.*

Polytopes are, in a sense, dual to polyhedral sets. They are defined to be the convex hull of a finite number of points. Let $X = \{x_1, x_2, \ldots, x_n\}$; then the set $P = \text{conv}X$ is a *polytope*. Faces of dimension 0 are called vertices; the set of vertices is denoted by vert$X$.

The following theorem is another classical result from the theory of convex sets. A proof can be found in Grünbaum (1967, Theorem 3.1.3).

Theorem 2 *A set P is a polytope iff it is a bounded polyhedral set.*

In Section 13.1.1 P^n_{LO} was described as a polytope, defined as the convex hull of the π_σ. Its description as a polyhedral set, which, according to Theorem 2 exists, is equivalent to finding necessary and sufficient conditions for $p \in P^n_{LO}$. Each finite set of necessary and sufficient linear inequalities of a polytope must contain the facet defining inequalities. Conditions in excess of these are redundant and can be omitted.

A prominent role is played by the theory of convexity in optimization theory; techniques of integer programming have proved useful in handling binary choices in general and will also be used in the sequel. For reference Schrijver (1986) is recommended.

The cutting plane algorithm and branch and bound techniques are mentioned in Section 13.2. For a concise exposition in connection with P^n_{LO}, Reinelt (1985) is recommended.

In Section 13.3 the following characterization of vertices of polyhedral sets employed in linear optimization is needed.

Theorem 3 *Let A be an $m \times n$ matrix of real numbers, $m < n$, and $b \in \mathbf{R}^m$. Then $x_0 \in \mathbf{R}^n$ is a vertex of the polyhedral set $P = \{x; Ax = b, x \geq 0\}$ iff the column vectors belonging to positive components of x_0 are linearly independent.*

This theorem follows easily from Schrijver (1986, p. 104, Eq.(23)) and is contained in somewhat disguised form in Chvátal (1983, Theorem 18.2).

Next the notation used in connection with the linear ordering polytope is summarized. If $\mathbf{A}$ denotes an $m \times n$-matrix, then $A(i,j)$ will be the i,j entry of this matrix, and $\mathbf{A}$ is equivalent to $(A(i,j))$.

The set of permutations on $\{1, \ldots, n\}$ is denoted by S_n and, for all $\sigma \in S_n$, $(\pi_\sigma(i,j))$ defined in (13.2) is an $n \times n$-matrix denoted by π_σ.

Definition 2 *Let $p = (p(i,j)) \in \mathbf{R}^{n^2}$. Then,*

a) $\Pi_n = \{\pi_\sigma; \sigma \in S_n\}$,

b) $P^n_{LO} = \text{conv}\Pi_n$,

c) $d_{ijk}(p) = p(i,j) + p(j,k) - p(i,k)$,

d) $T_n = \{p; 0 \leq p(i,j), d_{ijk}(p) \leq 1, p(i,i) = 1, p(i,j) = 1 - p(j,i)\}$,

e) $F_{ij}^{\alpha} = P_{LO}^{n} \cap \{p; p(i,j) = \alpha\}$ *for* $\alpha = 0, 1$, *called* trivial facet,

f) $F_{ijk}^{\alpha} = P_{LO}^{n} \cap \{p; d_{ijk}(p) = \alpha\}$ *for* $\alpha = 0, 1$, *called* triangle facet.

Furthermore, the inequalities $0 \leq p(i,j) \leq 1$ *and* $0 \leq d_{ijk}(p) \leq 1$ *are called* trivial *and* triangle *inequalities, respectively.*

The sets Π_n, P_{LO}^n, T_n are subsets of $\mathbf{R}^{n^2}$. Their actual dimension is, however, much lower than n^2. The diagonal entries of the elements are of no importance and are arbitrarily set to 1. It is helpful to fix them in this way because then matrix algebra can be used; this is convenient in places. Furthermore, the elements of T_n and of P_{LO}^n satisfy the $\binom{n}{2}$ linear restrictions $p(i,j) + p(j,i) = 1$ for $i \neq j$.

The following theorem collects some well-known results.

Theorem 4 $\dim P_{LO}^n = \dim T_n = \binom{n}{2}$,
$\dim F_{ij}^{\alpha} = \binom{n}{2} - 1$ *for* $\alpha = 0, 1$,
$\dim F_{ijk}^{\alpha} = \binom{n}{2} - 1$ *for* $\alpha = 0, 1$.

Proofs are given in Suck (1992). Further references can be found there and in Fishburn (1992). Other facets are known, they play, however, a minor role in the investigations of this paper.

Theorem 5 *A vertex of* T_n *is a vertex of* P_{LO}^n *iff its components are all zero or one.*

This theorem can be rephrased as follows: If a 0,1-matrix p satisfies $d_{ijk}(p) = 0$ or 1 and $p(i,j) = 1 - p(j,i)$ (and $p(i,i) = 1$), then $p \in \Pi_n$. The considerations of Section 13.2 are based on this result: Working in T_n is analytically easy and, if confined to the convex hull of the subset of 0,1-vertices of T_n, is always operating in P_{LO}^n.

13.2 Outline of an Algorithm

By a well-known algorithm attributed to Birkhoff (cf. Brualdi 1982), a doubly stochastic matrix X, i.e., $X(i,j) \geq 0$, and $\sum_i X(i,j) = \sum_j X(i,j) = 1$, can effectively be expressed as a convex combination of the permutation matrices P_σ ($P_\sigma(i,j) = 1$ if $\sigma(i) = j$ and 0 otherwise). Roughly speaking, this algorithm proceeds as follows: Choose a permutation $\sigma \in S_n$ such that $X(i, \sigma(i)) \neq 0$ for $i = 1, \ldots, n$, and let $\lambda_\sigma := \min_i X(i, \sigma(i))$. Then, if X' is defined by $X' := X - \lambda_\sigma P_\sigma$, the matrix $X_1 := (1 - \lambda_\sigma)^{-1} X'$ is again doubly stochastic, and the number of zero entries has increased by at least one. Proceeding with X_1 in the same way eventually leads to a permutation matrix, and from the sequence of λ_σ obtained in this process a convex representation of X can be calculated.

In geometric terms, each iteration of the algorithm consists of a projection of the doubly stochastic matrix on a facet of the polytope Ω_n of all doubly stochastic matrices.

An attempt to mimic this procedure in P_{LO}^n faces the difficulty that our knowledge of the facial structure of P_{LO}^n is rather limited. Nevertheless, it seems worthwhile to pursue this approach a bit further. If projections on trivial and triangle facets lead to a 0,1-matrix, then a convex combination has been found; if not, the final matrix may have only few fractional entries, and there may be a way to proceed from such a "wrong solution" to a correct one. This is essentially the idea of cutting planes and branch and bound methods in integer optimization.

Proposition 1 *If, for* $p \in P_{LO}^n$,

$$p = \sum_{\sigma \in S_n} \lambda_\sigma \pi_\sigma \tag{13.4}$$

is a convex representation, then, for all σ *such that* $\lambda_\sigma > 0$,

$$\begin{aligned} \pi_\sigma(i,j) = 1 \quad &\text{implies } p(i,j) > 0, \\ d_{ijk}(\pi_\sigma) = 1 \quad &\text{implies } d_{ijk}(p) > 0. \end{aligned}$$

Proof: Obvious.

Proposition 1 suggests exploiting the following value for a Birkhoff–like procedure: For $\sigma \in S_n$ and $p \in P_{LO}^n$ let

$$\mu_\sigma := \min\left\{\min_{i,j}\{p(i,j); \pi_\sigma(i,j) = 1\}, \min_{i,j,k}\{d_{ijk}(p); d_{ijk}(\pi_\sigma) = 1\}\right\}. \tag{13.5}$$

Proposition 2 *If* $1 > \mu_\sigma$, *then for*

$$p_1 := \frac{p - \mu_\sigma \pi_\sigma}{1 - \mu_\sigma} \tag{13.6}$$

it follows that

$$p_1(i,j) + p_1(j,i) = 1, \tag{13.7}$$

$$p_1(i,j) = \begin{cases} \dfrac{p(i,j)}{1-\mu_\sigma} & \text{if} \quad \pi_\sigma(i,j) = 0, \\[2ex] \dfrac{p(i,j) - \mu_\sigma}{1-\mu_\sigma} & \text{if} \quad \pi_\sigma(i,j) = 1, \end{cases} \tag{13.8}$$

$$d_{ijk}(p_1) = \begin{cases} \dfrac{d_{ijk}(p)}{1-\mu_\sigma} & \text{if} \quad d_{ijk}(\pi_\sigma) = 0, \\[2ex] \dfrac{d_{ijk}(p) - \mu_\sigma}{1-\mu_\sigma} & \text{if} \quad d_{ijk}(\pi_\sigma) = 1. \end{cases} \tag{13.9}$$

Proof: (13.7) and (13.8) are immediate. Also, (13.9) is easily seen to hold, since d_{ijk} is linear. □

Proposition 3 p_1 *satisfies all trivial and all triangle inequalities (i.e.,* $p_1 \in T_n$*).*

Proof: By (13.5), $p(i,j) - \mu_\sigma \pi_\sigma(i,j) \geq 0$ for all i,j, whence $p_1(i,j) \geq 0$. Thus, from (13.7) $p_1(i,j) \leq 1$. Also $d_{ijk}(p_1) \geq 0$ is obvious from (13.5) and (13.9). In case $d_{ijk}(\pi_\sigma) = 1$, $d_{ijk}(p_1) \leq 1$ follows immediately. If $d_{ijk}(\pi_\sigma) = 0$, then $d_{ikj}(\pi_\sigma) = 1$ and $\mu_\sigma \leq d_{ikj}(p) = 1 - d_{ijk}(p)$. Thus, by (13.9), $d_{ijk}(p_1) < d_{ijk}(p)(1-(1-d_{ijk}(p)))^{-1} = 1$. □

Proposition 4 *If $\langle a, x\rangle \leq a_0$ is a facet defining inequality and π_σ is on that facet (i.e., $\langle a, \pi_\sigma\rangle = a_0$) and $\langle a, p\rangle \leq a_0$, then $\langle a, p_1\rangle \leq a_0$.*

Proof: Simple estimation of $\langle a, p_1\rangle$ using (13.6).

Proposition 5 *If p is in a face F of P^n_{LO}, then, for some $\pi_\sigma \in F$, the value μ_σ defined in (13.5) is positive, and $p_1 \in \operatorname{aff} F$.*

Proof: If $p \in F$, then in (13.4) $\lambda_\sigma > 0$ only if $\pi_\sigma \in F$, and this holds for at least one permutation. Thus, by Proposition 1, for some $\pi_\sigma \in F$, $\mu_\sigma > 0$. Let $\langle a, x\rangle = a_0$ be the equation of a hyperplane containing F. Then, $\langle a, p\rangle = \langle a, \pi_\sigma\rangle = a_0$, whence $\langle a, p_1\rangle = a_0$. Since this holds for all hyperplanes containing F, it follows that $p_1 \in \operatorname{aff} F$. □

Proposition 6 *Let $G_1, G_2, \ldots, G_m$ be a collection of m distinct trivial or triangle facets. If $p \in \bigcap_{\nu=1}^m G_\nu$, then the value μ_σ defined in (13.5) is positive for at least one $\pi_\sigma \in \Pi_n \cap \bigcap_{\nu=1}^m G_\nu$. If, for such a permutation, $\mu_\sigma < 1$, then p_1 constructed according to (13.6) is in $\bigcap_{\nu=1}^m G_\nu \cap G_{m+1}$, where $G_{m+1} := T_n \cap \operatorname{aff} G$, with G being F^0_{ij} if the minimum in (13.5) is attained by $p(i,j)$, or F^0_{ijk} if it is attained by $d_{ijk}(p)$.*

Proof: Since $\bigcap_{\nu=1}^m G_\nu$ is a face of P^n_{LO}, for some $\pi_\sigma \in \bigcap_{\nu=1}^m G_\nu$ the value μ_σ is greater than zero according to Proposition 5. Thus, for such a σ,

$$p(i,j) = \alpha \quad \text{iff} \quad \pi_\sigma(i,j) = \alpha \quad \text{for } \alpha = 0, 1,$$

and

$$d_{ijk}(p) = \alpha \quad \text{iff} \quad d_{ijk}(\pi_\sigma) = \alpha \quad \text{for } \alpha = 0, 1.$$

In view of (13.8) and (13.9) it follows from these two equivalences that

$$p_1(i,j) = \alpha \quad \text{iff} \quad p(i,j) = \alpha \quad \text{for } \alpha = 0, 1,$$

and

$$d_{ijk}(p_1) = \alpha \quad \text{iff} \quad d_{ijk}(p) = \alpha \quad \text{for } \alpha = 0, 1,$$

yielding $p_1 \in \bigcap_{\nu=1}^m G_\nu$ since $p \in \bigcap_{\nu=1}^m G_\nu$.

However, suppose the minimum in (13.5) is attained by (i,j). Then, $p(i,j) = \mu_\sigma$ and $\pi_\sigma(i,j) = 1$, whence by (13.8) $p_1(i,j) = 0$, i.e., $p_1 \in F^0_{ij}$. Similarly, the case that μ_σ is defined by $d_{ijk}(p)$ leads to $p_1 \in F^0_{ijk}$. Denoting by G the respective facet (i.e., F^0_{ij} or F^0_{ijk}), it follows from Proposion 5 that $p_1 \in \bigcap_{\nu=1}^m G_\nu \cap \operatorname{aff} G$.

However, by Proposition 3, $p_1 \in T_n$; thus, the Proposition follows, defining $G_{m+1} := T_n \cap \text{aff}G$. □

The idea of the algorithm is now easily formulated:

1. Determine σ_1 such that $\mu_{\sigma_1} > 0$.
2. If p_1 is a 0, 1-matrix, the algorithm is completed; if not, repeat Step 1 with p_1 in place of p.

This procedure will eventually terminate. One possibility is that after some iterations the resultant p_m is a 0, 1-matrix, in which case the required representation has been found. The coefficients of a convex representation (13.4) can be recursively determined from the μ_{σ_i}. Note that by Proposition 6 each iteration increases the number of trivial and triangle inequalities which p_i satisfies with equality by at least one.

Another possibility is that no permutation can be found such that (13.5) gives a positive μ_σ. In the latter case, the matrix p_m which resulted from the last iteration of applying (13.6) is no longer in P^n_{LO}. A cutting plane procedure can now be applied. It consists of looking for violated inequalities. Here the techniques resemble those described by Grötschel et al. (1985) or, in more detail, by Reinelt (1985). For P^n_{LO} facet-defining inequalities different from the trivial and triangle facets are known. If one of those inequalities is violated by p_m (denote it by $\langle a, x\rangle \leq a_0$), we determine the largest index i such that $\langle a, p_i\rangle \leq a_0$ and $\langle a, p_{i+1}\rangle > a_0$. Next determine the solution in λ of $\langle a, (1-\lambda)^{-1}(p - \lambda\pi_{\sigma_i})\rangle = a_0$. If λ^* is this solution and $p^* = \frac{p-\lambda^*\pi_{\sigma_i}}{1-\lambda^*}$, then the procedure is resumed with p^* instead of p_{i+1}. From Proposition 5 we know that in the following iterations σ can always be chosen in such a way that all ensuing matrices will be on the hyperplane $\langle a, x\rangle = a_0$.

Checking all known facet-defining inequalities is a time consuming job. By Proposition 4 an inequality which π_{σ_i} satisfies with equality cannot be violated by p_i. Thus, such an inequality need not be checked.

There is, however, still the possibility that p_m violates an unknown inequality. This would mean that a cutting plane cannot be found. In this case one should resort to some elaborated "branch and bound"-like techniques or use "brute force" and hope for a result without enumerating too large a portion of S_n. Clearly, in this way the algorithm will always end. However, to find an adequate procedure which can handle matrices up to $n = 80$, say – the problem size Reinelt (1985) could treat successfully – remains an open problem.

13.3 The set of convex representations of $p \in P^n_{LO}$

For each $p \in P^n_{LO}$ there are in general very many convex representations. In this section, the set of all possible convex combinations of vertices of P^n_{LO} which yield a given p is described drawing on Theorem 1. The number $d+1$ of vertices needed to represent it is minimal; however, for special elements there may well be representations with fewer elements provided p is in the affine hull of less then $d+1$ vertices.

An instance of such a situation arises if p is on the boundary of the polytope because then it is an element of a polytope of dimension $k < d$. But even if p is in the (relative) interior of P^n_{LO}, this situation can arise. Any k–dimensional set of vertices whose affine hull intersects the interior gives rise to cases of "sparse" representations. On the other hand, inner points can be represented by a convex combination which employs all vertices.

The purpose of the present section is to investigate the structure of the convex representations of a given point in P^n_{LO}. It begins with two definitions.

Definition 3 *Let $p \in P^n_{LO}$. Then,*

$$K_p = \left\{ \lambda \in \mathbb{R}^{n!};\ \sum_{\sigma \in S_n} \lambda_\sigma = 1,\ \ \lambda_\sigma \geq 0,\ \ p = \sum_{\sigma \in S_n} \lambda_\sigma \pi_\sigma \right\}.$$

Definition 4 *Let $\lambda \in K_p$. Then λ is a* Carathéodory representation *of p iff for*

$$M_p(\lambda) = \{\sigma \in S_n; \lambda_\sigma > 0\} \text{ and } V_p(\lambda) = \{\pi_\sigma \in \Pi_n; \sigma \in M_p(\lambda)\},$$

the condition

$$\dim V_p(\lambda) = |M_p(\lambda)| - 1$$

is satisfied.

It is easily seen that Definition 4 is equivalent to

$$p \notin \operatorname{conv}(V_p(\lambda) - \{\pi_\sigma\}) \text{ for all } \sigma \in M_p(\lambda).$$

Theorem 6 *For all $p \in P^n_{LO}$, K_p is a polytope satisfying*

$$\dim K_p \leq n! - \binom{n}{2} - 1.$$

Proof: The elements $\lambda \in K_p$ satisfy $\binom{n}{2} + 1$ equations:

$$\sum_{\sigma \in S_n} \lambda_\sigma \pi_\sigma(i,j) = p(i,j) \text{ for } i < j \quad \text{and} \quad \sum_{\sigma \in S_n} \lambda_\sigma = 1.$$

The solutions of this system of linear equations form an affine subspace of $\mathbb{R}^{n!}$ of dimension $n! - \operatorname{rank}\Psi$, where Ψ is the following matrix

$$\Psi = \begin{pmatrix} \pi_1(1,2) & \pi_2(1,2) & \cdots & \pi_{n!}(1,2) \\ \pi_1(1,3) & \pi_2(1,3) & \cdots & \pi_{n!}(1,3) \\ \vdots & \vdots & \ddots & \vdots \\ \pi_1(n-1,n) & \pi_2(n-1,n) & \cdots & \pi_{n!}(n-1,n) \\ 1 & 1 & \cdots & 1 \end{pmatrix}. \tag{13.10}$$

In (13.10), the elements of Π_n are enumerated in an arbitrary but fixed order. Denoting $\overline{\lambda} := (\lambda_1, \ldots, \lambda_{n!})^T$ and $\overline{p} := (p(1,2), p(1,3), \ldots, p(n-1,n), 1)^T$, the system of equations above can be rewritten as

$$\Psi\overline{\lambda} = \overline{p}. \tag{13.11}$$

Since Ψ is a $(\binom{n}{2} + 1) \times n!$ matrix, its rank is at most $\binom{n}{2} + 1$. The solutions of (13.11) together with the conditions $\lambda_\sigma \geq 0$ form a polyhedral set which is contained in the unit cube because of $\lambda_\sigma \leq 1$. By Theorem 2 it is therefore a polytope. □

Remark. Although the dimension of the solution space of (13.11) can be shown to equal $n! - \binom{n}{2} - 1$, the dimension of K_p might be smaller. Take for example $p = \pi_\sigma$ for some $\sigma \in S_n$. In this case K_p consists of one point only.

Theorem 7 $\text{vert}K_p = \left\{\lambda \in \mathbb{R}^{n!}; \lambda \textit{ is a Carathéodory representation}\right\}$.

Proof: From Definition 4 we infer that λ is a Carathéodory representation iff the elements of $V_p(\lambda)$ are affinely independent. This in turn is equivalent to linear independence of the column vectors of Ψ corresponding to $\pi_i \in V_p(\lambda)$. By definition of $M_p(\lambda)$ and $V_p(\lambda)$, the vectors $\binom{\pi_i}{1}$ with $\pi_i \in V_p(\lambda)$ are linearly independent. Furthermore, they are in one–to–one correspondence to the positive components of λ. Therefore, they comprise the set of vertices of K_p by Theorem 3. □

This section finishes by describing how one can, in principle, find the vertices of K_p. According to Theorem 7, all Carathéodory representations need to be constructed. To this end, all subsets S of Π_n must be determined which satisfy

$$\dim S > \dim(S - \{x\}) \;\; \forall x \in S \quad \text{and} \quad p \in \text{aff}S. \tag{13.12}$$

Thus, conv$S$ is a simplex containing p in its relative interior. S being known, the representation is easily constructed by picking an arbitrary point in S and projecting p on the "opposing" facet of conv$S$, then iterating the procedure with the image of this projection. Section 13.2 tried to salvage as much as possible from this purely theoretical procedure. The main difficulty is that S is not given, so one begins with a "skillfully" chosen π_σ in the hope that it can be extended to a set S with the properties (13.12). If p is in a face of P^n_{LO}, then π_σ must also be in this face. But even if p is an inner point and π_σ is a member of a Carathéodory representation, not knowing the other elements of S can result in a disastrous "overshooting", i.e., p is projected onto a hyperplane corresponding to a trivial or triangle facet with the image unfortunately not being in P^n_{LO}. The ray emanating from π_σ through p has pierced a hidden facet. This difficulty can perhaps be avoided if an inner point is projected, not on the boundary, but on a subspace spanned by vertices which intersects the interior.

13.4 Discussion

Thurstone's law of Comparative Judgement was mentioned in Subsection 13.1.1 as a scaling technique which determines a random utility representation under the hypothesis of normal distributions. Thurstone's procedure always gives a result, and if no model testing is applied, is not really satisfactory.

Section 13.2 also dealt with the determination of a representation; however, in close connection with considering its existence. The procedure described here will not find a "solution" if none exists. However, if it is agreed that the $p(i,j)$ are fallible, then "small" violations of facet–defining inequalities might be tolerable, but how large is "small"? Statistical methods are needed to decide this question. In that case error minimizing techniques such as Thurstone's are superior. A discussion of this issue, important as it might be, is outside the scope of the present paper.

The word "outline" in the title of Section 13.2 was deliberately chosen because the procedure described there is not an algorithm in the strict sense of the word. It can, however, be reformulated in a true algorithmic way. But consideration should be given to two problems before implementing it. The first is the determination of the permutation σ satisfying $0 < \mu_\sigma < 1$. The second is to find an economic way of dealing with fractional entries in p_m, where m is the index at which the process of extracting π_{σ_i} ends. It is likely that most procedures dealing with the latter problem will stop short of enumerating the whole set S_n. Clearly, this is unsatisfactory, but there may be a way, similar to branch and bound techniques in optimization theory, to reduce the number of bad cases (in the above sense) in such a way that "real world" paired–comparison matrices can be efficiently represented for moderately large n.

The technique employed in Section 13.3 to characterize K_p is very general and is applicable to any convex polytope. Therefore, it can also be used to answer the question raised by Brualdi (1982) in connection with the Birkhoff algorithm (cf. Brualdi (1982) p. 197-198) for the polytope Ω_n. Clearly, the Carathéodory representations defined *mutatis mutandis* for Ω_n form the set of vertices of the polytope of convex representations of a doubly stochastic matrix.

A question of considerable interest (see Marley, 1990, or Colonius, 1984, Ch. 6 and Ch. 7.2) is the determination of an independent random utility representation. By this is meant a random vector $(U_1, \ldots, U_n)$ such that U_i, U_j are stochastically independent for $i \neq j$. In terms of the investigations of Section 3 this amounts to characterizing the subset I_p of K_p consisting of all $\lambda \in K_p$ such that λ admits an independent representation. Obviously, the usual way of calculating random variables satisfying (13.1), described for example in Suppes et al. (1989, Theorem 17.9), generates highly dependent variables. Colonius (1984) gives some results for the case $n = 3$, but apparently there are no generalizations for larger n, let alone a solution.

13.5 References

Brøndsted, A. (1983) *An introduction to convex polytopes.* New York: Springer.

Brualdi, R.A. (1982) Notes on the Birkhoff algorithm for doubly stochastic matrices. *Canadian Mathematical Bulletin, 25 (2),* 191-199.

Chvátal, V. (1983) *Linear programming.* New York: Freeman.

Colonius, H. (1984) *Stochastische Theorien des individuellen Wahlverhaltens.* Berlin: Springer.

Fishburn, P.C. (1992) Induced binary probabilities: A status report. *Mathematical Social Sciences, 23,* 67-80.

Grötschel, M., Jünger, M., & Reinelt, G. (1985) Facets of the linear ordering polytope. *Mathematical Programming, 33,* 43-60.

Grünbaum, B. (1967) *Convex polytopes.* New York: Wiley.

Heyer, D., & Niederée, R. (1992) Generalizing the concept of binary choice systems induced by rankings: One way of probabilizing deterministic measurement structures. *Mathematical Social Sciences, 23,* 31-44.

Luce, R.D., & Suppes, P. (1965) Preference, utility, and subjective probability. In R.D. Luce, R.R. Bush, & E. Galanter (Eds.) *Handbook of mathematical psychology* (pp. 252-410). New York: Wiley.

Leung, J., & Lee, J. (1990) Reinforcing old fences gives new facets. *Technical Report No. 90-22.* Department of Operations Research, Yale University, New Haven, CT.

Marley, A.A.J. (1990) A historical and contemporary perspective on random scale representations of choice probabilities and reaction times in the context of Cohen and Falmagne's (1990, Journal of Mathematical Psychology, 34) results. *Journal of Mathematical Psychology, 34,* 81-87.

McLennan, A. (1990) Binary stochastic choice. In J.S. Chipman, D. McFadden & M.K. Richter (Eds.) *Preferences, uncertainty, and optimality* (pp. 187-202). Boulder: Westview Press.

Reinelt, G. (1985) *The linear ordering problem: Algorithms and applications.* Heldermann Verlag: Berlin.

Reinelt, G. (1991) A note on small linear ordering polytopes Schwerpunktprogramm der Deutschen Forschungsgemeinschaft: Anwendungsbezogene Optimierung und Steuerung. *Augsburg Report No. 329.*

Schrijver, A. (1986) *Theory of linear and integer programming.* New York: Wiley.

Suck, R. (1992) Geometric and combinatorial properties of the polytope of choice probabilities. *Mathematical Social Sciences, 23,* 81-102.

Suppes, P., Krantz, D.M., Luce, R.D., & Tversky, A. (1989). *Foundations of measurement (Vol. II). Geometrical, threshold, and probabilistic representations.* New York: Academic Press.

14

Collective Choice Probabilities

Ákos Münnich[1]

ABSTRACT This paper studies probabilistic collective choice behavior. The concept of the composition rule for forced choice probabilities is determined in terms of the choice probabilities of choosing a stimulus from a finite set of stimuli that is "composed" of the unforced elementary response probabilities. This study starts, as usual, from the individual level of choice and extends our composition rule to the case of collective choice.

14.1 Introduction

Consider a set of *stimuli* (I) and a set of *subjects* (S). Let us suppose that in a trial situation (the exact specification of the trial situation is not important at present) each subject is faced with some stimuli (one or more). The outcomes can be represented by the dichotomous random variables X_{si} (called *unforced elementary response*), where $s \in S$ denotes the subject and $i \in I$ the stimulus. For the sake of simplicity, let the possible values of X_{si} be 1 and 0. The distribution of X_{si} can be specified as follows:

$$p(s,i) = P(X_{si} = 1), \tag{14.1}$$
$$q(s,i) = 1 - P(X_{si} = 1); \tag{14.2}$$

and $\#A$ will signify the number of elements in the finite set A.

A typical trial is as follows: Given a finite set $A \subset I$ of stimuli (we call it the *choice set*), the collective (a finite set of subjects, denoted by C) has to choose some elements according to specific instructions. The outcome on each trial is the subset of A chosen by the collective. If the instructions allow the collective to choose any subset of A, this is called an *unforced choice trial.* If the instructions restrict the sets of A which might be chosen to a proper subset of all subsets of A, then it is called a *forced choice trial.*

In what follows, we are concerned with the *choice probability*, denoted by $p(C, i, A)$, of a stimulus i being chosen from the choice set A by a finite collective of subjects $C \subset S$.

---

[1]Kossuth Lajos University, 4032 Debrecen, Egyetem tér 1, Hungary; e-mail: H2754MUN @ELLA.HU
The author thanks Prof. Dr. Gerhard Fischer and the organizing committee of the 22nd EMPG Meeting for their support of this work. The final version of this work has been supported in part by National Science Foundation grant OTKA-F4796. Moreover, the author thanks the editors, the referees of the manuscript, Prof. Dr. Robert Mokken, Prof. Dr. R. Duncan Luce, and Prof. Dr. A. A. J. Marley for their constructive comments.

Probabilistic individual choice behavior has been studied by Luce (1959), French (1986), Marley (in press), and Münnich (1991), among others. The study of the social or collective level of choice behavior started with deterministic models, e.g., Arrow (1959/1963), Sen (1970,1977), Bezembinder (1989); but Arrows' approach, based on some plausible conditions (their probabilistic versions are formulated in Theorem 6), culminated in an impossibility theorem. There are many papers dealing with probabilistic versions of collective choice, e.g., Williamson and Sargent (1967), Intrilligator (1973), Fishburn (1975), Fishburn and Gehrlein (1977), French (1986), Pattanaik and Peleg (1986), Van Blokland-Vogelesang (1990), and Marley (1990, in press). My approach, unlike the others mentioned, is based explicitly on individuals' "unforced preferences" (given by X_{si}), instead of individual orderings, or individual choice probabilities. One might say that this approach starts "one step before" the individual orderings or choice probabilities of the stimuli.

14.2 The individual level of choice

This section focuses on the individual level of choice, supposing the collective to be a single subject. In this case, a composition rule is defined for expressing $p(s,i,A)$ in terms of the unforced elementary response probabilities $p(s,j)$, where $s \in S$ and $j \in A$. This composition rule can be regarded as an axiom, and in some general cases it can be tested directly. The composition rule is, as will appear, based on classical conditional probability.

For $A \subset I$, $s \in S$, and $i \in A$, let

$$\varphi(s,i,A) = p(s,i)\left(\prod_{j \neq i} q(s,j)\right). \tag{14.3}$$

The function $\varphi(s,i,A)$ can be regarded as the "likelihood" of jointly "accepting" stimulus i and "rejecting" the other stimuli in A. It is worth noting that the random variables X_{si} need not be independent (at least in this paper), because their joint distribution does not enter into the argument.

Definition 1 *Let $A \subset I$ finite and $s \in S$ be given. Then, for $i \in A$,*

$$p(s,i,A) = \begin{cases} \dfrac{\varphi(s,i,A)}{\sum\limits_{j \in A} \varphi(s,j,A)} & \text{if} \quad \sum\limits_{j \in A} \varphi(s,j,A) > 0, \\[2ex] \dfrac{1}{\#A} & \text{if} \quad \sum\limits_{j \in A} \varphi(s,j,A) = 0, \end{cases} \tag{14.4}$$

will be called the composition rule for the forced choice probabilities of choosing stimulus i from the choice set A by subject s.

A much more general definition of this composition rule was given by Münnich (1991); here only certain properties are mentioned.

Theorem 1 (Pareto principle) *If* $p(s,i) \geq p(s,j)$, *then*

$$p(s,i,A) \geq p(s,j,A).$$

Theorem 2 (Luce's axiom) *If* $B \subset A$ *and* $p(s,j,A) > 0$ *for all* $j \in A$, *then*

$$p(s,i,A) = p(s,i,B) \sum_{j \in B} p(s,j,A).$$

Recall that Luce (1959) showed that the choice probabilities $p(s,i,A)$ can be expressed in terms of the pair–comparison probabilities $p(s,i,\{i,j\})$, $j \in A$. The present formulation is a little different from thc original; I dcsribc an "urn" model to illustrate its meaning.

For simplicity, let $p(s,i,\{i,j\})$ be denoted by $p_{ij}(s)$.

Theorem 3 (Luce, 1959) *If* $\#A \geq 3$ *and* $p(s,j) > 0$ *for all* $j \in A$, *then*

$$p(s,i,A) = \frac{\prod_{\substack{k \in A \\ k \neq i}} p_{ik}(s)}{\prod_{\substack{k \in A \\ k \neq i}} p_{ik}(s) + \sum_{\substack{h \in A \\ h \neq i}} p_{hi}(s) \prod_{\substack{k \in A \\ k \neq i \\ k \neq h}} p_{ik}(s)}. \tag{14.5}$$

Proof: See Luce (1959, p.16).

Example 1 *Let* $A = \{i,j,k\}$ *and* $B = \{i,j,k,l\}$, *and let* $p_{mn}(s)$ *be denoted by* p_{mn} *for simplicity. Then,*

$$p(s,i,A) = \frac{p_{ij}p_{ik}}{p_{ij}p_{ik} + p_{ij}p_{ki} + p_{ik}p_{ji}},$$

$$p(s,i,B) = \frac{p_{ij}p_{ik}p_{il}}{p_{ij}p_{ik}p_{il} + p_{ij}p_{ki}p_{il} + p_{ij}p_{ik}p_{li}}.$$

Individual "urn" model: Envisage $m+1$ different colors, $c_1, c_2, \ldots, c_{m+1}$, and m urns, $u_1, u_2, \ldots, u_m$, and let $a = m+1$. Let colored balls be placed in the urns in such a way that urn u_b contains balls of colors c_a and c_b . Let the probability of drawing a ball of color c_a at random from u_b be p_{ab}, and suppose that $0 < p_{ab} < 1$. Suppose now that one ball is chosen independently from each urn. Let E_{ab} be the event that a ball of color c_a is taken from urn u_b. Define the indicator function of E_{ab} by

$$\chi_{E_{ab}} = \begin{cases} 1 & \text{if } E_{ab} \text{ occurs,} \\ 0 & \text{otherwise.} \end{cases} \tag{14.6}$$

Let the random variable ξ_a be defined by

$$\xi_a = \sum_{b=1}^{m} \chi_{E_{ab}} \,. \tag{14.7}$$

Then the value of ξ_a is the number of the balls of color c_a independently chosen from the m urns. Suppose that, for every $i \in A$, there is an associated color c_i, and that these colors are different for different stimuli. Theorem 4 shows that the forced choice probability $p(s, i, A)$ can be represented as the probability of choosing a ball of color c_i from the m urns described above. According to the composition rule, the Luce–type choice probabilities are the same as $P(\xi_i = m \mid \xi_i \geq m-1)$. To check this assertion, compare (14.5) and (14.8). As a consequence, the Luce–type choice model may be taken to be an "almost sure choice" model of individual choice behavior.

Theorem 4 *Using the notation defined in the individual "urn" model,*

$$P(\xi_a = m \mid \xi_a \geq m-1) = \frac{\prod_{b=1}^{m} p_{ab}}{\prod_{b=1}^{m} p_{ab} + \sum_{d=1}^{m} p_{da} \prod_{b \neq d} p_{ab}} \,. \tag{14.8}$$

Proof: Suppose, for simplicity, that $a = m+1$; then,

$$P(\xi_a = m \mid \xi_a \geq m-1) = \frac{P(\xi_a = m)}{P(\xi_a \geq m-1)} = \frac{P(\xi_a = m)}{P(\xi_a = m) + P(\xi_a = m-1)} \tag{14.9}$$

and $P(\xi_a = m) = \prod_{b=1}^{m} p_{ab}$.

Since the events $E_{a1}, E_{a2}, E_{a3}, \ldots, E_{am}$ are independent,

$$P(\xi_a = m-1) =$$

$$= P(\overline{E}_{a1} E_{a2} \ldots E_{am}) + P(E_{a1} \overline{E}_{a2} \ldots E_{am}) + \ldots + P(E_{a1} E_{a2} \ldots \overline{E}_{am}) =$$

$$= p_{1a} p_{a2} p_{a3} \cdots p_{am} + p_{a1} p_{2a} p_{a3} \cdots p_{am} + \ldots + p_{a1} p_{a2} p_{a3} \cdots p_{ma} \,. \tag{14.10}$$

This completes the proof. □

Definition 2 *Introduce the odds $o(s,i)$ for $s \in S$, $i \in I$, and $0 < p(s,i) < 1$ as follows,*

$$o(s,i) = \frac{1 - p(s,i)}{p(s,i)} \,. \tag{14.11}$$

The reciprocal of $o(s,i)$ will be denoted by $o^{-1}(s,i)$, and the notation $o(i)$ will be used instead of $o(s,i)$ for simplicity.

The next theorem will prove useful in the following section when the individual composition rule is extended to the case of collective choice.

Theorem 5 *If* $0 < p(s,j) < 1$ *for all* $j \in A$*, then, for* $i \in A$*,*

$$p(s,i,A) = \frac{\prod\limits_{\substack{j \neq i \\ j \in A}} o(s,j)}{\sum\limits_{k \in A} \prod\limits_{\substack{h \neq k \\ h \in A}} o(s,h)}, \tag{14.12}$$

and

$$p(s,i,A) = \frac{o^{-1}(s,i)}{\sum\limits_{k \in A} o^{-1}(s,k)}. \tag{14.13}$$

Proof: The proof of (14.12) is straightforward dividing (14.4) by $\prod\limits_{k \in A} p(s,k)$. Similarly, dividing (14.4) by $\prod\limits_{k \in A} (1 - p(s,k))$, gives (14.13). □

Example 2 *Let* $A_1 = \{i,j\}$ *and* $A_2 = \{i,j,k\}$*; then,*

$$\begin{aligned}
p(s,i,A_1) &= \frac{o(j)}{o(i) + o(j)}, \\
p(s,i,A_2) &= \frac{o(j)o(k)}{o(i)o(j) + o(j)o(k) + o(i)o(k)}, \\
p(s,i,A_1) &= \frac{o^{-1}(i)}{o^{-1}(i) + o^{-1}(j)}, \\
p(s,i,A_2) &= \frac{o^{-1}(i)}{o^{-1}(i) + o^{-1}(j) + o^{-1}(k)}.
\end{aligned}$$

14.3 The collective level of choice

In this section the individual forced choice model is extended to the collective forced choice trial situation. The approach is motivated by the individual "urn" model. The main aim is to give a plausible definition for $p(C,i,A)$ in terms of the unforced single response probabilities $p(s,j)$, where $j \in A$, $s \in C \subset S$, $\#C = n$.

The collective "urn" model: As before, there are $m+1$ different colors, $c_1, c_2, \ldots, c_m$, and, for simplicity, $a = m+1$. Moreover, let there be n by m urns, as follows,

$$\begin{array}{c}
u_{11}, u_{12}, \ldots, u_{1m} \\
u_{21}, u_{22}, \ldots, u_{2m} \\
\cdots\cdots\cdots\cdots \\
\cdots\cdots\cdots\cdots \\
\cdots\cdots\cdots\cdots \\
u_{n1}, u_{n2}, \ldots, u_{nm}
\end{array}$$

where u_{rt} is the tth urn in the rth row.

Again, as in the individual "urn" model, colored balls are placed in the urns in such a way that u_{rt} contains balls of colors c_a and c_t. Denote the number of balls of color c_a in urn u_{rt} by $f(r,a)$ and of color c_t by $f(r,t)$. Denote the probability of choosing a ball of color c_a from urn u_{rt} by p^r_{at} , which is defined as

$$p^r_{at} = \frac{f(r,a)}{f(r,a)+f(r,t)}. \tag{14.14}$$

This much is equivalent to n "individual urn models", one corresponding to each of the n rows of urns.

Now consider m additional urns, $u_1, u_2, \ldots, u_m$, and place balls in these urns in such a way that urn u_b $(b = 1, 2, \ldots, m)$ contains a set of balls equivalent to the contents of urns $u_{1b}, u_{2b}, \ldots, u_{nb}$. Denote the number of balls of color c_a in urn u_t $(t = 1, 2, \ldots, m)$, by $f(a)$, and the number of balls of color c_t by $f(t)$. Then,

$$f(a) = \sum_{r=1}^{n} f(r,a), \quad \text{and} \quad f(t) = \sum_{r=1}^{n} f(r,t). \tag{14.15}$$

Denote the probability of choosing a ball of color c_a from urn u_t by p_{at}. It is obvious that

$$p_{at} = \frac{f(a)}{f(a)+f(t)}. \tag{14.16}$$

The "individual urn model" applies also to the joint urns, and this application leads to the following development.

Substitute a stimulus for each different color, stimulus i_b for color c_b, and the subjects of the given collective for the rows of urns, so that subject s_r corresponds to row r and the individual choice probability $p(s_r, i_a, \{i_a, i_t\})$ corresponds to the probability p^r_{at} of choosing a ball of color c_a from urn u_{rt}.

If we apply the pair–comparison probabilities p_{at} to (14.5) or (14.8), we obtain forced choice probabilities for the collective in terms of the "joined urns".

The questions now are: How can the choice probabilities $p(s_r, i_a, \{i_a, i_t\})$ and $p(C, i_a, \{i_a, i_t\})$ be related to the corresponding numbers $f(r,a)$, $f(r,t)$? What is the meaning of $f(r,t)$ in the "collective urn model"? Note that no restrictions were placed on the number of balls in the urns, so they can be regarded as "subjective utilities of preference" for the colors (or stimuli). What is the meaning of the "subjective utility of preference" in the "collective urn model"? The "basic information" resides in the unforced elementary response probabilities $p(s,i)$, and the objective is to express the "subjective utility of preference" of the stimuli in these terms.

Suppose that, with respect to stimulus i, subject s can be in one of two non-overlapping states, the state "do not prefer i" and the alternative "prefer i". Imagine the following process:

Step 0: Subject s is in the state "do not prefer i" with probability $q(s,i)$.

Step 1: Subject s remains in the state "do not prefer i" with compound probability $q^2(s,i)$, and the process continues; or the subject's state changes to "prefer i" with compound probability $p(s,i)\,q(s,i)$, and the process stops.

⋮

Step n: Subject s remains in the state "do not prefer i" with compound probability $q^{n+1}(s,i)$, and the process continues; or the subject's state changes to "prefer i" with compound probability $p(s,i)\,q^n(s,i)$, and the process stops.

The mean value of the "length" of the process (denoted by $E(s,i)$) can be treated as the "subjective utility" of preferring stimulus i, and $E(s,i)$ may be substituted for $f(s,i)$ in the "collective urn model".

After some elementary computations it can be seen that

$$E(s,i) = \sum_{k=0}^{\infty} k\, p(s,i)\, q^k(s,i) = p(s,i)\frac{q(s,i)}{(1-q(s,i))^2} = \frac{1-p(s,i)}{p(s,i)}. \tag{14.17}$$

Hence the suggestion is to apply (14.5),(14.8),(14.12),(14.13),(14.15), and (14.16) as follows:

Case 1: let $f(r,t) = o(s_r, i_t)$, $r = 1,2,\ldots,n$ and $t = 1,2,\ldots,m$;

Case 2: let $f(r,t) = o^{-1}(s_r, i_t)$, $r = 1,2,\ldots,n$ and $t = 1,2,\ldots,m$.

Definition 3 *Let $A \subset I$ and $C \subset S$ ($\#C \geq 2$) be given.*

Case 1: Let $0 < p(s,j)$ for all $j \in A$, then, for $i \in A$,

$$p_\nu(C,i,A) = \begin{cases} \dfrac{\prod\limits_{\substack{j \in A \\ j \neq i}} \nu(C,j)}{\sum\limits_{k\in A} \prod\limits_{\substack{h \in A \\ h \neq k}} \nu(C,h)} & \text{if } \exists k \in A,\ s \in C:\ p(s,k) < 1, \\ \dfrac{1}{\#A} & \text{otherwise,} \end{cases} \tag{14.18}$$

where

$$\nu(C,j) = \sum_{s\in C} o(s,j) \quad (j \in A)\,. \tag{14.19}$$

The rule given by (14.18) is called the ν-composition rule for a collective forced–choice trial situation.

Case 2: Let $p(s,j) < 1$ for all $j \in A$, then, for $i \in A$,

$$p_\mu(C,i,A) = \begin{cases} \dfrac{\mu(C,i)}{\sum\limits_{k\in A} \mu(C,k)} & \text{if } \exists k \in A,\ s \in C:\ 0 < p(s,k), \\ \dfrac{1}{\#A} & \text{otherwise,} \end{cases} \tag{14.20}$$

where

$$\mu(C,j) = \sum_{s \in C} o^{-1}(s,j) \quad (j \in A)\,. \tag{14.21}$$

The rule given by (14.20) is called the μ-composition rule for collective forced-choice trial situations.

14.4 Some decision and measurement theoretical properties of the collective choice probabilities

In this final section, it is shown that the collective composition rules satisfy some well-known requirements, namely, the conditions for an acceptable model of social choice given by Arrow (1959/1963) and also Luce's choice axiom.

Definition 4 *For $C \subset S$, and $s \in C$, let $R(s)$, $R^*_\nu(C)$, $R^*_\mu(C)$ be binary relations on the set of stimuli I defined as follows:*

(i) $$(i,j) \in R(s) \quad \text{iff} \quad p(s,i) \geq p(s,j)\,; \tag{14.22}$$

(ii) if $p(s,j) \neq 0$ for $s \in S$ and $j \in A$, then

$$(i,j) \in R^*_\nu(C) \quad \text{iff} \quad p_\nu(C,i,\{i,j\}) \geq p_\nu(C,j,\{i,j\})\,; \tag{14.23}$$

(iii) if $p(s,j) \neq 1$ for $s \in S$ and $j \in A$, then

$$(i,j) \in R^*_\mu(C) \quad \text{iff} \quad p_\mu(C,i,\{i,j\}) \geq p_\mu(C,j,\{i,j\})\,. \tag{14.24}$$

Lemma 1 *For $C \subset S$, $s \in C$, and $A \subset I$, $R(s)$ is a weak order on A, that is, $R(s)$ satisfies these conditions of transitivity and comparability:*

(i) transitivity: for every $i,j,k \in A$ such that $(i,j) \in R(s)$ and $(j,k) \in R(s)$, it holds that $(i,k) \in R(s)$;

(ii) comparability: for every $i,j \in A$, it holds that $(i,j) \in R(s)$ or $(j,i) \in R(s)$, or both hold.

Proof: Obvious.

Theorem 6 *For every $C \subset S$, $s \in C$, $A \subset I$,*

(i) $R^*_\nu(C)$ is a weak order on A;

(ii) $R^*_\nu(C)$ is defined for any number of stimuli;

(iii) $R^*_\nu(C)$ is defined whenever $R(s)$ is defined for all $s \in C$;

(iv) $R^*_\nu(C)$ satisfies the "independence of irrelevant alternatives" condition, that is, $(i,j) \in R^*_\nu(C)$ does not depend any $k \in A$, $k \neq i,j$;

(v) $R^*_\nu(C)$ *satisfies the "Pareto principle", that is, if* $(i,j) \in R(s)$ *for all* $s \in C$, *then* $(i,j) \in R^*_\nu(C)$;

(vi) *If for every* $s \in C$, $i,j \in A$, $(i,j) \in R(s)$, *there is* $s^* \in C$ *such that* $p(s^*,j) \geq p(s,i) \geq p(s,j) > p(s^*,i)$, *then* $R^*_\nu(\{s,s^*\})$ *satisfies the "non dictatorship" condition, that is: for every* $s \in C$ *there is* $s^* \in C$ *such that* $(i,j) \in R(s)$ *implies* $(i,j) \notin R^*_\nu(\{s,s^*\})$.

Proof: (i) Let $i,j,k \in A$ such that $(i,j) \in R^*_\nu(C)$ and $(j,k) \in R^*_\nu(C)$. Then, from (14.18) we have $\nu(C,k) \geq \nu(C,j) \geq \nu(C,i)$, and then $(i,k) \in R^*_\nu(C)$.
(ii)–(iv) Trivial.
(v) If $(i,j) \in R(s)$, $s \in C$, then $p(s,i) \geq p(s,j)$ and $o(s,j) \leq o(s,i)$, $s \in C$, hence $\nu(C,i) \leq \nu(C,j)$, which implies $(i,j) \in R^*_\nu(C)$.
(vi) Let $i,j \in A$, and $s,s^* \in C$ such that $p(s^*,j) \geq p(s,i) \geq p(s,j) > p(s^*,i)$; then $(i,j) \in R(s)$, but $(i,j) \notin R^*_\nu(\{s,s^*\})$. □

Theorem 7 *For every* $C \subset S$, $s \in C$, $A \subset I$, *and if* $p(s,j) < 1$ *for* $s \in S$ *and* $j \in A$, *then (i)–(vi) in Theorem 6 hold for* $R^*_\mu(C)$ *as well.*

Proof: Similar to the proof of Theorem 6.

Theorem 8 (Luce's choice axiom) *If* $B \subset A$, $C \subset S$ *and*

(i) $p_\nu(C,j,A) > 0$ *for all* $j \in A$, *then*

$$p_\nu(C,i,A) = p_\nu(C,i,B) \sum_{j \in B} p_\nu(C,j,A)\ ; \tag{14.25}$$

(ii) $p_\mu(C,j,A) < 1$ *for all* $j \in A$, *then*

$$p_\mu(C,i,A) = p_\mu(C,i,B) \sum_{j \in B} p_\mu(C,j,A)\ . \tag{14.26}$$

Proof: The result derives from (14.18) and (14.20). □

14.5 References

Arrow, K.J. (1963). *Social choice and individual values.* New York: John Wiley.

Bezembinder, T. (1989) Social choice theory and practice. In Ch. Vlek & G. Cvetkovich (Eds.), *Social decision methodology for technological projects* (pp. 15–37).

Fishburn, P.C. (1975). A probabilistic model of social choice: Comment. *Review of Economic Studies, 42*, 297–301.

Fishburn, P.C., & Gehrlein, W.V. (1977). Towards a theory of elections with probabilistic preferences. *Econometrica, 45*, 1907–1924.

French S. (1986). *Decision theory: An introduction to the mathematics of rationality.* Chichester: Ellis Horwood Limited.

Intrilligator, M.D. (1973). A probabilistic model of social choice. *Review of Economic Studies, 40*, 665–681.

Luce, R.D. (1959). *Individual choice behavior.* New York: Wiley.

Marley, A.A. (1990). Aggregation theorems and multidimensional stochastic choice models. *Theory and Decision, 30*, 245–272.

Marley, A.A. (in press). Aggregation theorems and the combination of probabilistic rank orders, In D. E. Critchlow, M. A. Flinger, & J. S. Verducci (Eds.), *Probability models and data analysis for ranking data.* New York: Springer–Verlag.

Münnich, Á. (in press). Probabilistic choice systems, *Proceedings of the* 7th *Meeting of the Psychometric Society*, Trier, July 1991.

Pattanaik, P.K., & Peleg, B. (1986) Distribution of power under stochastic social choice rules. *Econometrica, 54*, 909–921.

Sen, A.K. (1970). *Collective choice and social welfare.* San Francisco: Holden–Day.

Sen, A.K. (1977). Social choice theory: A re–examination. *Econometrica, 45*, 53–89.

Van Blokland–Vogelesang, A.W. (1990). *Unfolding and group consensus ranking for individual preferences.* Doctoral thesis. Leiden: University of Leiden.

Williamson, O.E., & Sargent, T.J. (1967). Social choice: A probabilistic approach. *Economic Journal, 77*, 797–813.

15

Random Walk Models of Digit Comparison: Some Problems and a Solution

Wolfgang Schwarz[1]

ABSTRACT Poltrock (1989) presented a general random walk model of cognitive distance effects (Moyer & Dumais, 1978) and derived approximate predictions for digit comparisons. The present article points out some problems inherent in his approach and formulates an improved model. The model predictions are fitted to data from a digit comparison experiment involving a speed-accuracy manipulation. The model permits an exact analysis at the level of the individual digit pairs and offers an answer to some open questions such as the processing speed conditional on the response given, and the cognitive representation of the digit one.

15.1 Introduction

In an influential paper, Moyer & Landauer (1967) have shown that the time required to choose the larger of two digits decreases with the absolute difference between the digits. Subsequently, similar cognitive distance effects in mental comparisons have been demonstrated with many other stimulus materials which can be ordered along some psychological dimension (Moyer & Dumais, 1978). A possible interpretation of these findings is that humans process information by converting discrete information into cognitive analog representations, much as elementary sensory coding processes are known to act. For some time, there was no precise quantitative formulation of these qualitative ideas. Usually some deterministic ad hoc equation was fitted to the RT as a function of the digits presented, a procedure which is unable to account for many psychologically interesting features such as the speed-accuracy tradeoff or the detailed probabilistic structure of the data.

In a highly original recent paper, Poltrock (1989, see also Link, 1990) proposed an explicit random walk model of digit comparison processes. In a typical digit comparison task, one digit is presented to the left and another to the right of the fixation point, and the subject has to press a response button on the side of the larger digit. Basically, Poltrock assumes that a serial decision process accumulates evidence concerning the digits over successive time intervals. During each step of this process, each digit induces a replication of its cognitive analog

[1]Psychologisches Institut der FU Berlin, Habelschwerdter Allee 45, W—1000 Berlin 33, FRG
I would like to thank D. Laming and an anonymous reviewer for their constructive comments.

representation. The overall evidence for one of the two decisions 'left digit larger' or 'right digit larger' conveyed in a single step of the process is equivalent to the noisy difference between the, possibly already noisy, internal representations of the digits. This evidence is accumulated until one of two fixed barriers is reached. The answer given by the subject depends on which barrier it is, and the RT on the number of steps which have been taken in reaching it.

Poltrock carried out three extensive digit comparison experiments, varying the speed–accuracy tradeoff, and fitted some general linear relations between response probabilities and mean RT to his data. He concluded that his model accounted for the major features of the data such as the distance effect and the relation between speed and accuracy. Perhaps the most impressive finding is the fact that, as the model required, all stimulus–dependent parameters were reasonably constant for all experiments, while speed–accuracy manipulations and response bias effects were represented by variations of the barriers and of response execution times only.

The aim of the present article is (Section 15.2) to point out some problems with Poltrock's approach. In Section 15.3 an improved model is formulated and used to illustrate the application of some related results recently derived by the author (Schwarz, 1990, 1991a). In Section 15.4 the model is fitted to some example data from a digit comparison experiment; in Section 15.5, the relation of the present model to diffusion models in continuous time is discussed and some further possible applications are outlined.

15.2 Some problems with Poltrock's (1989) model

Random walk models of choice RT are usually analyzed by means of Wald's approximation, neglecting the excess of the process over the absorbing barrier (for mathematical details see Townsend & Ashby, 1983; Luce, 1986; Schwarz, 1990, 1991a). The approximation requires the mean and the standard deviation of a single step to be small relative to the barriers. Under this assumption, it is possible to derive parameter estimates based on a log–odds analysis of the response probabilities and certain approximately linear relationships between mean RT and response probabilities, which can be used for an approximate regression analysis (cf. eqs.(4–6) in Poltrock, 1989). A problem here is that often no bounds for the error of Wald's approximation are known, while the log–odds are extremely sensitive to even a slight numerical imprecision – especially if the error proportions are near to zero, as is often the case for relatively easy tasks such as digit comparisons. Considering this strategy of analysis, Luce (1986, p. 346) has pointed out that it 'can be grossly misleading' to approximate a strongly nonlinear stochastic model by strictly linear regression methods.

For our purposes, this fact is perhaps best illustrated as follows. For his speeded digit comparison Experiment 2, Poltrock (1989, Table 7) presented results of an approximate random–walk regression analysis yielding a correlation of r = +0.63 between predicted and observed RTs for the 56 ordered pairs of the digits 2 − 9. From this and other comparable results he concluded that his model accounted for the major features of the data. However, if the parameter estimates from his Table 7 are resubstituted into his eq. (6) in order to obtain probability predictions for individual digit pairs, and if these predictions are evaluated by a conventional χ^2 test for the individual binomial proportions, the result is $\chi^2 \approx 9081$. In calculating this statistic, only the digit-pairs with an expected error frequency of at least two were considered; these pairs provided 49 proportions to be predicted with eight parameters estimated from the data, yielding df = 41. Thus, although the model reproduces all the qualitative features of the data such as the distance effect, the fit is not satisfactory, a fact already noted by Poltrock (p. 151–152). It is not clear whether this is attributable to the error inherent in Wald's approximation or to intrinsic psychological inadequacies of the assumptions of the model.

To illustrate the psychological significance of these issues, note that, in order to obtain reasonable parameter estimates from Wald's approximation, Poltrock had to discard *post hoc* all the 16 pairs involving the digit 1. This immediately suggests the following question: is the digit 1 processed as a 'cognitive unit', in a manner qualitatively different from all the other digits, or does the failure of the model for pairs including the digit 1 simply reflect the formal limitations of Wald's approximation; being thus without psychological significance?

In principle, one of the most interesting aspects of random walk models of choice RT is their ability to make detailed predictions concerning the tradeoff between speed and accuracy of cognitive processes. Are errors, for example, faster than correct responses, and does the cognitive distance effect hold also for incorrect RTs? These aspects are overlooked by the type of analysis employed by Poltrock (1989), which uses unconditional mean RT only, lumping together both correct and error responses. It would be desirable to obtain exact conditional mean response time predictions, depending on the correctness or incorrectness of the response uttered by the subject.

15.3 The Laplace model for digit comparison processes

The main assumption is that the cognitive analog representations of the single digit i are exponentially distributed with rate λ_i. Under Poltrock's general model, each step of the decision process discriminating between the digits i and j will be distributed as the difference of two exponential random variables with rates λ_i and λ_j; i.e., the accumulation of information is driven by a step size density g_{ij} of the Laplace type:

$$g_{ij}(x) = \frac{\lambda_i \lambda_j}{\lambda_i + \lambda_j} \exp(\lambda_j x) \qquad \text{for } x \leq 0,$$

$$g_{ij}(x) = \frac{\lambda_i \lambda_j}{\lambda_i + \lambda_j} \exp(-\lambda_i x) \qquad \text{for } x \geq 0.$$

For the purpose of illustration, let the left digit (i) be 7, and the right digit (j) be 3, and assume that during each interval of time a noisy analog representation of each digit is generated internally, a representation which is exponentially distributed with rates λ_7 and λ_3 , respectively. The evidence accumulated during the decision process is the difference of these representations. The mean analog representation of digit 7 will usually be larger than the mean representation of digit 3; i.e., $1/\lambda_7 > 1/\lambda_3$. On the average, the process will tend towards the upper barrier at a and initiate the correct response 'left digit larger'. Occasionally, however, it will also be absorbed at the lower barrier at $-b$, at least if λ_3 and λ_7 are not too different and the barriers not too widely spaced, so that temporary 'sampling errors' of the analog representations will sometimes determinate the response.

For this model, the following exact predictions can be deduced from results derived by Schwarz (1990). The probability of the response 'left digit larger' is

$$p_{ij} = \frac{1 - \psi_{ij}}{\phi_{ij} - \psi_{ij}},$$

where $\phi_{ij} = \exp[(\lambda_i - \lambda_j)\, a]\, \frac{\lambda_i}{\lambda_j}$ and $\psi_{ij} \equiv \exp[(\lambda_j - \lambda_i)\, b]\, \frac{\lambda_j}{\lambda_i}$.

Let $\mathbf{N}_{ij}$ be the discrete first passage time, i.e. the number of steps of the accumulation process required to reach one of the barriers for the first time. Then the corresponding expected first passage times, conditional on the absorbing barrier (a or $-b$), can be shown to be

$$\mathrm{E}_a\,[\mathbf{N}_{ij}] = \frac{(a + 1/\lambda_j)\,\phi_{ij}\,p_{ij} - (b + 1/\lambda_i)\,\psi_{ij}\,(1 - p_{ij})}{(1/\lambda_i - 1/\lambda_j)\,(\psi_{ij} - 1)} + \frac{\psi_{ij}}{\psi_{ij} - 1}\,\mathrm{E}\,[\mathbf{N}_{ij}], \quad (15.1)$$

$$\mathrm{E}_{-b}\,[\mathbf{N}_{ij}] = \frac{(a + 1/\lambda_j)\,\phi_{ij}\,p_{ij} - (b + 1/\lambda_i)\,\psi_{ij}\,(1 - p_{ij})}{(1/\lambda_i - 1/\lambda_j)\,(\phi_{ij} - 1)} + \frac{\phi_{ij}}{\phi_{ij} - 1}\,\mathrm{E}\,[\mathbf{N}_{ij}], \quad (15.2)$$

In (15.1)–(15.2),

$$\mathrm{E}\,[\mathbf{N}_{ij}] = \frac{(a + 1/\lambda_i)\,p_{ij} - (b + 1/\lambda_j)\,(1 - p_{ij})}{1/\lambda_i - 1/\lambda_j}$$

denotes the unconditional first passage time. If each step of the accumulation process has duration $\mathbf{D}$ with expectation μ_D, and the duration $\mathbf{M}$ of all non-decision processes has expectation μ_M, then the first passage times are related to the observable mean reaction times by the equation

$$\mathrm{E}_r[\mathrm{RT}_{ij}] = \mu_D\, \mathrm{E}_r[\mathbf{N}_{ij}] + \mu_M, \qquad r = a, -b.$$

In applying this model to data from digit comparison experiments, the rate parameters λ_i, the barriers a and b, and the two parameters μ_D and μ_M have to be estimated. Each additional speed–accuracy level requires two further barrier parameters, while, of course, all other parameters (a possible exception is μ_M, see Poltrock, 1989) should remain constant. All rate and barrier parameters are unique up to a scale factor.

As a simple illustration of the results, the Laplace model has been fitted to the data of Poltrock's (1989) Table 7 already discussed in Section 2. If all the digits $1-9$ (72 ordered pairs) are included and λ_1 is set to 1, then the minimum χ^2– estimates of the mean analog representations $1/\lambda_i$ of digits given in brackets are: $1.190\,(2), 1.272\,(3), 1.324\,(4), 1.487\,(5), 1.620\,(6), 1.618\,(7), 1.880\,(8)$, and $1.949\,(9)$, while the barriers are equal to $\mathrm{a} = 11.05$ and $\mathrm{b} = 8.82$. The corresponding value of χ^2 is 744. It was computed in exactly the same manner as for Poltrock's estimates and has $\mathrm{df} = 70 - 10 = 60$. If only the 56 proportions excluding the digit 1 are considered, then $\chi^2 = 572$ and $\mathrm{df} = 46$, as compared to $\chi^2 = 9081$ for only $\mathrm{df} = 41$ from Poltrock's (1989) estimates. It should be noted, however, that Poltrock did not use a minimum χ^2 estimation to determine his parameters, and it is possible that his model could achieve a lower χ^2 than 9081. The Laplace model can thus improve the fit considerably, but the absolute values of χ^2 are still much too large. Neither the estimated values of the parameters nor the resulting fit give any hint that comparisons involving digit 1 are qualitatively different from those which do not. For pairs involving the digit 1, $\mathrm{E}\,[\mathbf{N}_{ij}]$ is usually quite small, which explains the error ($16\% - 20\%$) of Wald's approximation in these cases.

It should be recognized, however, that the meaning of these results is questionable, because the reaction times and error proportions, to which the Laplace model was applied, are averaged over six subjects who are already known, from the separate analysis of individual subjects, to require rather different parameter values (see Table 4 of Poltrock, 1989). For the Laplace digit–comparison model at least the behaviour of the model subject to the mean parameters will be different from the mean of its behaviour subject to the individual parameter sets, so that analyzing grouped data can be misleading. If each proportion is estimated from 392 data as in Poltrock (1989), the statistical power is considerable, and even slight deviations as, e.g., introduced by averaging over subjects, can cause significant, but misleading deviations. It therefore seems preferable to analyze data from individual subjects, as Poltrock did and as in the next section. Recently, the author has also derived exact conditional standard deviations for the Laplace model (Schwarz, 1991a). Among other things, these results permit calculation of the standard error of the mean conditional RT for individual subjects, which can be useful in evaluating the fit of the model.

15.4 Experiment and results

Method. The digits $2-9$ were used so that there were $8 \times 7 = 56$ ordered digit pairs. The experiment was run at two performance levels: speed vs. accuracy. Two sessions prior to the main experiment were used for practice of the subjects, and there were four warm–up trials at the beginning of each block. In addition, there were 30 practice trials prior to each session. For each condition, there were five sessions, each consisting of 14 blocks. During each block, each of the 56 ordered digit pairs was presented exactly once in random order. There were thus $5 \times 14 = 70$ replications of each ordered pair under each performance condition. From all 56 distributions, the largest three and the smallest two RTs were discarded, so that finally there remained 65 RTs for each ordered digit pair under each condition. The data for just one subject out of three are reported here; the subjects were similar in all important respects.

The stimuli were displayed on a video monitor using an Atari computer, and the display timing was synchronized with the video refresh cycle. The digits were specifically designed on a virtual square of 40×40 pixels (1.2 cm $\times$ 1.2 cm). The viewing distance was approximately 57 cm, so that each digit was seen within an angle of $1° \, 20'$. Each trial started with the presentation of the fixation point for 250 ms. After a further delay of 100 ms, the digit pair was presented and remained visible until one of the two buttons on an external keyboard was pressed. The next trial started three seconds later. Between two blocks there was an arbitrary pause during which the mean RT of the preceding block was displayed on the screen.

Results. Figure 15.1, filled circles, shows the empirical error probabilities under the speed condition, separately for each left digit as a function of the right digit. Under the accuracy condition the error frequencies were too low to justify a plot. Each point in Figure 15.1 is a binomial proportion based on $n = 65$ data, that is, the standard errors range approximately from 2 % to 6 %. The main effect clearly exhibited by all graphs is that the error frequency increases with decreasing distance of the left and right digit (the distance effect). A comparison of pairs like $2-3$ vs. $8-9$ shows that for a given fixed digit difference the error probability usually increases as the digits involved become larger (the so–called minimum effect). The open circles joined by lines show that the Laplace model, given the parameter values estimated from the data, generally produces a pattern reflecting the distance and the minimum effect. If λ_2 is set to 2, then the minimum χ^2 estimates of the mean analog representations $1/\lambda_i$ of the digits given in brackets are: 2.45 (3), 2.81 (4), 3.13 (5), 3.42 (6), 3.54 (7), 3.75 (8), and 3.84 (9), while the minimum χ^2 estimates for the barriers are a = 94.52 and b = 113.88 under the speed condition, and a = 150.82 and b = 204.98 under the accuracy condition.

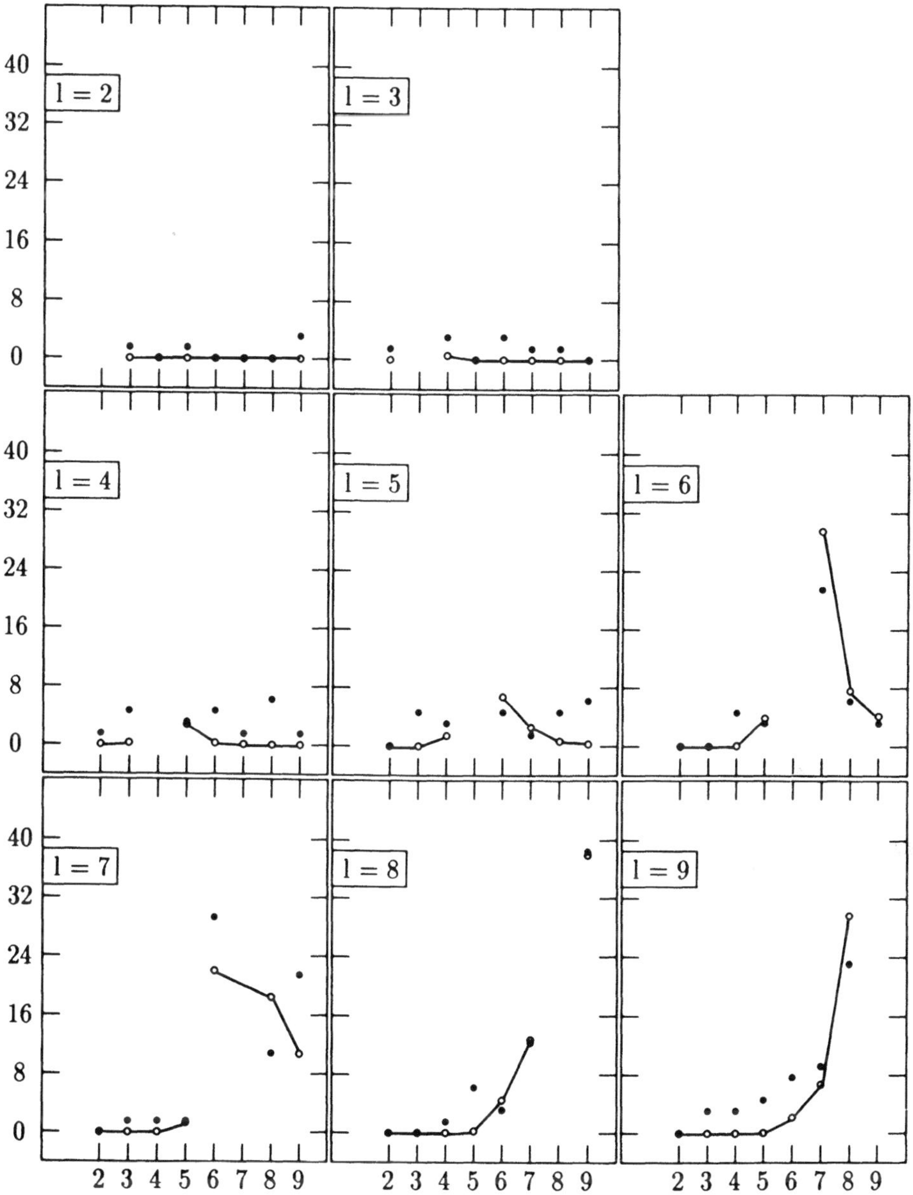

FIGURE 15.1. Percentage of errors (ordinates) under speed condition as a function of the right digit (abscissae), separately for each left digit (1). Filled circles: data, $n = 65$ per point. Open circles: Laplace model.

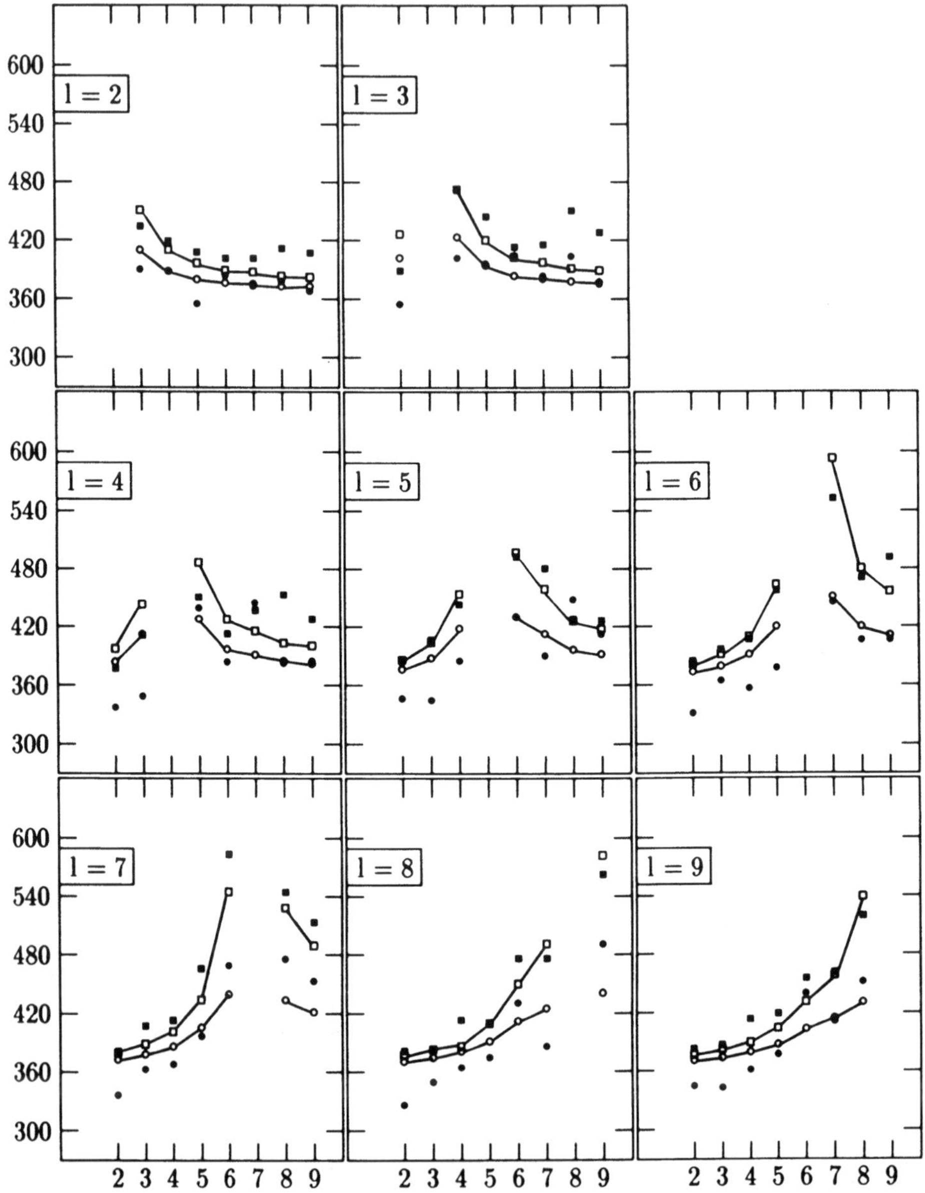

FIGURE 15.2. Mean correct RTs (ordinates) as a function of the right digit (absicissae), separately for each left digit (1). Filled symbols: data, $n = 65 \times$ P(correct) per point. Open symbols: Laplace model. Circles: speed. Squares: accuracy.

The distance effect is explained by the monotone increase of the mean analog representations $1/\lambda_i$ of the digits; it also implies an increasing distance $1/\lambda_i - 1/\lambda_j$ for increasing digit differences $(i-j)$. Consequent on the negative acceleration of this increase, the distance 2 to 3 is much larger than the distance 8 to 9, so that the minimum effect results. As was to be expected, the barriers are smaller under the speed condition. The difference of the barrier estimates indicates a general response bias towards the answer 'left digit larger' and explains the asymmetry of the response proportions; compare, e.g., the error frequencies of the pairs $8-9$ (39%) and $9-8$ (23%).

Turning to the RT predictions, the mean duration of a single comparison step (μ_D) was estimated to be 0.194 ms, corresponding to 5.15 comparisons per ms. The mean μ_M of all non–decision processes, restricted to be equal under both conditions, was found to be 360 ms. Allowing μ_M to depend on the speed vs. accuracy instruction did not improve the overall fit significantly. Descriptively, the correlation between all 112 predicted and observed mean correct RTs is +0.87.

Figure 15.2 shows observed and predicted mean correct RTs for both conditions. The standard errors of the means range from 4 to 10 ms, approximately. The speed vs. accuracy instruction was clearly effective. The distance effect is present in each individual graph. The minimum effect shows up in steeper curves for large digits, compare, e.g., the vertical ranges of the graphs for $l = 2$ and $l = 9$. The Laplace model correctly explains the general speed manipulation without altering the base time μ_M, and the pattern of the predicted mean RTs conforms to the distance and to the minimum effect. The deviations generally concern the few empirical points which run counter to these general effects. In theory, the whole experiment generates 2×56 probability predictions and $2 \times 2 \times 56$ mean conditional RT predictions for the separate ordered digit pairs, i.e. a total of 336 independent data which are predicted by a total of 13 parameters: 7 rates λ_i, 4 barriers, μ_D, and μ_M. Of course, many digit pairs gave almost no errors. However, even considering probability and correct mean RT predictions only, there are still 168 independent data in Figures 15.1 and 15.2. To obtain an approximate evaluation of the overall fit of the Laplace model, χ^2 was calculated for the response proportions as in Section 15.2, and for the mean reaction times from the squares of the $z-$scores of these means (Schwarz, 1991a, p. $260-261$). A value of $\chi^2 = 506$ was obtained with $df = 118$, which shows that the Laplace model as it stands cannot account for the quantitative structure of the data. One reason is a lack of stationarity in the data, which showed up in small but significant effects between the individual blocks and sessions of the experiment.

Finally, the error data from the speed condition address the question whether the distance effect holds for incorrect responses also. From the parameters estimated above, it is easy to derive corresponding theoretical conditional expectations under the Laplace model, which in fact does predict a general distance effect for incorrect responses also. The empirical mean incorrect RTs for the distances given in brackets are: 363, (1), 331 (2), 314 (3), 296 (4), 294 (5), 286 (6) and 253 (7) ms. Thus, there is a clear general distance effect for error responses, confirming the prediction of the Laplace model.

15.5 Discussion

The approximate predictions of the general random walk model of digit comparison presented by Poltrock (1989) accounts qualitatively for all gross features of relevant data published so far, and the Laplace model described above makes exact and detailed (conditional) predictions as shown in Figures 15.1 and 15.2. Now, is this due to the specific properties of the Laplace model, or simply to the fact that only for this model are exact results known? By invoking the central limit theorem, the discrete random walk for most digit pairs can be approximated by a continuous diffusion process with normal and independent increments (the Wiener process) between by two absorbing barriers. The fit of this model to the data to is similar to the one presented above. This is not surprising, because the mean individual steps $1/\lambda_i - 1/\lambda_j$ are generally small relative to the barriers. In view of this finding, a further step can be taken to validate the general approach: what would be the effect of presenting one of the digits subject to a certain delay? The basic assumption in modeling this modification is that the barriers are reflecting as long as only one digit is shown and they are absorbing after the presentation of the second digit. For the Laplace model, this generalization leads to considerable formal complications, while for the diffusion model the calculations are somewhat involved, but still feasible. Note that even in the presence of reflecting barriers the accumulation process can still move towards the response barrier corresponding to the first digit presented, and that this movement will increase with increasing delay. Thus, this mechanism provides a model of the time-dependent priming effects induced by the information presented first. The mathematical details of the diffusion theory of priming effects has recently been worked out, and the fit of preliminary results of relevant experiments have proved to be quite encouraging (Schwarz, 1991b).

15.6 References

Link, S. (1990). Modeling imageless thought: The relative judgment theory of numerical comparisons. *Journal of Mathematical Psychology, 34,* 2–41.

Luce, R.D. (1986). *Response times.* New York: Oxford University Press.

Moyer, R.S., & Dumais, S. T. (1978). Mental comparison. In G. H. Bower (Ed.), *The psychology of learning and motivation* (Vol. 12). New York: Academic Press.

Moyer, R.S., & Landauer, T. K. (1967). Time required for judgements of numerical inequality. *Nature, 215,* 1519–1520.

Poltrock, S.E. (1989). A random walk model of digit comparison. *Journal of Mathematical Psychology, 33,* 131–162.

Schwarz, W. (1990). Stochastic accumulation of information in discrete time: Comparing exact results and Wald approximations. *Journal of Mathematical Psychology, 34,* 229–236.

Schwarz, W. (1991a). Variance results for random walk models of choice reaction time. *British Journal of Mathematical and Statistical Psychology, 44,* 251–264.

Schwarz, W. (1991b). Diffusion models for speed–accuracy tradeoff and priming effects in digit comparison tasks. Paper presented at the 22nd meeting of the European Mathematical Psychology Group, Vienna, 9.-12. Sept. 1991.

Townsend, J.T., & Ashby, F. G. (1983). *Stochastic modeling of elementary psychological processes.* Cambridge: Cambridge University Press.

16

A Diffusion Model For Intersensory Facilitation of Reaction Time

Adele Diederich [1]

ABSTRACT An experiment was conduced to test for intersensory facilitation of reaction time with stimuli from three modalities (visual, auditory, tactile). Stimuli from two or three modalities with different intensity levels were presented with different interstimulus intervals in a simple reaction time experiment. A diffusion model, called the "Ornstein-Uhlenbeck Process with Drift", with one absorbing and one reflecting boundary, is presented to explain the observed facilitation of reaction time. Although no closed form is known for the diffusion equations for the case of an arbitrary boundary, numerical methods based on limits of discrete-time Markov chains provide practical approximations.

16.1 Introduction

From time to time psychologists have been interested in the effects on detection or recognition performance, of presenting two or more signals (targets). Using reaction time as measure of information processing time it can be observed that a response is faster when two or more targets are presented than when only one is. There are essentially two different approaches that have been suggested to explain the phenomenon: *coactivation models* and *separate activation models* (Miller, 1982). Separate activation models assume that processing in different sensory channels is parallel and independent. Presenting multiple stimuli (e.g., light, tone, and vibration) produces separate activation in each of the corresponding channels. In any channel activation might built to the level at which it can produce a response. That means that a response is produced by the signal whose processing is completed first. These models are also called "race models" since a race between several response activation processes takes place, and the response time is determined by the winner. Separate activation models predict faster average reactions to multiple stimuli compared to single stimuli since the average of the minimum of the processing time in different channels is smaller than the average processing time in any one channel, provided processing time is assumed to be a random variable. In accordance with earlier findings (Miller, 1982; Diederich & Colonius, 1987) the data presented in this paper show facilitation above the level predicted by the separate activation model (see also Diederich, 1992b).

An alternative approach to explaining intersensory facilitation is coactivation.

[1]C.v.Ossietzky Universität Oldenburg, Fachbereich Physik, 2900 Oldenburg, Germany

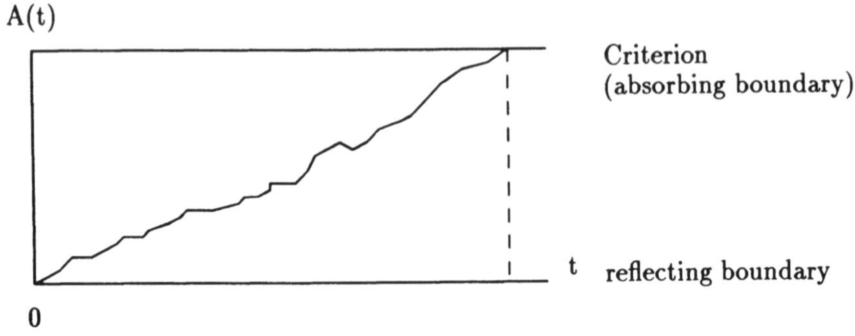

FIGURE 16.1. Information accumulation process when one stimulus only is presented

Coactivation models assume that interaction between different sensory modalities is possible. When multiple stimuli are presented activation of different channels can be combined to satisfy a single criterion for response initiation. Coactivation models predict faster average reaction to multiple stimuli compared to single stimuli because the combined activation reaches this criterion sooner.

Here I develop a specific coactivation model, a diffusion model. (For an alternative approach in terms of Poisson superposition mechanisms see Diederich, 1992a.) Diffusion models are stochastic processes with continuous-time and continuous-state space. Most research with these models concerns choice reaction time. Ratcliff (1978) used a diffusion process in his theory of memory retrieval. Recently, Busemeyer and Townsend (1989); (see also Busemeyer & Goldstein 1990, Busemeyer & Townsend, 1991) developed a dynamic stochastic model of decision making – a so-called decision-field theory – which can be described by diffusion processes. Diffusion models have also been applied in simple reaction time research, though to a much lesser extent. Pacut (1980, 1982) applied a diffusion model to escape reaction and escape learning data; Pacut and Tych (1982) applied it to avoidance reaction latency and avoidance learning. Recently, Smith (1990) applied a diffusion model to simple visual reaction time. In the following we introduce a diffusion model for intersensory facilitation.

16.2 The Diffusion Model for Intersensory Facilitation

Suppose that, as soon as a signal is presented to the subject, the nervous system accumulates very small amounts of activation stochastically and continuously over time. A response is initiated as soon as a preset criterion level is reached. Figure 16.1 illustrates the process.

Activation A grows nonlinearly towards an asymptote. After the signal is switched off, the amount of activation is assumed to decay to its starting level. This decay is in accordance with the neuronal activity dynamics studied in neurobiology (e.g., Ricciardi (1977)). It is assumed that activation never drops

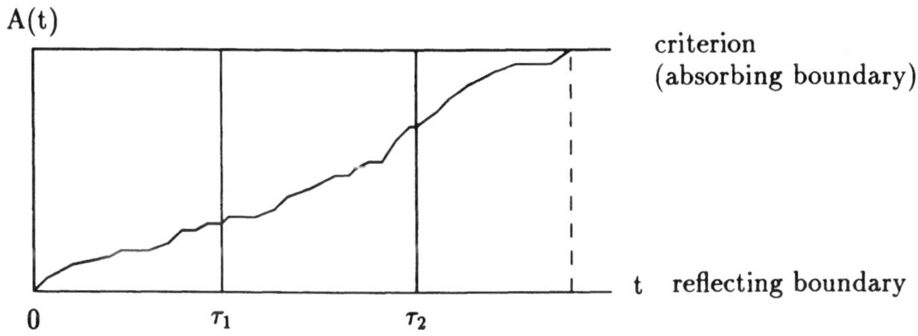

FIGURE 16.2. Information accumulation process when three different stimuli are presented, τ_1 and τ_2 msec apart

below its initial level; that is, the process is reflected on the lower boundary. This process can be described by a diffusion process called the "Ornstein-Uhlenbeck process with drift" with one reflecting and one absorbing boundary. Its drift, $\mu(x)$, is $\alpha - \beta x$, and its diffusion coefficient, $\sigma^2(x)$, is σ^2, where β and σ are arbitrary positive constants. The paramter α is the force directed towards the criterion, whereas β represents a restoring force directed towards the origin and of a magnitude proportional to the distance from it. While the diffusion coefficient is independent of the state of information accumulation, the drift rate changes with location x. Assume that in an experiment three different stimuli are presented, τ_1 and τ_2 msec apart. The information accumulation process for this situation is illustrated in Figure 16.2.

A tactile stiumuls, say, is presented at time $t = 0$, and the process starts with drift $\mu(x) = \alpha_T - \beta_x$ and diffusion coefficient σ^2. At time $t = \tau_1$, when the second (visual) stimulus is presented, the process is located somewhere between the reflecting barrier and the absorbing boundary. The drift coefficient now changes since more information is available; $\mu(x)$ is assumed to become $(\alpha_T + \alpha_V) - \beta x$, with $\alpha_V > \alpha_T$. Finally, at time $t = \tau_2$, when the third (auditory) stimulus is presented, the process still is located somewhere between the reflecting boundary and the absorbing boundary. The drift coefficient is now $\mu(x) = (\alpha_T + \alpha_V + \alpha_A) - \beta x$, with $\alpha_A > \alpha_V$. Call the parameters $\alpha_T, \alpha_V, \alpha_A$, representing the force directed toward the criterion, "intensity parameters". The diffusion coefficient is supposed to be the same for the whole process. Note that the diffusion proccss is time-homogeneous within each time interval $[0, \tau_1), [\tau_1, \tau_2)$, and $[\tau_2, \infty)$, but non-homogeneous across the interval $[0, \infty)$.

Of special interest, of course, is the time at which the stochastic process reaches the upper boundary, when a response is initiated. Consider a single state x and the probability distribution of the time T at which $X(t)$ enters this state for the first time. In this model the criterion state is an absorbing state and, once entered, is never left. Clearly T is a random variable taking values in the interval $[0, \infty)$ and is called the "first passage time". The backward or

Kolmogorov equation is appropriate for calculating the probability distribution of the first passage time since the objective is to determine the first passage time distribution as a function of the initial state x_0 for a fixed final state. In the case of an unrestricted process, with the boundaries at $\pm\infty$, starting at time $t_0 = 0$ and with $\alpha = 0$, the solution for the backward equation as a transition probability function for the Ornstein-Uhlenbeck process is a normal distribution with mean $x_0\exp(-\beta t)$ and variance $\sigma_2(1-\exp(-2\beta t)/2\beta$. Unfortunately, no solution is known in closed form for the case of arbitrary boundaries, e.g., one absorbing boundary and one reflecting boundary (see Ricciardi, 1977, p. 128; Bhattacharya & Waymire, 1990, p. 389). But there are numerical methods based on the limits of discrete-time Markov chains that provide practical approximations. These will be used in what follows (for a related approach, see Busemeyer & Townsend, 1992).

The presentation of a stimulus evokes an information accumulation process. This process is represented by a variable, A, called the state of information accumulation or activation. The state of activation changes over time, depending on how much information is available, producing a new state of activation at each moment in time, a stochastic process denoted by $A(t)$. As soon as $A(t) \geq c$, a criterion set by the subject, the response is initiated. It is assumed that activation changes by only very small amounts during small time intervals. Let $h, h > 0$, denote a very small time interval, and $\Delta A(t) = A(t) - A(t-h)$ the change of activation during the interval h. $\Delta A(t)$ may be a very small step equal to $\delta \equiv \sigma\sqrt{h}$. The parameter δ is called the step size, and the parameter σ^2 is called the diffusion coefficient for the discrete-time and -state Markov model. Since we assume a time-homogeneous Markov chain the transition probabilities are determined as

$$\begin{aligned} Prob[A(t) &- A(t-h) = -\delta \mid A(t-h) = x, A(t-2h) = x_1, \ldots] \\ &= Prob[\Delta A(t) = -\delta \mid A(t-h) = x] \\ &\equiv p(x), \end{aligned} \tag{16.1}$$

$$\begin{aligned} Prob[A(t) &- A(t-h) = \delta \mid A(t-h) = x, A(t-2h) = x_1, \ldots] \\ &= Prob[\Delta A(t) = \delta \mid A(t-h) = x] \\ &= 1 - p(x) \equiv q(x). \end{aligned} \tag{16.2}$$

Note that the transition probability functions $p(x)$ and $q(x)$ depend on the state x but not on the time t. The drift coefficient, $\mu(x)$, and the diffusion coefficient, σ^2, for the discrete-time Markov chain can be defined by the first and second moments of $\Delta A(t)$, conditioned on the previous state of activation, as follows:

$$\mu(x) = \frac{E[\Delta A(t) \mid A(t-h) = x]}{h} = \frac{\delta[q(x) - p(x)]}{h}, \tag{16.3}$$

$$\sigma^2 = \frac{E[\Delta A(t)^2 \mid A(t-h) = x]}{h} = \frac{\delta^2[q(x) + p(x)]}{h}. \tag{16.4}$$

The drift and diffusion coefficients are functions of the time unit h which is taken to be a fixed constant. Its value is chosen a priori to be as close to zero as is needed to make the discrete-time Markov chain model approximate the continuous-time Markov process to the desired accuracy. The transition probabilities, $p(x)$ and $q(x)$, can be written as functions of the drift and diffusion coefficients, $\mu(x)$ and σ^2, by solving for $p(x)$ and $q(x)$ in the two equations above:

$$\mu(x) = \frac{\delta(q(x) - p(x))}{h},$$
$$q(x) = \frac{h\mu(x)}{\delta} + p(x);$$

assuming $p(x) = 1 - q(x)$ and $\delta = \sigma\sqrt{h}$ yields

$$p(x) = \frac{1}{2}\left[1 - \frac{\sqrt{h}}{\sigma}\mu(x)\right], \tag{16.5}$$

$$q(x) = \frac{1}{2}\left[1 + \frac{\sqrt{h}}{\sigma}\mu(x)\right]. \tag{16.6}$$

In order to obtain $0 \leq p(x), q(x) \leq 1$, $\mu(x)/\sigma$ must be between $-1/\sqrt{h}$ and $1/\sqrt{h}$. The stochastic process obtained so far can be defined by a stochastic difference equation, i.e.,

$$A(t+h) = A(t) + \mu(A(t))h + \epsilon(t)\sqrt{h},$$

or

$$\Delta A(t+h) = \mu(A(t))h + \epsilon(t)\sqrt{h}, \tag{16.7}$$

where $\epsilon(t)$, a random variable, is a stochastic disturbance term with mean 0 and variance σ^2 for small h. Remembering that the drift coefficient of the Ornstein-Uhlenbeck process is $\mu(x) = \alpha - \beta x$, where α reflects the force directed towards the criterion, and β a restoring force directed towards the origin and of a magnitude proportional to the distance, leads to the following stochastic difference equation for changes in activation,

$$\Delta A(t+h) = [\alpha - \beta A(t)]h + \sigma\epsilon(t)\sqrt{h}. \tag{16.8}$$

To summarize the arguments so far: the objective is to describe a simple reaction time process where a single stimulus is represented by an Ornstein-Uhlenbeck stochastic process. The Ornstein-Uhlenbeck process is a continuous-time, continuous-state processes with Markov property (continuous-time Markov process). Since, in this special case, there are difficulties in determining the distribution of the first passage time, the continuous process is approximated by a discrete-time, discrete-state Markov process (a discrete-time Markov chain) with the time interval and the step size tending to zero in the limit. The transition probabilities for the discrete-time Markov chain are expressed in terms of the drift and diffusion coefficient.

To calculate the first passage time distribution and the expected value and the variance of the time the approximating process will take to reach the boundary, it is necessary to determine the transition probability matrix for the Markov chain. This requires a prior definition of the state space of the process. Let S be a set of states of activation of the information accumulation process. The total number of states can be expressed as a function of the step size δ and the criterion c by setting $c = k\delta$. Consequently,

$$S = \{0, \delta, 2\delta, \ldots, k\delta\},$$

and the cardinality of S is $k + 1 = m$. Since the process is assumed never to skip states in its evolution, the transition matrix can be characterized in the following way

$$p_{i,j} = \begin{cases} p((i-1)\delta) & \text{for } j = i - 1, \\ q((i-1)\delta) & \text{for } j = i + 1, \\ 0 & \text{otherwise,} \end{cases}$$

where p and q are as defined earlier in (16.5) and (16.6). Note that the transition probabilities depend on the state which the process occupies.

It is convenient to represent the transition matrix $\mathbf{T}$ in canonical form. Let the m-state Markov chain consist of r recurrent states and $(m - r)$ transient states. The transition matrix $\mathbf{T}$ can be put in the form

$$\mathbf{T} = \left[\begin{array}{c|c} \mathbf{0} & \mathbf{P} \\ \hline \mathbf{Q} & \mathbf{R} \end{array}\right],$$

where $\mathbf{P}$ is an $r \times r$ matrix with the transition probabilities among the recurrent states for its elements; $\mathbf{Q}$ is an $(m-r) \times (m-r)$ substochastic matrix (with at least one row sum less than 1) with probabilities of transition only among the transient states for its elements; and $\mathbf{R}$ is an $(m-r) \times r$ matrix whose elements are the probabilities of the one-step transition from the $(m-r)$ transient states to the r recurrent states. Clearly, $\mathbf{P}$ is an 1×1 matrix containing the number 1 since there is just one absorbing boundary, the criterion c. The matrix $\mathbf{Q}$ is an $(m-1) \times (m-1)$ matrix containing the transition probabilities p_{ij} stated above, except for the last row where $p_{1,2} = 1$, and the other elements zero, since there is one reflecting boundary. Finally, $\mathbf{R}$ is an $(m-1)$ vector containing the transition probabilies from the transient to the absorbing state. The first element is q_{m-1} and zero. To see how the process can be mapped onto the transition matrix, consult Figure 16.3.

Consider next the starting position of the process. At time $t = 0$ the process is set in motion either by starting at a fixed state i_0, called the *initial state*, or by random location in the state space according to a probability distribution $\mathbf{Z}$,

	Z											
m		0	0	0	0	$\cdots$	0	0	0	0	1	P
m-1	0	0	0	0	0	$\cdots$	0	0	p_{m-1}	0	q_{m-1}	
m-2	0	0	0	0	0	$\cdots$	0	p_{m-2}	0	q_{m-2}	0	
						$\cdots$	p_{m-3}	0	q_{m-3}	0	0	
$\vdots$						$\cdots$						
		0	0	p_4	0	q_4	$\cdots$					R
3	0	0	p_3	0	q_3	$\cdots$	0	0	0	0	0	
2	0	p_2	0	q_2	0	$\cdots$	0	0	0	0	0	
1	1	0	1	0	0	$\cdots$	0	0	0	0	0	
		1	2	3		$\cdots$			m-2	m-1	m	

FIGURE 16.3. Transition matrix for the information accumulation process when one stimulus is presented

called the initial distribution. In the former case, **Z** is the distribution concentrated at the state i_0, i.e., $Z_j = 1$ if $j = i_0$, $Z_j = 0$ if $j \neq i_0$. In the latter case, the probability is z_i that at time zero the process will be in state i, where $0 \leq z_i \leq 1$ and $\sum_i z_i = 1$. Thus the initial distribution **Z** is a $(m - r)$ vector containing the initial probability distribution over the transient states. It is assumed that the process starts at time $t = 0$ with probability 1, so that **Z** is a $(m - 1)$ vector with $z_{m-1} = 1$, and the other elements zero.

Equations for mean reaction time can be derived using standard methods developed in Markov chain theory (e.g., Bhat, 1984; Bhattacharya & Waymire, 1990).

Consider the probability of entering an absorbing state and leaving the class of transient states. It should be noted that, if the process enters the absorbing state, it does not thereafter leave it. Let $f_{ij}^{(n)}$ be the probability that starting from transient state i, the process enters the absorbing state j in n steps. Noting that a transition from j to i is not possible, the number of steps can be taken to be proportional to the time needed for a first passage from i to j. If T_{ij} represents this random variable, its distribution may be written as $f_{ij}^{(n)}$, given by

$$P(T_{ij} = n) = f_{ij}^{(n)}.$$

Let $\mathbf{F}^{(n)}$ be the matrix with elements $f_{ij}^{(n)}$. A useful theorem (see, e.g., Bhat 1984, p.79) states that

$$\mathbf{F}^n = \mathbf{Q}^{n-1}\mathbf{R},$$

where **Q** and **R** are as defined earlier. Starting at a particular transient state, we have

$$\mathbf{F} = \mathbf{Z}'\mathbf{Q}^{n-1}\mathbf{R}.$$

Thus, the probability distribution of the time to reach the criterion c starting at time $t = 0$, $T_{o,c} \equiv T$, is

$$Prob[T = t] = \mathbf{Z}'\mathbf{Q}^n\mathbf{R},$$

where $t = (n+1)h$. The p-th moment for the distribution of times to reach the boundary is therefore

$$E[T^p] = h^p\mathbf{Z}'\sum_n n^p\mathbf{Q}^{n-1}\mathbf{R}, \qquad n = 1,\ldots,\infty. \tag{16.9}$$

In particular, the mean time and the second moment are

$$E[T] = h\mathbf{Z}'[\mathbf{I}-\mathbf{Q}]^{-2}\mathbf{R},$$

and

$$E[T^2] = h^2\mathbf{Z}'[2(\mathbf{I}-\mathbf{Q})^{-3} - (\mathbf{I}-\mathbf{Q})^{-2}]\mathbf{R},$$

respectively, where $\mathbf{I}$ is the identity matrix.

In the experiment below three different stimuli are presented, τ_1 and τ_2 msec apart. The first part of the process, when only one stimulus is present, is exactly as described above. The process starts at time $t = 0$ with probability 1. At time $t = \tau_1$, when the second stimulus is presented, the process may be somewhere between the absorbing and the reflecting boundary. The location of the process at that point in time is the starting position for a new process that starts with a different drift coefficient at time $t = \tau_1$. Clearly, the location now follows a probability distribution over all transient states. At time $t = \tau_2$, when the third stimulus is presented, the process may be anywhere between the absorbing and the reflecting boundary. Again, a new process starts with a different drift coefficient at time $t = \tau_2$.

16.3 Expected Value and Variance of the Reaction Time for the Diffusion Model

The observed reaction time is assumed to be additively composed of the stimulus processing time, denoted by T, plus all processes following or preceding it (e.g., temporal uncertainty of stimulus onset). The durations of these residual processes, which may include motor preparation and response execution components, are represented by a random variable, R. In general, T and R need not be stochastically independent, and to test the model at the level of the entire distribution functions, the bivariate distribution for (R,T) has to be specified. However, at the level of expectations,

$$E[RT] = E[T+R] = E[T] + E[R]. \tag{16.10}$$

If T and R are stochastically independent, the variance of the observable RT will be

$$Var[RT] = Var[T] + Var[R]. \tag{16.11}$$

For the sake of parsimony, we assume below that R is constant, $R = r$. Now equations (16.10) and (16.11) simplify to

$$E[RT] = E[T] + r,$$
$$Var[RT] = Var[T].$$

The corresponding expected value $E[T]$ is

$$\begin{aligned} E[T] &= h\Bigg(\mathbf{Z}'\sum_{i=1}^{n_1} i\mathbf{Q}_1^{i-1}\mathbf{R}_1 + \mathbf{Z}'\mathbf{Q}_1^{n_1}\sum_{i=n_1+1}^{n_2} i\mathbf{Q}_2^{(i-(n_1+1))}\mathbf{R}_2 \\ &\quad + \mathbf{Z}'\mathbf{Q}_1^{n_1}\mathbf{Q}_2^{n_2-n_1}\sum_{i=n_2+1}^{\infty} i\mathbf{Q}_3^{(i-(n_2+1))}\mathbf{R}_3\Bigg), \end{aligned} \tag{16.12}$$

where $\mathbf{Q}_1$, $\mathbf{Q}_2$, and $\mathbf{Q}_3$ are the transition matrices with transient states for the first, second, and third part of the process, respectively; n_1 and n_2 are $\tau_1/h - 1$ and $\tau_2/h - 1$, respectively; $\mathbf{R}_1$, $\mathbf{R}_2$, and $\mathbf{R}_3$ are the vectors that contain the transition probability from the transient to the absorbing state for the first, second, and third part of the process; $\mathbf{Z}'\mathbf{Q}_1^{n_1}$ the initial defective distribution for the second part, and $\mathbf{Z}'\mathbf{Q}_1^{n_1}\mathbf{Q}_2^{n_2-n_1}$ the initial defective distribution for the third part of the process. Setting the transition matrices $\mathbf{T}_1 = \mathbf{T}_2 = \mathbf{T}_3$ implies $\mathbf{Q}_1 = \mathbf{Q}_2 = \mathbf{Q}_3$ and $\mathbf{R}_1 = \mathbf{R}_2 = \mathbf{R}_3$. This leads to a diffusion process where only one stimulus is presented. The variance is

$$Var[T] = E[T^2] - E[T]^2,$$

where

$$\begin{aligned} E[T^2] &= h^2\Bigg(\mathbf{Z}'\sum_{i=1}^{n_1} i^2\mathbf{Q}_1^{i-1}\mathbf{R}_1 \\ &\quad +\mathbf{Z}'\mathbf{Q}_1^{n_1}\sum_{i=n_1+1}^{n_2} i^2\mathbf{Q}_2^{(i-(n_1+1))}\mathbf{R}_2 \\ &\quad +\mathbf{Z}'\mathbf{Q}_1^{n_1}\mathbf{Q}_2^{n_2-n_1}\sum_{i=n_2+1}^{\infty} i^2\mathbf{Q}_3^{(i-(n_2+1))}\mathbf{R}_3\Bigg). \end{aligned} \tag{16.13}$$

For computational purposes the equations for the expected value and the second moment can be greatly simplified:

$$\begin{aligned} E[T] = \quad & h\left[\mathbf{Z}'(\mathbf{I}-\mathbf{Q}_1)^{-1}\left((\mathbf{I}-\mathbf{Q}_1^{-1}(\mathbf{I}-\mathbf{Q}_1^{n_1}) - n_1\mathbf{Q}_1^{n_1}\right)\mathbf{R}_1\right. \\ & +\mathbf{Z}_2'(\mathbf{I}-\mathbf{Q}_2)^{-1}\left((\mathbf{I}-\mathbf{Q}_2)^{-1}(\mathbf{I}-\mathbf{Q}_2^{(n_2-n_1)}) + n_i\mathbf{I} - n_2\mathbf{Q}_2^{(n_2-n_1)}\right)\mathbf{R}_2 \\ & \left. + \mathbf{Z}_3'(\mathbf{I}-\mathbf{Q}_3)^{-1}\left((\mathbf{I}-\mathbf{Q}_3^{-1} + n_2\mathbf{I}\right)\mathbf{R}_3\right], \end{aligned} \tag{16.14}$$

$$
\begin{aligned}
E[T^2] = \; & h^2 \left[\mathbf{Z}' \left((2(\mathbf{I}-\mathbf{Q}_1)^{-3} - (\mathbf{I}-\mathbf{Q}_1)^{-2}) (\mathbf{I}-\mathbf{Q}_1^{n_1}) \right.\right. \\
& - \left(2(\mathbf{I}-\mathbf{Q}_1)^{-2} + n_1(\mathbf{I}-\mathbf{Q}_1^{-1}) \, n_1 \mathbf{Q}_1^{n_1}\right) \mathbf{R}_1 \\
& + \mathbf{Z}_2' \left((2(\mathbf{I}-\mathbf{Q}_2)^{-3} - (\mathbf{I}-\mathbf{Q}_2)^{-2}) \left(\mathbf{I}-\mathbf{Q}_2^{(n_2-n_1)}\right) \right. \\
& + (\mathbf{I}-\mathbf{Q}_2)^{-1} \left(2n_1(\mathbf{I}-\mathbf{Q}_2)^{-1} + n_1^2\mathbf{I}\right) \\
& - \left. \left(2(\mathbf{I}-\mathbf{Q}_2)^{-2} + n_2(\mathbf{I}-\mathbf{Q}_2)^{-1}\right) n_2 \mathbf{Q}_2^{(n_2-n_1)} \right) \mathbf{R}_2 \\
& + \mathbf{Z}_3' \left((2(\mathbf{I}-\mathbf{Q}_3)^{-3} - (\mathbf{I}-\mathbf{Q}_3)^{-2}) \right. \\
& \left.\left. + (\mathbf{I}-\mathbf{Q}_3)^{-1} \left(2n_2(\mathbf{I}-\mathbf{Q}_3)^{-1} + n_2^2\mathbf{I}\right)\right) \mathbf{R}_3 \right], \qquad (16.15)
\end{aligned}
$$

where $\mathbf{Z}_2' = \mathbf{Z}'\mathbf{Q}_1^{n_1}$ and $\mathbf{Z}_3' = \mathbf{Z}'\mathbf{Q}_1^{n_1}\mathbf{Q}_2^{(n_2-n_1)}$. The matrix computation can be further simplified by spectral analytic methods. A tridiagonal transition matrix $\mathbf{Q}$ is similar to a symmetric tridiagonal matrix, i.e., there exists a real valued matrix $\mathbf{D}$ such that $\mathbf{DQD}^{-1} = \mathbf{S}$, where $\mathbf{S}$ is a symmetric tridiagonal matrix (see Horn and Johnson, 1990). This fact guarantees that $\mathbf{Q}$ has $m-1$ linearly independent eigenvectors and $m-1$ real eigenvalues (*ibid.*). According to the *Frobenius-Perron* theorem (Bhattacharya & Waymire, 1990, Ch.3), all of the eigenvalues in $\mathbf{Q}$ are less than 1. Therefore, $\mathbf{Q}$ can be factorized as follows,

$$
\begin{aligned}
\mathbf{Q} &= \mathbf{P\Lambda P}^{-1}, \\
\mathbf{Q}^2 &= (\mathbf{P\Lambda P}^{-1})(\mathbf{P\Lambda P}^{-1}) = \mathbf{P\Lambda}^2\mathbf{P}^{-1}, \\
&\;\;\vdots \\
\mathbf{Q}^n &= \mathbf{P\Lambda}^n\mathbf{P}^{-1},
\end{aligned}
$$

where $\mathbf{P}$ is the matrix of linear independent eigenvectors, and $\mathbf{\Lambda}$ the diagonal matrix of real valued eigenvalues less than one in magnitude. Efficient computer programs exist for finding the eigenvalues of tridiagonal symmetric matrices.

16.4 Parameter Estimation

The parameters from the triple stimulus condition are estimated according to the diffusion model. Seven parameters need to be determined for each triple stimulus combination: the intensity parameters α_A, α_V, and α_T for the auditory, visual, and tactile stimuli; the parameter β, the force directed towards the origin; the residual constant r; the diffusion coefficient σ^2; and the boundary c as a size of transition matrix, i.e., m. The time constant h is not estimated from the data but chosen as close to zero as needed to approximate a continuous time process. For the data h is set to 1 msec. Considering (16.5) and (16.6) for $p(x)$ and $q(x)$, it is obvious that nothing is lost in setting $\sigma = 1$. Since the step size δ was defined as $\delta = \sigma\sqrt{h}$, $\delta = 1$ also. The criterion c was expressed in terms of the step size, i.e., $c = k\delta$, and the size of the transition matrix m was defined as $m = k+1$. Thus the matrix size is $c+1$. Parameter β is viewed as a function of the intensity parameters and the matrix size, i.e., $\beta = \gamma(1+\alpha)/m$, α being the corresponding intensity parameter or one of its combinations, and γ a number to be estimated between 0 and 1. A technical reason is to constrain the drift rates $\mu(x)$ to values between 0 and 1, since the drift rates function as probabilities in the transition matrix. From a psychological point of view this is reasonable because, with increasing intensity of a stimulus, the force directed to the origin should be larger. There remain six parameters to be estimated, $\alpha_A, \alpha_V, \alpha_T, r, m, \gamma$.

The experiment presented three different stimulus combinations to each subject (weak tactile stimulus and light and 70db tone, strong tactile stimulus and light and 70db tone, weak tactile stimulus and light and 90db tone), requiring five different intensity parameters: α_V for light, α_{A70} for the 70db tone, α_{A90} for the 90db tone, α_{T1} for the weak tactile stimulus, α_{T3} for the strong tactile stimulus. According to the model the relation between the intensity parameters should be

$$\alpha_{T1} < \alpha_{T3} < \alpha_V < \alpha_{A70} < \alpha_{A90}, \tag{16.16}$$

since the subjects responded fastest to the 90 db tone and slowest to the weaker tactile stimulus. The criterion c is supposed to be the same for all three stimulus conditions. For all three triple stimulus conditions it is necessary to estimate ten parameters from the data: the intensity parameters $\alpha_{A70}, \alpha_{A90}, \alpha_V, \alpha_{T1}$, and α_{T3}; three residual parameters r_1, r_2 and r_3, the matrix size m; and γ. The diffusion model was fitted to the triple stimulus data by minimizing the function

$$SSE = \sum_{j=1}^{n} \sum_{i=1}^{m} (M_{ij} - (E_{ij}[T] + r_j))^2,$$

using the optimization program GAUSS.OPTMUM. In this expression M_{ij} represents the observed mean RT for the ith stimulus interval combination (τ_1, τ_2) in the jth triple stimulus combination; $E_{ij}[T]$ represents the predicted stimulus processing time for the ith interstimulus interval combination in the jth triple stimulus combination; and r_j represents the residual time in the jth triple stimulus combination. With these estimated parameters the standard deviations of the RT for all triple stimulus conditions can also be predicted. Moreover, the same estimated intensity parameters give the matrix size m and the residual time r for each double stimulus condition and hence the expected mean RT as well as the standard deviations for those conditions.

16.5 Results

For all subjects the predicted order of the intensity parameters in (16.16) was observed. The fit of the means for the triple stimulus condition is very good for all subjects. But the predictions of the standard deviations for the triple stimulus

TABLE 16.1. Observed and predicted mean reaction times and standard deviations for triple stimulus condition T1VA70 (weak tactile stimulus, T1, followed by a visual stimulus, V, followed by a 70db tone, A70). The first and the second stimulus are τ_1 msec apart, the first und the third stimulus are τ_2 msec apart.

(τ_1, τ_2)	Mean RT	Prediction	StD	Prediction
(20,50)	171.85	173.40	13.20	11.51
(30,50)	177.22	177.84	13.82	10.02
(40,50)	179.14	181.33	15.48	9.46
(50,50)	185.60	183.95	19.42	9.85
(20,80)	177.53	178.89	16.24	18.64
(30,80)	186.71	185.57	17.74	17.95
(40,80)	191.95	191.85	14.25	17.50
(50,80)	198.47	196.79	15.10	17.43

TABLE 16.2. Observed and predicted mean reaction times and standard deviations for double stimulus condition VA70 (visual stimulus, V, followed by a 70db tone, A70, τ msec apart)

τ	Mean RT	Prediction	StD	Prediction
0	138.21	137.65	25.37	3.47
10	141.51	142.94	16.71	4.34
20	149.02	148.28	18.24	6.88
30	150.83	151.70	14.30	10.06
40	154.76	153.75	15.57	12.77
50	154.96	154.97	15.50	14.84

conditions are poor for all subjects under all conditions. Table 16.1 shows some of these results. The whole set of results can be found in Diederich (1992a).

With the same intensity parameters and the β parameter estimated from the triple stimulus condition, the matrix size m and the residual time r were estimated for each double stimulus condition (eight different conditions) and the expected mean RT predicted as well as the standard deviations. The fit for the means was almost perfect for all the double stimulus conditions for all subjects. But the predictions for the standard deviation for the double stimulus conditions was poor under all conditions. Table 16.2 and Table 16.3 give some of these results. (See Diederich, 1992a, for more detailed results.)

One reason for the poor fit of the standard deviation might be that the residual time r is not constant. It has been assumed that the observed reaction time

TABLE 16.3. Observed and predicted mean reaction times and standard deviations for double stimulus condition T1A90 (weak tactile stimulus, T1, followed by a visual stimulus, V, τ msec apart)

τ	Mean RT	Prediction	StD	Prediction
50	181.99	180.48	13.82	11.21
60	187.46	188.01	12.62	14.54
70	196.02	195.07	23.82	19.16
80	200.91	201.67	19.10	21.96
90	206.75	207.85	17.09	25.87
100	213.58	213.63	22.61	29.84

consists of two additive variables: the stimulus processing time and a residual time in which all remaining processes are lumped together. For simplicity, the residual time was taken to be a constant within a given condition. But the residual time was very large compared to the observed reaction time, and it is plausible that it subsumes processes that are not constant but contribute to the overall variance. However, simply adding a constant to the predicted variances does not solve the problem because the difference between the observed and predicted variance is not constant within a given stimulus configuration. Some variances are underpredicted as well as others overpredicted (see Table 16.1). The next step will be to abandon the assumption of independence between the stimulus processing time and the residual time or, more generally, to make an asumption about the bivariate distribution of S and R.

16.6 References

Bhat, U.N. (1984). *Elements of applied stochastic processes.* New York: Wiley.

Bhattacharya, R.N., & Waymire, E.C. (1990). *Stochastic processes with applications.* New York: Wiley.

Busemeyer, J.R., & Goldstein, W.M. (1992). Linking together different measures of preference: A dynamic model of matching derived from choice theory. To appear in *Organizational Behavior and Human Decision Processes.*

Busemeyer, J.R., & Townsend, J.T. (1989). Decision field theory: A dynamic-cognitive approach to decision making. *Purdue Mathematical Psychology Program Technical Report No. 89-7.*

Busemeyer, J.R., & Townsend, J.T. (1992). Fundamental derivations from decision field theory. *Mathematical Social Sciences, 23,* 155–182.

Diederich, A. (1992a). *Intersensory facilitation with multiple stimuli: Race, superposition, and diffusion models for reaction time.* Frankfurt: Verlag Peter Lang.

Diederich, A. (1992b). Probability inequalities for testing separate activation models of divided attention. *Perception & Psychophysics, 52(6)*, 714–716.

Diederich, A., & Colonius, H. (1987). Intersensory facilitation in the motor component? A reaction time analysis. *Psychological Research, 49*, 23–29.

Horn, R.A., & Johnson, Ch.R. (1990). *Matrix analysis*. Cambridge: Cambridge University Press.

Miller, J. (1982). Divided attention: Evidence for coactivation with redundant signals. *Cognitive Psychology, 14*, 247–279.

Pacut, A. (1980). Mathematical modelling of reaction latency: The structure of the models and its motivation. *Acta Neurobiologiae Experimentalis, 40*, 199–215.

Pacut, A. (1982). Mathematical modelling of reaction latency. Part II: A model of escape reaction and escape learning. *Acta Neurobiologiae Experimentalis, 42*, 379–395.

Pacut, A., & Tych, W. (1982). Mathematical modelling of reaction latency. Part III: A model of avoidance reaction latency and avoidance learning. *Acta Neurobiologiae Experimentalis, 42*, 397–420.

Ratcliff, R. (1978). A theory of memory retrieval. *Psychological Review, 85*, 59–108.

Ricciardi, L.M. (1977). *Lecture notes in biomathematics*. Berlin: Springer-Verlag.

Smith, P.L. (1990). *Attention, redundancy, and perceptual integration*. Paper presented at the 23rd Annual Mathematical Psychology Meeting, August 1990, University of Toronto, Canada.

Part III
Social Systems

17

Cooperation in Heterogeneous Populations

Karl Sigmund[1] and Martin Nowak[2]

ABSTRACT This paper deals with stochastic reactive strategies for the Iterated Prisoner's Dilemma. It considers populations of individuals meeting randomly, and noisy interactions. Both the analysis of monomorphic and heteromorphic populations show that the reciprocal strategy Tit For Tat acts like a pivot: it triggers an evolution towards cooperation, but is not the ultimate beneficiary of such an evolution.

17.1 Introduction

Ever since the early days of game theory, the Prisoner's Dilemma (PD) has been viewed as a major paradigm for discussing cooperation between selfish individuals in the absence of constraints. From 1979 onwards, the topic received a further strong impetus by Axelrod's computer tournaments (see Axelrod, 1984). In addition to theoretical work on the mathematical aspects of the game, a wide variety of research on its psychological, biological, political, and economical implications and applications has developed (see Axelrod and Dion, 1988).

The two players of a PD game each have a choice between two strategies, namely, to cooperate (C) or to defect (D). If both players cooperate, each receives as payoff a reward R which is larger than the payoff P (punishment) they receive if both defect. But a player defecting unilaterally obtains a payoff T (the temptation) which is still higher than R, while the opponent receives only S (the sucker's payoff) which is even lower than P. In addition to $T > R > P > S$, it is assumed that $T+S < 2R$, so that mutual cooperation is better than agreeing on a unilateral defection and sharing the payoffs. (In all the computer simulations, Axelrod's payoff values $T = 5, R = 3, P = 1$, and $S = 0$ are used, but other choices lead to similar results.) Since D dominates C (no matter what the co-player does, it is always better to defect), both players ought to opt for D: But then, their payoff P is smaller than the payoff R they would have obtained by both cooperating.

Many people experience difficulties in accepting defection as the "rational" solution (for an entertaining experiment, see chapter 30 of Hofstadter, 1985); but, to quote Luce and Raiffa (1957), "no, there appears no way around this dilemma.

[1]Department of Mathematics, University of Vienna, A-1090 Vienna, Strudlhofgasse 4, Austria
[2]Department of Zoology, University of Oxford, Oxford OX1 3PS, South Parks Road, England

We do not believe there is anything irrational or perverse about the choice of D and we must admit that if we were actually in this position we would make these choices."

One way of persuading oneself of the validity of this conclusion is to omit any appeal to "rationality" and to consider a ***population*** (biological or artificial) playing this game in randomly selected pairwise encounters. If one assumes that a successful strategy spreads (by inheritance, imitation, or infection, say), then necessarily defectors spread faster and end up by taking over the whole population.

Nevertheless, there is a sizeable amount of cooperation around us, and a tendency towards cooperation within us. How can this be explained? One possible answer would be that the PD is only of marginal relevance for the emergence of cooperation. But this suggestion can quickly be discarded. It seems, in fact, that most business transactions have the same structure as the PD: the players profit by cooperating in a joint enterprise (like trading or bartering), but could profit still more by not delivering their own full share. This remark leads to another possible explanation for cooperation: PD-like transactions are so common because they can be repeated. What we observe is frequently an Iterated Prisoner's Dilemma (or IPD), rather than a single-shot Prisoner's Dilemma.

If one assumes that there exists a constant probability w for repeating the encounter between the two players, one obtains a vastly more complex game. There is now an infinity of different strategies, and it can easily be shown that, if w (the *shadow of the future*, to use Axelrod's expression) is larger than $\frac{T-R}{T-P}$, there is no dominating strategy any more, (i.e., no strategy which is best regardless of what the other player does). In this case, the question of which strategy the other player is likely to choose becomes essential. In 1979, Axelrod put the diverging opinions on this matter to the test by inviting game theorists to submit their strategies (in the form of programs) to a computer tournament. The rest is history: the winner of the round-robin contest was Tit For Tat (or TFT), submitted by A. Rapoport. It consists in playing C in the first round and then doing whatever the other player has done in the previous round: first cooperate, then imitate, to put it in a nutshell.

The tournaments' outcome was surprising, not only because of TFT's utter simplicity, but also because, as is easily seen, a TFT-player does never better than the co-player. Whoever uses TFT can never be ahead in the count (but never lags far behind either). The strong point of TFT is that it elicits cooperation from the co-player. By virtue of its transparency, it allows the co-player to dictate the evolution of the game and thereby manipulates this co-player to opt for cooperation.

Since the IPD is not a zero-sum game, the TFT-program can win the tournament although it achieves at best a draw against any other program. It earns a large payoff by cooperating with strategies which, against other strategies or their like, engage in fruitless defections. This implies that the full advantage of TFT is only displayed within a population. In further work, Axelrod kept investigating this context: first, by repeating the tournament with a considerably larger number of entries, and then by applying game dynamics, i.e., selection: the population of programs was submitted, generation after generation, to a round-robin contest, and the frequencies of the programs increased at a rate proportional to their payoff in the previous generation. Again, TFT did best.

This led to a heightened interest in the role of reciprocity in human and animal behaviour. However, Axelrod's simulations dealt with a highly selective sample of strategies submitted by sophisticated contestants, running its programs in an error-free world. These two features are rather artificial. In order to approximate more realistic scenarios, May (1987) suggested to "take stochasticities into account and investigate representative samples". The following paper is a step in this direction. It analyses a highly heterogeneous population of players under conditions where noise blurs the perception and implementation of the players' moves. This approach shows, among other things, that TFT is a catalyst rather than a beneficiary of an emerging cooperation.

We start with a short account of game dynamics and a brief sample of biological applications. Then, we introduce stochastic reactive strategies and discuss the evolution both of monomorphic and polymorphic populations. Finally, we compare this with related work and sketch possible extensions.

17.2 Game Dynamics

One can compute the payoff A_n obtained in the n-th round by a player using strategy E against a player using strategy E', and hence the total payoff $A(E, E') = \sum_n A_n w^n$ (or, in the limiting case $w = 1$ of the infinitely iterated PD, $A(E, E') = \lim \frac{1}{n}(A_1 + ... + A_n)$). Consider now a large population where the strategies $E_1, ..., E_n$ exist with frequencies $x_1, ..., x_n$ (with $\sum_i x_i = 1$). The average payoff for strategy E_i in this population is given by $f_i(x) = \sum_j x_j A(E_i, E_j)$. According to the usual discrete game dynamics (see, e.g., Hofbauer and Sigmund, 1988), the frequency x_i' of E_i in the next generation is given by

$$x_i' = x_i \frac{f_i(x)}{\bar{f}},$$

where $\bar{f} = \sum_i x_i f_i(x)$ is the average payoff in the population. Thus the composition of the population changes from generation to generation. This composition in turn affects the average payoff for the strategies. Nowak and Sigmund (1989) show that with just three or four different strategies a rich spectrum of dynamic behaviour can be obtained, including bistability and limit cycles.

If w is sufficiently large, a population playing $AllD$ (i.e., always defecting) can be invaded by TFT players if they enter in a cluster exceeding a certain size, see Axelrod (1984). (TFT-players can make up among themselves more than they lose against defectors. They therefore increase in frequency, which in turn increases the frequency of contacts with other TFT-players and hence their advantage over *AllD*). Imagine that such a cluster is given by a family, for instance. A TFT-population, on the other hand, cannot be invaded by a cluster of defectors. *AllC*-players, however, do as well as the TFT-players in a TFT-population. Thus, there is no selection against *AllC*, so that it can spread by random drift. But once a sizeable portion of the population consists of individuals who do not retaliate, *AllD* players can invade again.

Stochastic events can threaten a TFT-population in a more direct way too: any mistake gives rise to a long sequence of alternating defections. In a heterogeneous population, this effect is less pronounced: when Axelrod repeated his tournament with an error rate of 1%, TFT still finished first. But in a homogeneous population of retaliators, even rare errors are very costly, at least if w is large. It is obviously better to be generous and to forgive, at least occasionally, a defection by the co-player. One such generous strategy is Tit For Two Tats (or $TFTT$), which defects only if the co-player defected in the previous *two* rounds. (This strategy would have won the first of Axelrod's tournaments had it been used: but when it was entered in the second tournament, it finished only twenty-first.)

In a mixed population of TFT and $TFTT$, all players do equally well if there are no mistakes. In a noisy environment, however, $TFTT$ has a considerable advantage. But if a minority of the population acquires the knack to cooperate and defect in turns, then it can systematically exploit $TFTT$, wheras TFT holds its own. On the other hand, a small group of players using Suspicious Tit For Tat (or $STFT$), which defects on the first move and then repeats whatever the other did in the previous round, will hurt TFT considerably, while $TFTT$ escapes almost unharmed.

17.3 Reciprocation in Animal Behaviour

The Prisoner's Dilemma is a favoured game of experimental psychologists (see, e.g., Selten and Stoecker, 1986). But while human communities offer many examples of IPDs, these are usually influenced by factors like social pressure, law enforcement, ideologies, etc., which alter the payoffs in a complex way. Biological populations pose less problems in that respect. Some of the leading sociobiologists, like Hamilton (see Axelrod and Hamilton, 1981) and Trivers (1985), have studied the aspects of cooperation and reciprocation in animal behaviour extensively. Here we list a few recent well documented cases:

(a) Predator inspection. If sticklebacks notice a pike in their vicinity, they tend to approach (up to a certain distance) in order to obtain information on this new enemy. The risk of such an approach is considerably reduced if sticklebacks approach in pairs. As long as the sticklebacks approach together or alternate in taking the lead, they cooperate. In an ingenious experiment, Milinski (1987) had a single stickleback confronted with a dummy pike, and used a mirror to make the stickleback think that it had a companion. Depending on the inclination of the mirror, this fake companion kept either abreast or lagged behind by a few inches consistently. In the former case, the deluded stickleback usually approached the dummy much closer, which suggests that its strategy is based on reciprocity.

(b) Grooming. The so-called cleaner fishes attend to the teeth of much larger fish. This is to their mutual benefit, but the larger fish could be tempted to cap their profit by swallowing the cleaner. The stability of the interaction seems to be due to the probability of a repetition; indeed, such grooming has only been observed among fish with a fixed home-range offering a reliable meeting place (see Axelrod and Hamilton, 1982).

(c) Egg-sharing. Sea-basses are equipped with male and female sex organs. In the course of a mating, they switch their sexual roles up to ten times in a row. Since eggs are more expensive than sperm, sea-basses wait, after producing some eggs, until their partner produces a few on its own (Fischer, 1980). By dividing the interaction into several steps, the sea-basses thus change a single-shot PD into an IPD.

(d) Food-sharing. Vampire bats who found no victim may, upon returning to their cave, be fed by their luckier brethren, who regurgitate some of their stomach's content, in what seems to be a case of reciprocal altruism (Wilkinson, 1984).

(e) Wife-sharing. One young male baboon starts to quarrel with the overlord of an estrous female, while his comrade uses the digression to mount her sneakily. On the next occasion, the two young males exchange their roles (see Packer, 1979).

(f) Mutual assistance. Vervet monkeys living in a group can recognize each other's vocalizations. Their support calls were tape-recorded and later played by a hidden loudspeaker. A monkey is more likely to respond if it has been groomed by the caller a short while ago (see Cheney and Seyfarth, 1982). Interestingly, responsiveness within a family is higher, but remains unaffected by recent grooming – acts of kindness appear to elicit less response among kin.

17.4 Homogeneous Populations

What is common to these real-life examples is that the decision whether or not to cooperate depends on what the other player has been doing, but not in a strictly deterministic way. Individuals do not obey a clear-cut rule, rather they exhibit a stronger or weaker tendency to play C. This appears to be a fuzzy program. To model it, we consider stochastic reactive strategies defined by triplets (y, p, q), where y is the probability to play C in the first round, while p and q are the conditional probabilities to play C after an opponent's C and, D respectively, in the previous round. These strategies correspond to points in the unit cube. *AllC* is the corner point $(1,1,1)$, *AllD* is the point $(0,0,0)$, TFT is the point $(1,1,0)$, and $STFT$ the point $(0,1,0)$. These strategies, however, are not properly stochastic: they correspond to extremal points of our strategy set. As soon as we take noise into account, we are in the interior of the cube.

It is clear that these (y, p, q)-strategies form only a small part of the huge set of all possible strategies for the IPD. For instance, $TFTT$ does not belong to this class. But a drastic restriction is necessary whenever one wants to consider representative samples; otherwise, such samples would have to be far too large and therefore unsuited for analysis. Furthermore, the class of strategies considered is sufficiently broad to contain highly diverse specimens, covering a wide range of cooperative, defective, reciprocal, or fully random behaviours; as a consequence, this set of strategies allows for complex dynamical extensions.

If a player using strategy $E = (y, p, q)$ is matched against a player using strategy $E' = (y', p', q')$, then the probability of the former to cooperate in the n-th round determines the co-player's probability to cooperate in the following round and hence also the own level of cooperation in the $(n+2)$-nd round. This echo-effect determines recursive relations which imply that the probability of the first player choosing C converges (for $n \to \infty$) to

$$c(E, E') = \frac{q + (p-q)q'}{1 - (p-q)(p'-q')}.$$

In particular, the probability to cooperate with a copy of oneself converges to

$$c(E, E) = \frac{q}{1 - p + q}.$$

If (p', q') lies on the line from (p, q) to $(1, 0)$, then $c(E, E) = c(E', E') = c(E, E') = c(E', E)$, and hence E and E' have the same asymptotic level of cooperativity. If (p', q') lies above this line, then E' is more cooperative, both against E and against a copy of itself. Hence this line divides the (p, q)-space into one part where cooperativity is higher (which includes *AllC*) and one part where it is lower (which includes *AllD*).

It is straightforward to compute the payoff $A(E, E')$, see Nowak and Sigmund (1989b). For the sake of simplicity, we shall only consider the limiting case $w = 1$ (but emphasize that our results essentially apply whenever w is large). We then can forget about y, the probability to cooperate in the first move, which only affects the initial phase and plays no role in the long run. We then obtain as payoff for strategy $E = (p, q)$ against $E' = (p', q')$ the value

$$A(E, E') = (R+P-T-S)c(E, E')c(E'E) + (S-P)c(E, E') + (T-P)c(E', E) + P.$$

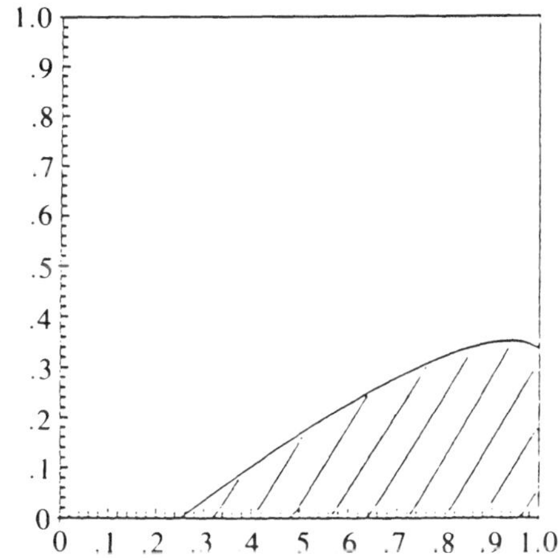

FIGURE 17.1. The square of (p, q)-strategies (see text). Point (1.0) corresponds to *TFT*; in its neighborhood (the cooperation-rewarding zone), a population can be invaded by mutants with a higher p- or q-value.

This allows us to analyse the evolution of populations which are monomorphic in the sense that the great majority of players settles on one fixed strategy E'. We may characterise those strategies E which do better, against E', than E' itself (see Nowak, 1990). A small minority using such a strategy E can invade the E'-population. It turns out that the space of our (p, q)-strategies, i.e., the unit square, is divided into two regions (see Fig.17.1). In the so-called C-region, which contains $TFT = (1, 0)$, an E'-population can be invaded by strategies E having a higher level of cooperation, but not by any strategies with a lower level. In the D-region, which contains $AllC = (1, 1)$ and $AllD = (0, 0)$, the converse holds. If E' is on the boundary line between these two regions, no E does better than E' in an E'-population, and only those E on the line through E' and $(1, 0)$ do equally well, i.e., have a chance to invade by neutral drift. This shows that a tendency towards cooperation exists only in the C-region, i.e., in a neighborhood of TFT. It must be emphasized that this is not a tendency *towards* TFT: it pays most to deviate from the "consensus"-strategy E' played by the homogeneous population, not in the direction towards TFT (such a move would not affect payoff at all), but in an *orthogonal* direction.

In the limiting case of very small noise, the strategy $(1, q)$ with $q = \min[1 - \frac{T-R}{R-S}, \frac{R-P}{T-P}]$ can be shown to be optimal in the sense that it affords the highest payoff for a population adopting it, compared to all reactive strategies immune to invasion by less cooperative strategies (i.e., with lower p-and q-values), see Molander (1985) and Nowak and Sigmund (1990). This strategy, which is called Generous Tit For Tat (or $GTFT$), can be viewed as a mixture of TFT and $AllC$. It is grateful in the sense that it never forgets a good deed, and tolerant in the sense that it sometimes forgives a defection (but not always).

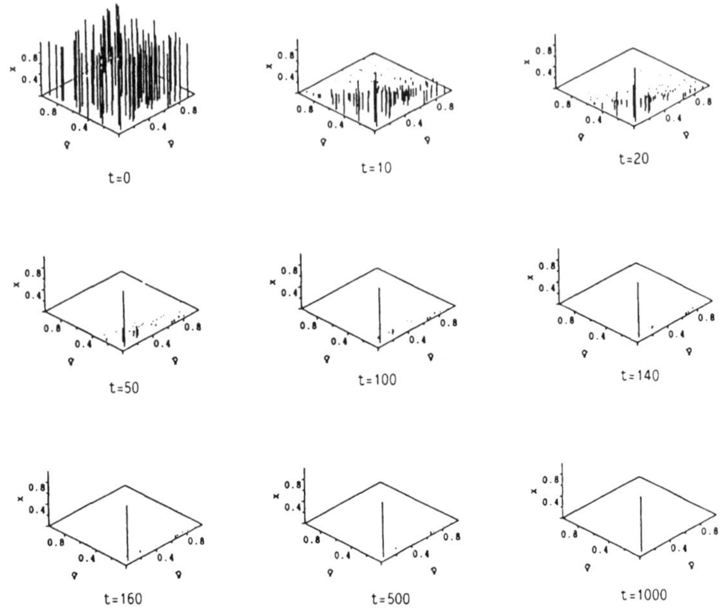

FIGURE 17.2. The evolution towards defection. We start out with 100 randomly chosen strategies, all equally frequent at the beginning. The defectors, close to the $(0,0)$-corner, increase in frequency and ultimately take over.

17.5 Heterogeneous Populations

If the population is heterogeneous, the evolution is considerably more complex. If there are two, three, or four different strategies in the population, the game dynamics can still be studied analytically (see Nowak and Sigmund 1989b). For still more diversified populations, one has to resort to computer simulations.

In Nowak and Sigmund (1992) we studied random samples consisting of hundred different (p,q)-strategies (which means that 10000 interactions have to be computed in each generation). If the strategies are uniformly distributed on the unit square, all initially equally frequent, the evolution tends in almost every case towards *AllD*. This means that the strategies closest to $(0,0)$ increase in frequency from generation to generation, while all others vanish (see Fig.17.2). Such a tendency is due to the fact that a large percentage of the initial random sample has high q-values and does not retaliate against exploiters. It pays to defect against inveterate "suckers". By the time these born victims are eliminated from the population, the defectors are so preponderant that cooperation cannot be established.

If the initial population contains a strategy which is *very* close to TFT, however, the evolution takes a remarkably different turn (see Fig. 17.3). The first

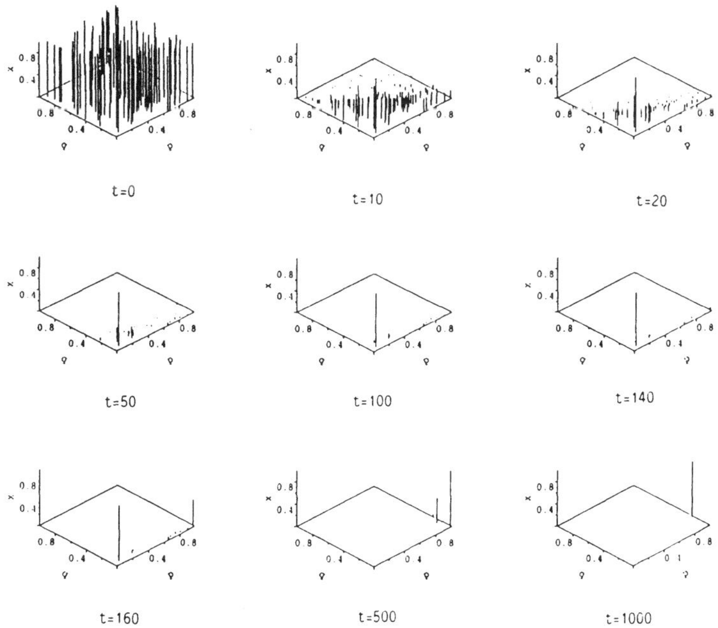

FIGURE 17.3. This time, a strategy close to TFT (with $p = 0.99$ and $q = 0.01$) has been added by hand. At first, defectors thrive, but when the "Suckers" who do not retaliate are driven from the population, the TFT-like strategy takes over. This is in turn superseded by a more generous strategy.

phase is almost indistiguishable from the runs described previously. The strategies close to *AllD* grow rapidly, feasting on their diet of "suckers". All reciprocating strategies close to $(1,0)$ seem to vanish. But when the victims are so reduced that the defectors can no longer exploit them, the small band of reciprocators returns upon the scene. The fortune of the exploiters takes a decided turn for the worse. Their frequencies dwindle. Now it is the TFT-like strategies who dominate the population. But their hegemony is transient, too. For, when the defectors are exterminated, the reciprocators give way to more generous strategies and vanish also. When the evolution finally grinds to a halt, after several hundred generations, the population consists almost exclusively of strategies close to $GTFT$. Even if we allow for random fluctuations, for instance by introducing occasionally a small percentage of another strategy, this will not alter the final equilibrium.

It should be noted that this turn towards cooperation can only be triggered by a strategy that is extremely near to the TFT-corner $(1,0)$. In general, a random sample of 100 strategies *uniformly* distributed over the unit square will not contain such a strategy. In the simulations, it can either be introduced by hand, or by choosing a probability distribution which places more weight upon the boundaries of the range $]0,1[$ of p- and q-values. The most plausible candidate for such a distribution has the familiar U-shaped density function

$$f(p) = \frac{1}{\pi\sqrt{p(1-p)}},$$

which is a particular beta-distribution playing a considerable role in statistics. Such a weighting of the values close to all or nothing (i.e. to 0 and 1) makes sense, since a change in, say, 1% in their vicinity can be much more relevant for the future evolution than a change from 50% to 49%. Whenever random samples according to such a distribution are used, a dramatic turn of the tide in favour of cooperation is obtained: while the exploiters seem to be unstoppable at first, and after ten generations only *AllD*-like strategies appear to be present, a minority of reciprocators survives and acts like a lever to twist the momentum away from defection. After fifty generations, all q-values larger than 10% have gone into hiding, and the battle between *AllD* and *TFT* takes a determined shift in the latters' favour. But after a hundred generations, when only p-values larger than 95% remain on stage, higher q-values re-enter and generosity starts to pay. In such an evolution, *TFT*-like strategies act as catalysers. They need be present only in a trifling amount in order to trigger the emergence of cooperation. During the transient phase of the reaction, their concentrations grow, but then vanish again.

The simulation in polymorphic populations confirms the result obtained by an analysis of the monomorphic scenario: *TFT* is the pivot rather than the aim of evolution. One way to visualize this pivotal role of *TFT* is to ask when an *AllD* population can be invaded by a cluster of E-strategists, for a given size $d \in]0,1[$ of the cluster. In a dimorphic population where d is the relative frequency of $E = (p,q)$ and $1-d$ the relative frequency of *AllD*, the amount of E will grow if and only if

$$dA(E,E) + (1-d)A(E, AllD) > dA(AllD, E) + (1-d)A(AllD, AllD).$$

For any given value d, this defines a neighborhood U_d of *TFT* in the strategy square. If d decreases, the neighborhood shrinks to the corner point $(1,0)$. Thus *TFT* is the strategy that can invade defectors in a cluster of minimal size.

17.6 Discussion

There are several directions in which these results could be extended. In particular, the range of strategies ought to be enlarged. The (p,q)-strategies depend only on the co-player's move in the previous round. It would be more realistic to admit strategies with a memory of several rounds, taking the own moves into account. In principle, this is straightforward: such a process is simply a Markov chain. With a memory of length 2, for instance, there are 16 relevant histories.

Each strategy is a 16-tuple $p_0, ..., p_{15}$ of probabilities to play C after a given history. This yields a 16×16 transition matrix, which has a unique left eigenvector $\pi_0, ..., \pi_{15}$ corresponding to the eigenvalue 1 which is stochastic (i.e., $\pi_i \geq 0$ and $\sum \pi_i = 1$). This describes the probabilities for cooperation between the two players in the stationary state, and hence the asymptotic payoff. The problem with this approach is less the computation of the eigenvector (the 16×16-matrix has many 0-entries), but the fact that the dimensionality of the strategy space is so huge that samples have to be impossibly large in order to represent all possibilities.

This problem occurs even if we consider only stochastic strategies with memories of lenght 1 (but taking the own move into account). Even if one uses U-shaped distributions for the sample, it may be that the TFT-corner is underrepresented and cannot exert its policing function. Some simulations lead towards $AllD$, while others display the turn towards cooperation as described in the previous section.

A very interesting approach is due to Lindgren (1991), who considered strategies which are less stochastic than ours, but have longer memories (2 or 3). More precisely, the probabilities to play C, given this or that history, are either ϵ or $1 - \epsilon$, where ϵ is a small number corresponding to the noise. This means that the strategies are "intended" to be deterministic, but may suffer with a small probability ϵ from a mistake. Lindgren's simulations exhibit highly complex evolutionary histories, frequently leading to meta-stable states displaying coexistence, and even symbiosis. An interesting result is the emergence of a large class of strategies which (a) cooperate if both players have cooperated in the previous two rounds, defected in the previous two rounds, or simultaneously switched from D to C in the last round; and (b) defect if they, or the co-player, have unilaterally switched from C to D. Such strategies handle mistakes in a surprising, but efficient way: if one player accidentally plays D, both players play two rounds of D and then resume playing C. Such strategies have already been proposed by Micko et al. (1977); they are difficult to exploit systematically, but do not do very well against TFT.

Lindgren's model extends simulations by Axelrod (1987), who used a genetic algorithm to simulate the evolution of a population of strategists whose memory has a length of 3. Such algorithms use random mutations and recombinations to introduce new strategies and submit this diversity to culling by selection. Axelrod also observed evolutionary histories which first lead toward defection and later turn toward cooperation.

It would be interesting to see whether increasing the memory capacity or the ability to randomize is more advantageous in playing the IPD. Preliminary investigations seem to show that the situation is bistable. On one hand, (p, q)-strategies cannot invade a meta-stable population of the type described by Lindgren; on the other hand, strategies of that type cannot invade a population of

GTFT-players. In order to be efficient under conditions of noise and uncertainty, a strategy will probably need both to keep track of the last few rounds and to react in an unpredictable way to isolated defections.

17.7 References

Axelrod R. (1984). *The evolution of cooperation.* New York: Basic Books Inc. (reprinted by Penguin, Harmondsworth)

Axelrod, R. (1987). The evolution of strategies in the iterated prisoner's dilemma. In D. Davis (Ed.), *Genetic algorithms and simulated annealing* (pp. 32-43). London: Pitman.

Axelrod, R., & Dion, D. (1988). The further evolution of cooperation, *Science, 242,* 1385-1390.

Axelrod, R., & Hamilton, W.D. (1981). The evolution of cooperation. *Science, 211,* 1390-1396.

Cheney, D.L., & Seyfarth, D. (1982). Recognition of individuals within and between groups of free-ranging vervet monkeys. *American Zoologist, 22,* 519-529.

Fischer, E.A. (1980). The relationship between mating system and simultaneous hermaphroditism in the coral reef fish. *Hypoplectrus nigricans. Animal Behaviour, 28,* 620-633.

Hofbauer, J., & Sigmund, K. (1988). *The theory of evolution and dynamical systems.* Cambridge: Cambridge University Press.

Hofstadter, D. (1985). *Metamagical themas: Questing for the essence of mind and pattern.* New York: Basic Books.

Lindgren, K. (1991) Evolutionary phenomena in simple dynamics. In Langton et al. (Eds.), *Artificial Life II* (pp. 295-312), Proceedings of the Santa Fe Institute Studies, Addison-Wesley.

Luce, R.D., & Raiffa, H. (1957). *Games and decisions.* New York: Wiley.

May, R.M. (1987). More evolution of cooperation. *Nature, 327,* 15-17.

Micko, H.-Ch., Brückner, G., & Ratzke, H. (1977). Theories and strategies for prisoner's dilemma. In W.F. Kempf & B.H. Repp (Eds.), *Mathematical Models for Social Psychology* (pp. 214-276). Bern: Huber.

Milinski, M. (1987). Tit for tat in sticklebacks and the evolution of cooperation. *Nature, 325,* 434-435.

Molander, P. (1985). The optimal level of generosity in a selfish, uncertain environment. *Journal of Conflict Resolution, 29,* 611-618.

Nowak, M. (1990). An evolutionarily stable strategy may be inaccessible, *Theoretical Population in Biology, 142,* 237-241.

Nowak, M., & Sigmund, K. (1989a). Oscillations in the evolution of reciprocity. *Journal of Theoretical Biology, 147,* 21-26.

Nowak, M., & Sigmund, K. (1989b). Game dynamical aspects of the prisoner's dilemma. *Journal of Applied Mathematics and Computation, 30,* 191-213.

Nowak, M. & Sigmund, K. (1990). The evolution of reactive strategies in iterated games. *Acta Applicandae Mathematicae, 20,* 247-265.

Nowak, M., & Sigmund, K. (1992). Tit for tat in heterogeneous populations. *Nature, 353,* 250-253.

Packer, C. (1979). Reciprocal altruism in *Papio anubis. Nature, 265,* 441-443.

Trivers, R. (1985). *Social evolution.* Menlo Park: Benjamin-Cummings.

Selten, R., & Stoecker, R. (1986). End behaviour in sequences of finite prisoner's dilemma supergames. *Journal of Economical Behavior and Organization, 7,* 47-70.

Wilkinson, G.S. (1984). Reciprocal food sharing in the vampire bat. *Nature, 308,* 181-184.

Part IV

Measurement and Psychometrics

18

An Axiomatization for Bilinear Models

Uta Wille[1]

18.1 Bilinear Models and Measurement Theory

In the social and behavioural sciences, different kinds of mathematical models are used to describe, to explain, and to predict social or individual behaviour. Social scientists often describe observed relations by quantitative models combining the underlying attributes. In the case of bilinear models, dependencies between attributes are expressed by sums of products, i.e., by equations of the form

$$C = \sum_{i=1}^{n} A_i \cdot B_i,$$

where C, A_i, B_i are variables modelling the attributes under study. A prominent area where bilinear models are used is the theory of rational choice; in particular, the subjective expected utility principle (SEU-principle) is expressed in this way.

The formulation of such quantitative models presupposes that the values of the attributes are numbers or can be represented by numbers. Often it is difficult to see what attributes have numerical measurements and what numerical manipulations of the measured values are meaningful. Such questions are treated by measurement theory.

According to measurement theory, a bilinear model can only be justified if it is based on an empirical model that can be verified by data. Moreover, a bilinear model representing an empirical model is only meaningful if the sums of products are invariant under different numerical representations. This paper concentrates on the following question:

> *Under what assumptions can an empirical model be represented by a bilinear model?*

Suprisingly, there have been only a few attempts to treat the problem of representation and meaningfulness for bilinear models. Orth (1985, 1987, 1988) published a few articles on this subject in the "Zeitschrift für Sozialpsychologie". He showed that the variables occurring in a sum of products have to be measured on the level of ratio scales to yield a meaningful numerical statement. But requirements for the existence of a representation by bilinear models (i.e., an axiomatization in the sense of measurement theory) were not discussed. Such an

[1]Department of Mathematics of the Justus-Liebig-Universität Gießen, Arndtstr. 2, D-6300 Gießen Germany

axiomatization is desirable and important because it makes an empirical verification of bilinear models possible. Moreover, it provides a deeper understanding of the underlying empirical situation since the assumptions involved are expressed by axioms. The purpose of this paper is to report a basic result for such an analysis of bilinear models by methods of measurement theory. First, a mathematical formulation of the problem will be given. Then, an axiomatization for bilinear models is described and justified by a general representation theorem. Finally, an interpretation of the axioms and the representation theorem is outlined.

18.2 Formulation of the Mathematical Problem

The measurement theory approach first requires an exact description of the interface between reality and mathematics, i.e., a formal model for the assumed empirical situation. Then the question arises: under which assumptions can this formal model be represented by a bilinear model? An answer is given by a set of axioms which can be used for a representation theorem.

What are suitable empirical models which allow representations by bilinear models? In most applications, the underlying data are given in the form of a table that reports, for the objects under study, their values with respect to given attributes. A formalization of such data tables can be taken from formal concept analysis (cf. R. Wille, 1982; R. Wille, 1987; Ganter and R. Wille, 1989), where methods of data analysis based on order and lattice theory have been developed.

Definition 1 *A* **many-valued context** *is a quadruple (G, M, W, I), where G, M, and W are sets, and I is a ternary relation between G, M, and W (i.e., $I \subseteq G \times M \times W$), such that $(g, m, v) \in I$ and $(g, m, w) \in I$ always imply $v = w$; the elements of G, M, and W are called* **objects, attributes,** *and* **attribute values,** *respectively. (G, M, W, I) is called* **complete** *if, for every $(g, m) \in G \times M$, there exists a value $w \in W$ such that $(g, m, w) \in I$. (Instead of $(g, m, w) \in I$, one often writes $m(g) = w$.)*

Since object/attribute/value tables occur most frequently, many-valued contexts (as formalizations of such tables) are useful empirical models. In particular, in our case, it is appropriate to choose complete many-valued contexts as empirical models. Thus, for the following, it is assumed that the data examined are given in the form of a complete many-valued context.

It is necessary to choose a numerical model in which sums of products can be expressed. In linear algebra, the basic type of sums of products is expressed by means of the scalar product. In its most general case, the scalar product occurs in the form

$$\varphi(\boldsymbol{x}) := a_1 \cdot x_1 + \ldots + a_n \cdot x_n,$$

where $\boldsymbol{x} := \boldsymbol{e}_1 \cdot x_1 + \ldots + \boldsymbol{e}_n \cdot x_n \in K^n$ and $\varphi := a_1 \cdot \boldsymbol{e}_1^* + \ldots + a_n \cdot \boldsymbol{e}_n^* \in (K^n)^*$

(K is an arbitrary field, $e_1, \ldots, e_n$, the canonical basis of the vector space K^n, $e_1^*, \ldots, e_n^*$ its dual basis in the dual space $(K^n)^*$, and $x_1, \ldots, x_n, a_1, \ldots, a_n$ are scalars from K). This setting motivates the following definition.

Definition 2 *Let V be an n-dimensional vector space over a field K and let V^* be the dual space of V. Then we call the complete many-valued context $\mathbb{K}(V) := (V, V^*, K, E)$, for which the relation E is defined by*

$$(v, \varphi, k) \in E \quad :\Leftrightarrow \quad \varphi(v) = k,$$

a **bilinear context** *of dimension n.*

This definition allows us to formulate the general problem of the representability by bilinear models as follows:

> *Under what assumptions is a complete many-valued context isomorphic to a bilinear context ?*

The question is how to define an addition and a multiplication on W such that W together with these operations forms a field and, furthermore, to find bijections $\alpha : G \longrightarrow W^n$ and $\beta : M \longrightarrow (W^n)^*$ such that the following condition holds:

$$(g, m, w) \in I \qquad \Leftrightarrow \qquad (\alpha(g), \beta(m), w) \in E.$$

The representation theorem stated in the next section describes conditions for the existence of such addition, multiplication, and bijections guaranteeing that the pair (α, β) establishes an isomorphism between the many-valued contexts (G, M, W, I) and $(W^n, (W^n)^*, W, E)$.

18.3 The Representation Theorem

Useful tools for the treatment of coordinatization problems are "frames" which were introduced by John von Neumann (1960), generalizing the idea of coordinate systems. In the case of a representation by bilinear contexts, a two-fold coordinatization has to be performed to create a vector space and its dual space. This is achieved by the definition of a "double frame" in a many-valued context. A double frame consists of elements $g_0, \ldots, g_n$ in G and attributes $m_0, \ldots, m_n$ in M that are suitably related. To simplify the definition, for every m_i and every g_i ($i \in \{0, \ldots, n\}$), the equivalence relations Θ_i^G and Θ_i^M on G and M are defined by

$$\begin{aligned} g\Theta_i^G h &\quad :\Leftrightarrow \quad m_i(g) = m_i(h), \\ m\Theta_i^M n &\quad :\Leftrightarrow \quad m(g_i) = n(g_i); \end{aligned}$$

furthermore, the abbreviations $\Psi_{ij}^G := \bigcap_{k \neq i,j} \Theta_k^G$ and $\Psi_{ij}^M := \bigcap_{k \neq i,j} \Theta_k^M$ shall be used.

Definition 3 *Let* $\mathbb{K} := (G, M, W, I)$ *be a many-valued context. Then we call an* $2(n+1)$*-tuple* $\Phi := (g_0, \ldots, g_n; m_0, \ldots, m_n)$ *in* $G^{n+1} \times M^{n+1}$ **a double frame** *of order n in* $\mathbb{K}$*, if there are two elements* $0 \neq 1$ *in* W *such that* $m_i(g_j) = \delta_{ij}$ *(i.e.,* $m_i(g_j) = 1$ *if* $i = j$ *and* $m_i(g_j) = 0$ *if* $i \neq j$*) for all* $i, j \neq 0$*, and if, for all* $i, j \in \{0, \ldots, n\}$ *with* $i \neq j$*, the following holds:*

$$\bigcap_{k \neq i} \Theta_k^G = id_G \quad \text{and} \quad \bigcap_{k \neq i} \Theta_k^M = id_M,$$

$$(P_{ij}^G) \qquad \forall g, k \in G \exists h \in G : g \Theta_i^G h \Psi_{ij}^G k,$$

$$(P_{ij}^M) \qquad \forall l, p \in M \exists m \in M : l \Theta_i^M m \Psi_{ij}^M p.$$

A double frame is called **surjective** *if* $\{m_i(g) \mid g \in G\} = W = \{m(g_i) \mid m \in M\}$ *for every* $i \in \{1, \ldots, n\}$*. By* o_G *we denote the object with* $m_i(o_G) = 0$ *for all* $i \in \{1, \ldots, n)$*, and by* o_M *the attribute with* $o_M(g_i) = 0$ *for all* $i \in \{1, \ldots, n\}$*.*

For a field K and a natural number n, $\Phi_{K,n} := (e_0, \ldots, e_n; e_0^*, \ldots, e_n^*)$ is a surjective double frame of the bilinear context $(K^n, (K^n)^*, K, E)$, where $e_1, \ldots, e_n$ is the canonical basis of the vector space K^n, and $e_1^*, \ldots, e_n^*$ its dual basis in the dual space $(K^n)^*$, $e_0 = e_1 + \ldots + e_n$, and $e_0^* = e_1^* + \ldots + e_n^*$.

For an understanding of the structure of a many-valued context $\mathbb{K} := (G, M, W, I)$, the following pairs of operators between the power sets of G and M are basic: For $r \in W \setminus \{0\}$, $A \subseteq G$ and $B \subseteq M$, we define

$$A \mapsto A^r := \{m \in M \mid m(g) = r \text{ for all } g \in A\},$$

$$B \mapsto B^r := \{g \in G \mid m(g) = r \text{ for all } m \in B\}.$$

By well-known results on Galois connections (cf. Birkhoff, 1967, p. 111 and pp. 121ff.), the sets $\mathfrak{U}_r(\mathbb{K}) := \{A \subseteq G \mid A = A^{rr}\}$ and $\mathfrak{I}_r(\mathbb{K}) := \{B \subseteq M \mid B = B^{rr}\}$ together with set inclusion form complete lattices in which $\bigwedge_{t \in T} X_t = \bigcap_{t \in T} X_t$ and $\bigvee_{t \in T} X_t = (\bigcup_{t \in T} X_t)^{rr}$ ($X_t \in \mathfrak{U}_r(\mathbb{K})$ or $X_t \in \mathfrak{I}_r(\mathbb{K})$ for all $t \in T$, respectively). The following conditions $B^G(r, s)$ and $B^M(r, s)$ ensure that, for $r, s \in W$, $\mathfrak{U}_r(\mathbb{K}) \subseteq \mathfrak{U}_s(\mathbb{K})$ and $\mathfrak{I}_r(\mathbb{K}) \subseteq \mathfrak{I}_s(\mathbb{K})$.

$$B^G(r, s) \qquad \forall m \in M \quad \exists n \in M : \quad \{m\}^r = \{n\}^s,$$

$$B^M(r, s) \qquad \forall g \in G \quad \exists h \in G : \quad \{g\}^r = \{h\}^s.$$

Representation Theorem 4 *Let n be an integer greater than 2. A complete many-valued context* (G, M, W, I) *is isomorphic to a bilinear context of dimension n if and only if* (G, M, W, I) *has a surjective double frame* $\Phi = (g_0, \ldots, g_n; m_0, \ldots, m_n)$ *of order n with respect to some distinguished elements* $0 \neq 1$ *in* W *such that the conditions* $B^G(r, s)$ *and* $B^M(r, s)$ *are fulfilled for all* $r, s \in W \setminus \{0\}$ *and the following five axioms are satisfied:*

(A0) *If* $m \in M$ *and* $r \in W \setminus \{0\}$*, then* $\{m\}^{01r} = \{m\}^0$ *and* $o_G \in \{m\}^0$*.*

(A1) *If* $m_j(g) = 0 = m_i(h)$ *and* $m_k(g) = 0 = m_k(h)$ *for all* $k \in \{1,\ldots,n\} \setminus \{i,j\}$, *where* $i,j \in \{1,\ldots,n\}$ *with* $i \neq j$, *then there is an attribute* $m \notin \{o_M, m_1, \ldots, m_n\}$ *with* $m(g) = m(h)$; *this can be chosen as* m_0 *if* $m_i(g) = m_j(h)$.

(A2) *If* $m_i(g) = m_i(h)$, $m_i(g') = m_i(h')$ *for some* $i \in \{1,\ldots,n\}$ *and* $m_k(g) = m_k(g')$, $m_k(h) = m_k(h')$ *for all* $k \in \{1,\ldots,n\} \setminus \{i\}$, *then every* m *in* M *satisfies*

$$m(g) = m(h) \quad \Rightarrow \quad m(g') = m(h').$$

(A3) *For all* $m, n \in M \setminus \{o_M\}$ *and for all* $s, t \in W\setminus\{0\}$ *with* $\{m\}^s \cap \{n\}^t = \emptyset$, *there is an* $r \in W\setminus\{0\}$ *with* $\{m\}^{s1r} = \{n\}^t$.

(A4) *For every* $g \in G \setminus \{o_G\}$ *and for every* $m \in M \setminus \{o_M\}$, *there is a value* $r \in W \setminus \{0\}$ *such that* $\{g\}^{rr} = \{g\}$ *and* $\{m\}^{rr} = \{m\}$, *respectively.*

18.4 Interpretation of the Representation Theorem

The representation theorem proved by U. Wille (1991) yields conditions under which a complete many-valued context (G, M, W, I) is isomorphic to a bilinear context $(W^n, (W^n)^*, W, E)$ of dimension $n > 2$. Because the axioms of the theorem express the assumptions made by using bilinear models, it is desirable to discuss the meaning of those axioms. This is the purpose of the final section.

If a complete many-valued context (G, M, W, I) is isomorphic to the bilinear context $(W^n, (W^n)^*, W, E)$ as in the representation theorem, an attribute value $m(g)$ is determined by

$$m(g) = \sum_{i=1}^{n} a_i \cdot x_i,$$

where the object g corresponds to the vector $(x_1, \ldots, x_n) \in W^n$ and the attribute m corresponds to the linear form described by the vector $(a_1, \ldots, a_n)$. Thus, the assumption of a bilinear representation for a complete many-valued context is connected with the interpretation of each value $m(g)$ as composition of the values of g with respect to m in different "states". To emphasize this interpretation, the value $m(g)$ is called an *aggregate value* composed by certain *state values* $m(g)_i$ of g with respect to m in the state i. The integer n can be interpreted as the number of states considered, the set of which shall be denoted by $Z := \{1, \ldots, n\}$. Let us summerize that, in the case of an isomorphism (α, β) from (G, M, W, I) onto $(W^n, (W^n)^*, W, E)$, an aggregate value $m(g)$ is composed by the state values

$m(g)_i := a_i \cdot x_i$, where $\alpha(g) = e_1 \cdot x_1 + \ldots + e_n \cdot x_n$ and $\beta(m) = a_1 \cdot e_1^* + \ldots + a_n \cdot e_n^*$, i.e.,

$$m(g) = \sum_{i=1}^{n} m(g)_i.$$

It should be emphasized that consideration is given not simply to objects and attributes, but to objects and attributes in certain states fixed in a given set Z of states.

The first assumption of the representation theorem is the existence of a surjective double frame $\Phi := (g_0, \ldots, g_n; m_0 \ldots, m_n)$ with respect to distinguished elements 0 and 1 of W. The value 0 can be interpreted as an *equilibrium value* while 1 can be understood as the *unit value* for the coordinatization which is elaborated on the base of the double frame Φ. In a natural way, Φ yields bijections $\alpha_G : G \longrightarrow W^n$ and $\alpha_M : M \longrightarrow W^n$ by

$$\begin{aligned} \alpha_G(g) &:= (m_1(g), \ldots, m_n(g)) \qquad (g \in G), \\ \alpha_M(m) &:= (m(g_1), \ldots, m(g_n)) \qquad (m \in M). \end{aligned}$$

The condition $m_i(g_j) = \delta_{ij}$ $(i, j \neq 0)$ suggests calling the objects $g_1, \ldots, g_n$ of Φ *unit objects*, and the attributes $m_1, \ldots, m_n$ of Φ *unit attributes*. The remaining object g_0 and attribute m_0 play a joining and unifying role for the "dimensions" established by $g_1, \ldots, g_n$ and $m_1, \ldots, m_n$, respectively. Therefore, g_0 will be called the *unifying object* and m_0 the *unifying attribute* of the double frame Φ.

Since the investigation of dependencies between attributes is an important subject in data analysis, it is also interesting to interpret double frames with respect to dependencies between attributes. A basic type of dependency is the functional dependency between attributes which is defined in the language of many-valued contexts as follows: In a complete many-valued context (G, M, W, I), a set Y of attributes *functionally depends* on a set X of attributes if, for $g, h \in G$, $x(g) = x(h)$ for all $x \in X$ implies $y(g) = y(h)$ for all $y \in Y$ (cf. Maier, 1983, pp. 42–70). In the case of bilinear contexts it can be proved that functional dependency coincides with linear dependency (cf. R. Wille, 1988).

Let $\Phi := (g_0, \ldots, g_n; m_0, \ldots, m_n)$ be a double frame of the complete many-valued context (G, M, W, I) which has the property that every two objects have different values with respect to at least one attribute and, conversely, every two attributes differ on at least one object (i.e., (G, M, W, I) does not have two equal rows or columns). It should be emphazised that this condition is not a serious restriction since we may simply identify equal rows and columns. Then, for $i \in \{0, \ldots, n\}$, the condition $\bigcap_{k \neq i} \Theta_k^G = id_G$ is equivalent to the property that every attribute of M functionally depends on $\{m_0, \ldots, m_n\} \setminus \{m_i\}$. Moreover, $m_i(g_j) = \delta_{ij}$ $(i, j \neq 0)$ ensures that m_i is not functionally dependent on $\{m_1, \ldots, m_n\} \setminus \{m_i\}$ for every $i \in \{1, \ldots, n\}$. Thus the unit attributes form a "basis" with respect to functional dependency in M. For objects the dual arguments hold. The remaining conditions (P_{ij}^G) and (P_{ij}^M) are more or less technical conditions which ensure that the many-valued context (G, M, W, I) is rich enough (cf. R. Wille & U. Wille, 1992). The conditions (P_{ij}^G) and (P_{ij}^M) guarantee, in particular the existence of o_G and o_M. The axiom $(A0)$ additionally requires that $m(o_G) = 0$ for every $m \in M$ and $o_M(g) = 0$ for every $g \in G$; o_M is called the *equilibrium attribute* and o_G the *equilibrium object*.

By axiom $(A1)$, for every two objects $g, h \in G$ with $g\Psi_{0i}^G g_i$ and $h\Psi_{0j}^G g_j$, there exists an attribute $m \notin \{o_M, m_1, \ldots, m_n\}$ with $m(g) = m(h)$. This again is a richness condition. The second part of $(A1)$, which requires

$$m_0(\alpha_G^{-1}(0, \ldots, \underset{i}{x}, \ldots, 0)) = m_0(\alpha_G^{-1}(0, \ldots, \underset{j}{x}, \ldots, 0))$$

for every $x \in W$ and $i, j \in \{1, \ldots, n\}$, is more interesting. A special consequence is $m_0(g_i) = m_0(g_j)$ for all $i, j \in \{1, \ldots, n\}$, i.e., $\alpha_M(m_0) = (c, \ldots, c)$ for an element c in W. Thus, axiom $(A1)$ ensures the unifying character of m_0.

As with axiom $(A1)$, axiom (A2) can be reformulated as follows:

$$m(\alpha_G^{-1}((a, x_2, \ldots, x_n))) = m(\alpha_G^{-1}((a, y_2, \ldots, y_n))),$$

implies

$$m(\alpha_G^{-1}((b, x_2, \ldots, x_n))) = m(\alpha_G^{-1}((b, y_2, \ldots, y_n))).$$

This shows that $(A2)$ has a similar content as the axiom $P2$ in the formulation of Savage's Expected-Utility-Theorem by Fishburn (1970), who gives the following interpretation: "$P2$ says that preferences between acts should not depend on those states that have identical consequences for the two acts" (Fishburn, 1970, p. 193). An analogous interpretation of $(A2)$ may be formulated as follows: $(A2)$ says that the equality of the aggregate values of objects with respect to a fixed attribute m should not depend on those states that have identical state values for the two objects with respect to m.

For a bilinear context $\mathbb{K}(V)$, the lattices $\mathfrak{U}_0(\mathbb{K}(V))$ and $\mathfrak{I}_0(\mathbb{K}(V))$ consist of all linear subspaces of V and V^*, respectively. For $r \neq 0$, $\mathfrak{U}_r(\mathbb{K}(V))$ and $\mathfrak{I}_r(\mathbb{K}(V))$ are the lattices of all non-linear affine subspaces of V and V^*, respectively; in particular, $\mathfrak{U}_r(\mathbb{K}(V)) = \mathfrak{U}_s(\mathbb{K}(V))$ and $\mathfrak{I}_r(\mathbb{K}(V)) = \mathfrak{I}_s(\mathbb{K}(V))$ for $r \neq 0 \neq s$. In the representation theorem these equalities are guaranteed for the many-valued context $\mathbb{K}$ by the conditions $B^G(r, s)$ and $B^M(r, s)$ for $r, s \in W \setminus \{0\}$. By a well-known result on Galois connections (see Birkhoff, 1967, p. 123), the mappings $\varphi_r : \mathfrak{U}_r(\mathbb{K}) \to \mathfrak{I}_r(\mathbb{K})$ and $\varphi_r^{-1} : \mathfrak{I}_r(\mathbb{K}) \to \mathfrak{U}_r(\mathbb{K})$ given by $\varphi_r(A) = A^r$ and $\varphi_r^{-1}(B) = B^r$ are inverse anti-isomorphisms between the complete lattices $(\mathfrak{U}_r(\mathbb{K}), \subseteq)$ and $(\mathfrak{I}_r(\mathbb{K}), \subseteq)$. If one interprets a pair (A, B) with $A^r = B$ and $B^r = A$ as a "natural" *concept* of $\mathbb{K}$ with *extension* A and *intension* B (see R. Wille, 1982), the anti-isomorphisms express in mathematical terms what in philosophy is known as the reciprocity law for concepts (cf. Wagner, 1973). According to this interpretation, the conditions $B^G(r, s)$ and $B^M(r, s)$ $(r, s \in W \setminus \{0\})$ yield a coincidence of *extensional structures*, namely $\mathfrak{U}_r(\mathbb{K}) = \mathfrak{U}_s(\mathbb{K}) =: \mathfrak{U}(\mathbb{K})$, and of *intensional structures*, namely $\mathfrak{I}_r(\mathbb{K}) = \mathfrak{I}_s(\mathbb{K}) =: \mathfrak{I}(\mathbb{K})$.

Axiom ($A3$) says that the extensional structure $\mathfrak{U}(\mathbb{K})$ is highly symmetric which also follows for the intensional structure $\mathfrak{J}(\mathbb{K})$. For all $r, s \in W \setminus \{0\}$, the composition $\varphi_s^{-1} \circ \varphi_r$ is an automorphism of $\mathfrak{U}(\mathbb{K})$. ($A3$) requires that any coatom $\{m\}^s$ of $\mathfrak{U}(\mathbb{K})$ can be mapped onto each disjoint coatom $\{n\}^t$ by a special automorphism $\varphi_r^{-1} \circ \varphi_1$, which can be geometrically understood as dilatation by the factor r fixing the equilibrium object o_G.

Assuming the conditions $B^G(r,s)$ and $B^M(r,s)$, axiom ($A4$) yields that $\{\{g\} \mid g \in G \setminus \{o_G\}\}$ is the set of atoms and $\{\{m\}^r \mid m \in M \setminus \{o_M\}\}$ is the set of coatoms of the lattice $(\mathfrak{U}(\mathbb{K}), \subseteq)$ for every $r \in W \setminus \{0\}$; dually, $\{\{m\} \mid m \in M \setminus \{o_M\}\}$ is the set of atoms and $\{\{g\}^r \mid g \in G \setminus \{o_G\}\}$ the set of coatoms of $(\mathfrak{J}(\mathbb{K}), \subseteq)$ for every $r \in W \setminus \{0\}$. Hence, ($A4$) is another richness condition because it requires that objects can be separated by attributes and attributes can be separated by objects.

The representation theorem establishes a general frame for the analysis of bilinear models. There might be data contexts for which all the conditions of the representation theorem can be verified so that there is a bilinear context as an adequate description of the data. But, in most cases, data are more or less incomplete. Then the question is whether the given data can be extended by "potential" objects, attributes, and values to constitute a complete many-valued context isomorphic to a bilinear context. Hence, for the application of bilinear models, further research should focus on the following problem:

> *Under what assumptions can a complete many-valued context* $\mathbb{K}$ *be embedded in a bilinear context such that the extensional and intensional structures of* $\mathbb{K}$ *are respected?*

18.5 References

Birkhoff, G. (1967). *Lattice theory* (3rd ed.). Providence, Rhode Island: A.M.S. Colloquium Publ.

Fishburn, P.C. (1970). *Utility theory for decision making.* New York: Wiley.

Ganter, B., & Wille, R. (1989). Conceptual scaling. In F.S. Roberts (Ed.), *Applications of combinatorics and graph theory to the biological and social sciences* (pp.139-167). New York: Springer.

Maier, D. (1983). *The theory of relational databases.* Rockville, Maryland: Computer Science Press.

Orth, B. (1985). Bedeutsamkeitsanalyse bilinearer Einstellungsmodelle. *Zeitschrift für Sozialpsychologie, 16,* 101-115.

Orth, B. (1987). Formale Untersuchungen des Modells von Fischbein & Aijzen zur Einstellungs–Verhaltensbeziehung: I. Bedeutsamkeit und erforderliches Skalenniveau. *Zeitschrift für Sozialpsychologie, 18,* 152–159.

Orth, B. (1988). Formale Untersuchungen des Modells von Fischbein & Aijzen zur Einstellungs–Verhaltensbeziehung: II. Modellmodifikation für intervallskalierte Variablen. *Zeitschrift für Sozialpsychologie, 19,* 31–40.

von Neumann, J. (1960). *Continuous geometry.* Princeton, New Jersey: Princeton University Press.

Wagner, H. (1973). Begriff. In H. Krings, H.M. Baumgartner, & C. Wild (Eds.), *Handbuch philosophischer Grundbegriffe* (pp. 191-209). München: Kösel.

Wille, R. (1982). Restructuring lattice theory: An approach based on hierarchies of concepts. In I. Rival (Ed.), *Ordered sets* (pp. 445-470). Boston-Dordrecht: Reidel.

Wille, R. (1987). Bedeutungen von Begriffsverbänden. In B. Ganter, R. Wille, & K.E. Wolff (Eds.), *Beiträge zur Begriffsanalyse* (pp. 161-211). Mannheim: B.I.-Wissenschaftsverlag.

Wille, R. (1988). Dependencies of many-valued attributes. In H.H. Bock (Ed.), *Classification and related methods of data analysis* (pp. 581-586). Amsterdam: North-Holland.

Wille, R., & Wille, U. (1992). On the controversy over Huntington's equations: When are such equations meaningful? *Mathematical Social Sciences* (to appear).

Wille, U. (1991). Eine Axiomatisierung bilinearer Kontexte. *Mitteilungen aus dem Mathematischen Seminar Giessen, 200,* 71-112.

19

On Item Parameter Estimation in Certain Latent Trait Models

J. Pfanzagl[1]

ABSTRACT The paper deals with the estimation of item difficulty parameters $(\delta_1, \ldots, \delta_m)$ if a subject with ability α produces the response vector $(x_1, \ldots, x_m) \in \{0,1\}^m$ with probability $\prod_{i=1}^m H(\alpha - \delta_i)^{x_i}\left(1 - H(\alpha - \delta_i)\right)^{1-x_i}$.
If $H = \Psi$ (the logistic distribution function), conditional maximum likelihood estimators are consistent under mild conditions on the ability parameters α_j, $j = 1, 2, \ldots$. If the ability parameters are a random sample from an unknown distribution, conditional maximum likelihood estimators are consistent and asymptotically efficient. Marginal maximum likelihood estimators share these properties, since they are asymptotically equivalent to conditional maximum likelihood estimators by a theorem of de Leeuw and Verhelst.
If $H = \Phi$, estimating $(\delta_1, \ldots, \delta_m)$ is impossible if the distribution of ability parameters is completely unknown, since the item difficulty parameters are not identifiable in this case.

19.1 Introduction

Assume that abilities of subjects and difficulties of tasks can be measured on scales such that the probability for a subject of ability α to respond positively to a task of difficulty δ, say $\Pi(\alpha, \delta)$, can be expressed as $H(\alpha - \delta)$, where $H : \mathbb{R} \to (0,1)$ is an increasing and continuous function fulfilling $\lim_{u \to -\infty} H(u) = 0$ and $\lim_{u \to \infty} H(u) = 1$. If such a representation is possible, the scale values α, δ and the function H are unique up to linear transformations: If there are other scales, related to the given scales by increasing transformations m_1 and m_2 such that

$$H^*\left(m_1(\alpha) - m_2(\delta)\right) = H(\alpha - \delta) \qquad \text{for all } \alpha, \delta \in \mathbb{R},$$

then

$$\begin{aligned} m_1(\alpha) &= a\alpha + b_1, \\ m_2(\delta) &= a\delta + b_2, \\ H^*(u) &= H\left(\frac{u - b_1 + b_2}{a}\right) \end{aligned}$$

(see Pfanzagl, 1971, p. 171, Theorem 11.4.1, or Falmagne, 1985, p. 192, Theorem 7.26).

---

[1]Mathematical Institute, University of Cologne, Weyertal 86, 5000 Cologne 41, Germany
The author wishes to thank R. Hamböker for checking the mathematical details, and the referees for suggesting improvements of the presentation.

The existence of scales for ability and difficulty such that $\Pi(\alpha,\delta) = H(\alpha-\delta)$ is sufficient and — under suitable regularity conditions on Π — even necessary for "specific objectivity" in the sense of Rasch. That means the existence of a function $U : (0,1)^2 \to \mathbf{R}$ such that $U\big(\Pi(\alpha',\delta),\Pi(\alpha'',\delta)\big)$ is independent of δ. (For details see Fischer, 1988, p. 94, Theorem 1.)

If we choose a special version of H, the scale values α and δ are unique up to a common translation. To make the scales unique, we may designate one item the "standard item" to which difficulty 0 is assigned. A natural side condition in choosing H is $H(0) = 1/2$, which means that the probability for a positive response is 1/2 if the subject's ability matches the item difficulty.

Our basic assumption can conveniently be written as follows: The probability of response $x \in \{0,1\}$ is $H(\alpha-\delta)^x\big(1-H(\alpha-\delta)\big)^{1-x}$ (where "1" denotes a positive, and "0" a negative response). We adopt the prevailing assumption of "local independence", which is nothing else than stochastic independence between the subject's responses to the test items; i.e., a subject of ability α, confronted with the standard item of difficulty $\delta_0 = 0$ and items with difficulties $(\delta_1,\ldots,\delta_m)$, produces the response vector $(x_0,x_1,\ldots,x_m) \in \{0,1\}^{1+m}$ with probability

$$\prod_{i=0}^{m} H(\alpha-\delta_i)^{x_i}\big(1-H(\alpha-\delta_i)\big)^{1-x_i}. \tag{19.1}$$

This defines a probability measure with support $\{0,1\}^{1+m}$ and parameters $(\delta_1,\ldots,\delta_m) \in \mathbf{R}^m$ and $\alpha \in \mathbf{R}$; it will be denoted by

$$H^*_{(\delta_1,\ldots,\delta_m),\alpha}. \tag{19.2}$$

In the following we discuss the problem of estimating the item difficulty parameters $(\delta_1,\ldots,\delta_m)$, based on the response vectors $(x_{j0},x_{j1},\ldots,x_{jm})$ of a large number of subjects $j = 1,\ldots,n$. The basic difficulty is this: with each subject j another — unknown — ability parameter α_j enters the model.

Any estimation procedure for $(\delta_1,\ldots,\delta_m)$ presumes the function H to be known. What the statistician would like to have is a model of the process underlying the response behavior that suggests what the function H should be. Then one could test this model before proceeding to the estimation of $(\delta_1,\ldots,\delta_m)$, based on this particular H. There are suggestions pointing to $H = \Phi$ (the normal distribution function), or $H = \Psi$ (the logistic distribution function), but none of these seems to be cogent. (See Gigerenzer, 1981, Section 7.5.4, or Falmagne, 1985, Chapter 5.) To estimate H from the data appears to be hopeless. Feller, as early as 1939, demonstrated that different functions H fit the same data (growth data, in his case) equally well. The same point is made by Chambers and Cox (1967).

In view of this unpleasant situation a different aspect comes into focus: Estimation of $(\delta_1,\ldots,\delta_m)$ becomes particularly simple for the logistic model, $H = \Psi$, with

$$\Psi(u) = \exp[u]/(1+\exp[u]). \tag{19.3}$$

Some authors (see, e.g., Roskam and Jansen, 1984, p. 296) are of the opinion that Rasch's specific objectivity implies the Rasch model. It does imply $\Pi(\alpha, \delta) = H(\alpha - \delta)$ for suitably chosen scales, but it does not imply $H = \Psi$ unless one adds further requirements such as the existence of a nontrivial sufficient statistic.

The interest of mathematical psychologists in the logistic model is, obviously, not reciprocated by specialists in this field. The "Handbook of the Logistic Distribution", edited by Balakrishnan (1992), a volume of 601 pages, devotes just 2 lines (on p. 475) to applications in psychology.

In Section 2 we present some results on the use of CML (conditional maximum likelihood) estimators which are applicable for $H = \Psi$ only. In Section 3, MML (marginal maximum likelihood) estimators are considered as an alternative.

Some readers will miss a discussion of more general models which include more (say discrimination- and guessing-) parameters, or which relax the assumption of local independence. One motivation for this omission is that certain clear-cut theoretical results hold true for the simple model (19.1), but not for its generalizations. A second motivation is that the advantage of some of these generalizations is questionable. Certainly, any more general model is more flexible and will fit the data better. However, this does not substantiate the more general model as a more accurate image of reality. This is true, a fortiori, if the generalization affects the operational significance of the original parameters.

19.2 Conditional maximum likelihood estimators for $(\delta_1, \ldots, \delta_m)$

This section contains, for the case $H = \Psi$, a condition on the sequence of abilities which implies consistency of CML estimators. If this condition is violated, consistent estimators do not exist at all (Theorem 1). The consistency result applies, in particular, if the abilities are a random sample (Theorem 2). If the distribution of abilities is completely unknown, CML estimators are asymptotically efficient (Theorem 3). Proposition 1 establishes that the method of CML estimators applies for $H = \Psi$ only.

Consider now the following more general problem: We have independent observations $x_1, \ldots, x_n$, with x_j distributed according to P_{ϑ,η_j} over some measurable space $(X, \mathcal{A})$. The problem is to estimate ϑ.

If applied statisticians think of estimators, they usually think of ML (maximum likelihood) estimators. ML estimators are known to have certain asymptotic optimum properties, but these results are restricted to the case where the nuisance parameters $\eta_1, \ldots, \eta_n$ are known, or unknown but identical. In our case, the nuisance parameters are the abilities of the subjects, and these are neither known nor identical.

With a nuisance parameter which varies from observation to observation, the

ML estimator is generally biased. This is also true in the special model (19.1) with $H = \Psi$. For two difficulty parameters ($m = 1$ in our notation) the sequence of ML estimators for δ_1 converges to $\delta_1/2$ (if it converges at all; see Andersen, 1973a, pp. 66–69, Example 2.6). As m, the number of items, increases, the bias presumably becomes smaller and smaller, perhaps at a rate $1/m$ (according to some empirical evidence mentioned in Wright and Douglas, 1977, p. 285), but nothing precise seems to be known. However, this matters little, since, in the case $H = \Psi$, CML estimators are available.

Consider now a family of probability measures $P_{\vartheta,\eta}$ with $\vartheta \in \Theta \subset \mathbf{R}^m$ and $\eta \in Y$ (some abstract parameter space), where the density of $P_{\vartheta,\eta}$, say $p(\cdot,\vartheta,\eta)$, can be written as

$$p(x,\vartheta,\eta) = q(x,\vartheta)p_0(S(x),\vartheta,\eta) \tag{19.4}$$

(i.e., there is a statistic S, not depending on ϑ, which is sufficient for each subfamily $\{P_{\vartheta,\eta} : \eta \in Y\}$). A representation of this type is not unique. However, if such a representation is at all possible, then p_0 can always be chosen to be a density of the distribution of S under $P_{\vartheta,\eta}$. For this "canonical" representation, one has

$$\int \frac{q(x,\vartheta)}{q(x,\vartheta_0)} P_{\vartheta_0,\eta}(dx) = 1 \qquad \text{for all } \vartheta \in \Theta,\ \eta \in Y.$$

Therefore, the function $\vartheta \to \int \log q(x,\vartheta) P_{\vartheta_0,\eta}(dx)$ attains its maximum for $\vartheta = \vartheta_0$. This is the ultimate reason why CML estimators, defined by maximization of $\vartheta \to \prod_1^n \log q(x_\nu,\vartheta)$, are consistent under certain regularity conditions.

The nuisance parameters do not show in the computation of the CML estimator. They play, however, a role for its consistency and its asymptotic distribution.

For $P_{\vartheta,\eta} = \Psi^*_{(\delta_1,\ldots,\delta_m),\alpha}$ (see (19.2)), the canonical factorization, now with probabilities rather than densities, is

$$\begin{aligned}\Psi^*_{(\delta_1,\ldots,\delta_m),\alpha}&(x_0,x_1,\ldots,x_m) = \\ &= Q_{(\delta_1,\ldots,\delta_m)}(x_0,x_1,\ldots,x_m)\hat{P}_{(\delta_1,\ldots,\delta_m),\alpha}\{\sum_{i=0}^m x_i\}.\end{aligned} \tag{19.5}$$

In this representation, $\hat{P}_{(\delta_1,\ldots,\delta_m),\alpha}\{k\}$ is the probability of $\sum_0^m x_i = k$, given by

$$\hat{P}_{(\delta_1,\ldots,\delta_m),\alpha}\{k\} = \Delta(\delta_1,\ldots,\delta_m;k)\exp[\alpha k] / \prod_{i=0}^m (1+\exp[\alpha-\delta_i]), \tag{19.6}$$

with

$$\Delta(\delta_1,\ldots,\delta_m;k) := \sum \exp[-\sum_{i=1}^m \delta_i y_i],$$

where the summation extends over all $(1+m)$–tuples $(y_0, y_1, \ldots, y_m) \in \{0,1\}^{1+m}$ with $\sum_0^m y_i = k$, and

$$Q_{(\delta_1,\ldots,\delta_m)}(x_0, x_1, \ldots, x_m) := \frac{1}{\Delta(\delta_1, \ldots, \delta_m; \sum_{i=0}^{m} x_i)} \exp\left[-\sum_{i=1}^{m} \delta_i x_i\right], \tag{19.7}$$

the conditional probability of $(x_0, x_1, \ldots, x_m)$, given $\sum_0^m x_i$.

If $\boldsymbol{x}_j = (x_{j0}, x_{j1}, \ldots, x_{jm}) \in \mathbf{R}^{1+m}$ denotes the response vector of subject j, the CML estimator, say $(\delta_1^{(n)}, \ldots, \delta_m^{(n)})$, is based on the maximization of

$$(\delta_1, \ldots, \delta_m) \rightarrow \prod_{j=1}^{n} Q_{(\delta_1,\ldots,\delta_m)}(x_{j0}, x_{j1}, \ldots, x_{jm}). \tag{19.8}$$

Remark. The standardization $\delta_0 = 0$ was arbitrary. It should, therefore, be mentioned that changing this standardization has no other effect than a common shift of the estimators. To allow for values $\delta_0 \neq 0$, we have to change our notation from $Q_{(\delta_1,\ldots,\delta_m)}$ to $Q_{(\delta_0,\delta_1,\ldots,\delta_m)}$ (where the summation in (19.7) and in the definition of $\Delta(\delta_1, \ldots, \delta_m; k)$ now extends from 0 to m). What is, in fact, estimable, are the differences $\delta_i - \delta_j$, $i, j = 0, 1, \ldots, m$. If we change our standardization, the differences between the CML estimators remain unchanged. This is an immediate consequence of

$$Q_{(\delta_0+c,\delta_1+c,\ldots,\delta_m+c)} = Q_{(\delta_0,\delta_1,\ldots,\delta_m)} \quad \text{for every } c \in \mathbf{R}.$$

Theorem 1: *(i) The CML estimator is strongly consistent, i.e.*

$$\lim_{n\to\infty} \delta_i^n(\boldsymbol{x}_1, \ldots, \boldsymbol{x}_n) = \delta_i \qquad \textit{for } i = 1, \ldots, m,$$

except for a set of sequences $(\boldsymbol{x}_j)_{j\in\mathbf{N}}$ *of* $\bigtimes_{j=1}^{\infty} \Psi^*_{(\delta_1,\ldots,\delta_m),\alpha_j}$*–measure zero if*

$$\sum_{j=1}^{\infty} \exp[-\mid \alpha_j \mid] = \infty. \tag{19.9}$$

(ii) If condition (19.9) fails, not even weakly consistent estimators for $(\delta_1, \ldots, \delta_m)$ exist.

Recall that the probability measure occurring in the definition of "strong consistency" is a probability measure over the infinite product space $X^{\mathbf{N}}$, describing the distribution of sequences consisting of infinitely many stochastically independent realizations $\boldsymbol{x}_j$, $j \in \mathbf{N}$. In the present context, the distribution of $\boldsymbol{x}_j$ is $\Psi^*_{(\delta_1,\ldots,\delta_m),\alpha_j}$, the product measure therefore $\bigtimes_{j=1}^{\infty} \Psi^*_{(\delta_1,\ldots,\delta_m),\alpha_j}$. If the probability measure is the same for every j (such as $\Psi^*_{(\delta_1,\ldots,\delta_m),M}$ in Theorem 2), the infinite product is denoted by $(\Psi^*_{(\delta_1,\ldots,\delta_m),M})^{\mathbb{N}}$.

Let D denote the set of all $(1+m)$-tuples in $\{0,1\}^{1+m}$, except for $(0,0,\ldots,0)$ and $(1,1,\ldots,1)$. Since $Q_{(\delta_1,\ldots,\delta_m)}(0,0,\ldots,0) = 1$ and $Q_{(\delta_1,\ldots,\delta_m)}(1,1,\ldots,1) = 1$, for every $\delta \in \mathbb{R}^m$, these two $(1+m)$–tuples contribute nothing to the estimation of $(\delta_1,\ldots,\delta_m)$. They occur in (19.8) as the factor 1. Condition (19.9) for the consistency of CML estimators is satisfactory in so far as it is equivalent to the condition that $\boldsymbol{x}_j \in D$ infinitely often for all sequences $\boldsymbol{x}_j \in \{0,1\}^{1+m}$, $j \in \mathbb{N}$, except for a set of measure zero under $\times_{j=1}^{\infty} \Psi^*_{(\delta_1,\ldots,\delta_m),\alpha_j}$. (See the Corollary in Pfanzagl, 1992.) If $\sum_1^\infty \exp[-|\alpha_j|] < \infty$, then (19.8) contains, as n tends to infinity, only a finite number of factors different from 1.

Since the standardization of the scale for α assigns difficulties $\delta_0 = 0$ to an *arbitrary* standard item, it ought to be mentioned that

$$\sum_{j=1}^{\infty} \exp[-|\alpha_j + c|] = \infty \quad \text{for } \textit{some } c \in \mathbb{R}$$

implies the same relation for *all* $c \in \mathbb{R}$. (This follows immediately from $|\alpha_j + c'| \leq |\alpha_j + c| + |c' - c|$.)

Part (i) of Theorem 1 is due to Hillgruber (1990, p. 35, Satz 2.3.26). It is obtained as an application of a more general result in Pfanzagl (1992), which also contains a proof of part (ii). Less satisfactory results have earlier been stated by Andersen; in (1973a, p. 159, Theorem 5.1) he asserts consistency under the condition that $\sup_{j\in\mathbb{N}} |\alpha_j| < \infty$, in (1973b, p. 128, Theorem 2) under the condition that $\sum_{i=0}^m x_{ji}$ attains each of the values $1,\ldots,m$ for infinitely many subjects $j \in \mathbb{N}$; in (1980, p. 247, Theorem 6.2) under the condition that $\delta_i \neq \delta_j \neq 0$ for $i \neq j \neq 0$ and that $\sum_{i=0}^m x_{ji}$ is in $\{1,\ldots,m\}$ (i.e. $\boldsymbol{x}_j \in D$) for infinitely many subjects $j \in \mathbb{N}$.

These conditions of Andersen, involving the infinite sequence of response vectors $\boldsymbol{x}_j$, $j \in \mathbb{N}$, are hard to interpret unless one knows that these conditions hold true for all sequences $\boldsymbol{x}_j$, $j \in \mathbb{N}$, except for a set of measure zero under $\times_{j=1}^{\infty} \Psi^*_{(\delta_1,\ldots,\delta_m),\alpha_j}$. This is the case if and only if condition (19.9) is fulfilled.

Moreover, Andersen's "consistency" results are on solutions to the conditional likelihood equation, not on the CML estimator. They can be read as results on CML estimators only if the solution of the conditional likelihood equation is unique. Andersen's condition for this uniqueness (1980, p. 248, Theorem 6.3) is not correct (see Fischer, 1981, p. 60). The (necessary and sufficient) condition for uniqueness, discovered by Fischer (1981, pp. 66–69, Theorems 3 and 4) is, however, fulfilled if n is sufficiently large for all sequences $\boldsymbol{x}_j$, $j \in \mathbb{N}$, except for a set of measure zero under $\times_{j=1}^{\infty} \Psi^*_{(\delta_1,\ldots,\delta_m),\alpha_j}$, provided condition (19.10) holds true.

A consistent sequence of CML estimators will be asymptotically normal under the usual regularity conditions. Nothing can be said, however, about any kind of asymptotic optimality of the CML estimator if α_j, $j \in \mathbb{N}$, is just a sequence of unknown numbers satisfying condition (19.9). The principal difficulty lies in the absence of any meaningful bounds for models with infinitely many unknown nuisance parameters. (For a discussion of this problem, see Pfanzagl, 1990b.)

The situation is different if the unknown nuisance parameters are, in fact, random; more precisely, if they are realizations from some (unknown) distribution.

Many statisticians are of the opinion that such an assumption is legitimate under general conditions. As a typical example we cite Hambleton and Swaminathan (1985, p. 29). "Suppose ... we are willing to make the assumption that the information regarding the ability of any one examinee is not different from that of any other examinee; i.e., the information on examinees is *exchangeable.* This assumption implies that the abilities ... may be considered as a random sample from a population." While the validity of this argument is questionable for sequences of nuisance parameters in general, it is not unreasonable to assume that the abilities are a random sample governed by some unknown distribution. (Making the same assumption for the item difficulties would be hard to justify, though.)

Going back to our general model (19.4): If the nuisance parameters η_j, $j = 1, 2, \ldots$, are realizations from some unknown distribution M over Y, the observations $x_1, \ldots, x_n$, being independent realizations from $P_{\vartheta,\eta_1}, \ldots, P_{\vartheta,\eta_n}$, can now be considered as realizations from one and the same probability measure $P_{\vartheta,M}$, defined by

$$P_{\vartheta,M}(A) := \int P_{\vartheta,\eta}(A)M(d\eta), \qquad A \in \mathcal{A}.$$

Here we presume that $(Y, \mathcal{B})$ is a measurable space and that $\eta \to P_{\vartheta,\eta}(A)$ is measurable for $A \in \mathcal{A}$ and $\vartheta \in \Theta$.

If this mixture model is accepted, the asymptotic theory for nonparametric families of probability measures (based on the concepts of "tangent space" and "canonical gradient") can be applied to obtain bounds for the asymptotic concentration of regular estimator–sequences for ϑ, and to verify that the estimator–sequence in question attains this bound asymptotically.

For the general model (19.4), the asymptotic efficiency of CML estimators is obtained in Pfanzagl and Wefelmeyer (1982, pp. 231–236). See also Pfanzagl (1990a, pp. 41/2, Remark 7.10). Under suitable regularity conditions, this holds true if the family of possible distributions of η is large, and the family of induced distributions $P_{\vartheta,\eta} * S$, $\eta \in Y$, is complete for every $\vartheta \in \Theta$.

These results on mixture models apply immediately to model (19.5) if we consider the abilities as a random sample from an unknown distribution, say $M|\mathbb{B}$, the Borel algebra over the real line, $\mathbb{R}$.

What about the consistency of CML estimators in the mixture model?

Condition (19.9) on the sequence $\alpha_j, j \in \mathbb{N}$, is certainly fulfilled if the sequence $|\alpha_j|, j \in \mathbb{N}$, is bounded. It also allows that, as n increases, more and more larger and larger values $|\alpha_j|$ occur — to a certain extent. That is also what happens if $\alpha_j, j \in \mathbb{N}$, is a sequence of independent realizations from some distribution M of abilities. Is the occurrence of larger and larger values $|\alpha_j|$, in this case, sufficiently moderate so that condition (19.9) will be satisfied, and under what conditions on M? The answer is quite satisfactory: Without any further condition on M, $M^{\mathbb{N}}$-almost all sequences $\alpha_j, j \in \mathbb{N}$, fulfill condition (19.9). This leads to the following

Theorem 2: *The CML estimators are strongly consistent in the mixture model, i.e.,*

$$\lim_{n\to\infty} \delta_i^{(n)}(\boldsymbol{x}_1, \ldots, \boldsymbol{x}_n) = \delta_i \qquad \text{for } i = 1, \ldots, m,$$

except for a set of sequences $(\boldsymbol{x}_j)_{j\in\mathbb{N}}$ *of* $(\Psi^*_{(\delta_1,\ldots,\delta_m),M})^{\mathbb{N}}$*-measure zero, for any probability measure* $M|\mathbf{B}$ *(where*

$$\Psi^*_{(\delta_1,\ldots,\delta_m),M}(x_0, x_1, \ldots, x_m) := \int \Psi^*_{(\delta_1,\ldots,\delta_m),\alpha}(x_0, x_1, \ldots, x_m) M(d\alpha). \;)$$

For a proof see the Corollary in Pfanzagl (1992).

Moreover, the family $\{\hat{P}_{(\delta_1,\ldots,\delta_m),\alpha} : \alpha \in \mathbf{R}\}$ is exponential, see (19.6), and therefore complete for every $(\delta_1, \ldots, \delta_m) \in \mathbf{R}^m$. Hence, the general result on the asymptotic efficiency of CML estimators for mixture models implies the following

Theorem 3: *Assume that the family $\mathcal{M}$ of possible prior distributions of α contains all probability measures over* $\mathbf{R}$ *with positive and continuous Lebesgue density. Then any sequence of CML estimators for δ is asymptotically optimal (in the sense that it is regular, and no other regular estimator-sequence can be more concentrated asymptotically about the true value $(\delta_1, \ldots, \delta_m)$).*

Observe that, in general, CML estimators are not asymptotically efficient any more if M is known to be a member of a certain *parametric* family $\mathcal{M}$.

The favorable results for CML estimators (in particular: their asymptotic efficiency) lead to the question whether there are other functions H, different from Ψ, for which the method of CML applies. Regrettably, this is not the case. The optimality result refers to mixture models based on (19.4). The basic ingredient of (19.4) is a sufficient statistic S. In (19.5) we have seen that $\{H^*_{(\delta_1,\ldots,\delta_m),\alpha} : \alpha \in \mathbf{R}\}$ admits for $H = \Psi$ the sufficient statistic $S(x_0, x_1, \ldots, x_m) = \sum_0^m x_i$.

The method of CML requires some sufficient statistic for the family $\{H^*_{(\delta_1,\ldots,\delta_m),\alpha} : \alpha \in \mathbf{R}\}$; it does not require a special form of this statistic (like $\sum_0^m x_i$). The sufficient statistic, S, has to be nontrivial, of course. That means: it has to provide *some* reduction of the data, i.e., there have to be at least two distinct $(1+m)$–tuples $(x_0, x_1, \ldots, x_m)$ and $(x'_0, x'_1, \ldots, x'_m)$ in $\{0,1\}^{1+m}$ such that $S(x_0, x_1, \ldots, x_m) = S(x'_0, x'_1, \ldots, x'_m)$. This cannot be the case unless $H = \Psi$.

Proposition 1: *Assume that for some $m \in \mathbf{N}$ there exists a nontrivial statistic S which is sufficient for every family $\{H^*_{(\delta_1,\ldots,\delta_m),\alpha} : \alpha \in \mathbf{R}\}$, $(\delta_1, \ldots, \delta_m) \in \mathbf{R}^m$. Then, $H(u) = \Psi(au + c)$ for some $a > 0$, $c \in \mathbf{R}$. Moreover, $S(x_0, x_1, \ldots, x_m) = \sum_0^m x_i$.*

Proof: We write $\boldsymbol{x} = (x_0, x_1, \ldots, x_m)$, if convenient. Since S is defined on the finite set $\{0,1\}^{1+m}$, we may assume w.l.g. that S maps $\{0,1\}^{1+m}$ into $\mathbf{R}$. If S is sufficient for $\{H^*_{(\delta_1,\ldots,\delta_m),\alpha} : \alpha \in \mathbf{R}\}$, the factorization theorem implies the existence of functions q and f such that, for $\boldsymbol{x} \in \{0,1\}^{1+m}$, $\alpha \in \mathbf{R}$, and $\delta \in \mathbf{R}^m$,

$$\prod_{i=0}^{m} H(\alpha - \delta_i)^{x_i} \big(1 - H(\alpha - \delta_i)\big)^{1-x_i}$$
$$= q\big(\boldsymbol{x}, (\delta_1, \ldots, \delta_m)\big) f\big(S(\boldsymbol{x}), (\delta_1, \ldots, \delta_m), \alpha\big).$$

Introducing $L(u) = H(u)/\big(1 - H(u)\big)$ this may be rewritten as

$$\prod_{i=0}^{m}\big(1 - H(\alpha - \delta_i)\big) \prod_{i=0}^{m} L(\alpha - \delta_i)^{x_i} = q(\boldsymbol{x}, (\delta_1, \ldots, \delta_m)) f\big(S(\boldsymbol{x}), (\delta_1, \ldots, \delta_m), \alpha\big).$$

If there exist $\boldsymbol{x}' \neq \boldsymbol{x}$ in $\{0,1\}^{1+m}$ with $S(\boldsymbol{x}') = S(\boldsymbol{x})$, we obtain

$$\prod_{i=0}^{m} \frac{L(\alpha - \delta_i)^{x'_i}}{L(\alpha - \delta_i)^{x_i}} = \frac{q(\boldsymbol{x}', (\delta_1, \ldots, \delta_m))}{q(\boldsymbol{x}, (\delta_1, \ldots, \delta_m))} \tag{19.10}$$

for all $\alpha \in \mathbf{R}$ and $\delta \in \mathbf{R}^m$.

Let $k := \sum_0^m \big(1_{\{1\}}(x_i) - 1_{\{1\}}(x'_i)\big)$. Assume w.l.g. that $x'_1 = 1$, $x_1 = 0$. Relation (19.10), applied with $\delta_1 = \delta \in \mathbf{R}$ and $\delta_i = 0$ for $i \neq 1$, yields

$$L(\alpha - \delta) = \frac{q(\boldsymbol{x}', (\delta, 0, \ldots, 0))}{q(\boldsymbol{x}, (\delta, 0, \ldots, 0))} L(\alpha)^{k+1} \qquad \text{for } \alpha, \delta \in \mathbf{R}. \tag{19.11}$$

Let $\lambda(u) := L(u)/L(0)$. Using relation (19.10) for arbitrary α and for $\alpha = 0$, we obtain

$$\lambda(\alpha - \delta) = \lambda(-\delta)\lambda(\alpha)^{k+1}, \qquad \text{for } \alpha, \delta \in \mathbf{R}.$$

Applied with $\delta = 0$ this yields $k = 0$, hence

$$\lambda(\alpha - \delta) = \lambda(-\delta)\lambda(\alpha).$$

Since λ is increasing, this implies (see Aczél, 1966, p. 38, Theorem 1) $\lambda(u) = \exp[au]$, hence $L(u) = L(0)\exp[au]$, with $a > 0$. This proves the assertion. □

The question about the existence of a nontrivial sufficient statistic for the family $\{H^*_{(\delta_1,\dots,\delta_m),\alpha} : \alpha \in \mathbf{R}\}$ must not be mixed up with the following different question: Assume that for item i there is an *item characteristic curve* $\alpha \to \pi_i(\alpha)$ which states the probability of response "1" by a subject with ability α. In case of local independence, the response vector $(x_1, \dots, x_m)$ has probability $\prod_1^m \pi_i(\alpha)^{x_i}(1-\pi_i(\alpha))^{1-x_i}$. If $\pi_i(\alpha) = \Psi(\alpha - \delta_i)$ (or, more generally, $\Psi(m(\alpha) - \delta_i)$ with some increasing function m), the statistic $\sum_1^m x_i$ is sufficient for the family of probability measures

$$\Big\{(x_1,\dots,x_m) \to \prod_{i=1}^m \pi_i(\alpha)^{x_i}\big(1-\pi_i(\alpha)\big)^{1-x_i} : \alpha \in \mathbf{R}\Big\}. \tag{19.12}$$

Starting with Birnbaum (1968), much effort has been devoted to the question whether the existence of a nontrivial sufficient statistic characterizes the logistic model. The most satisfying result seems to be that of Fischer (1968, pp. 95–100, with a valid proof given in 1974, pp. 193–203): If $\sum_1^m x_i$ is sufficient for the family (19.12), then $\pi_i(\alpha) = \Psi(m(\alpha) - \delta_i)$; in particular: the item difficulty can be expressed by a single number, δ_i. Our Proposition 1 starts from the assumption $\pi_i(\alpha) = H(m(\alpha) - \delta_i)$, or $\pi_i(\alpha) = H(\alpha - \delta_i)$, without loss of generality. Thanks to this more special assumption, a weaker condition on the sufficient statistic ("S nontrivial" instead of "$S(x_1,\dots,x_m) = \sum_1^m x_i$") is enough to deduce $H = \Psi$.

19.3 Marginal maximum likelihood estimators for $(\delta_1, \dots, \delta_m)$

This section recalls that, in the case $H = \Psi$, MML estimators for item difficulty parameters are asymptotically equivalent to CML estimators if the distribution of abilities is totally unknown. Consequently, they share the favorable asymptotic properties of CML estimators. In the case $H = \Phi$, the difficulties are not identifiable if the distribution of abilities is totally unknown. Hence they are not estimable at all.

Proposition 1 in Section 19.2 makes clear that, among mixture models based on (19.1), only the one with $H = \Psi$ lends itself to the method of CML estimators. If the use of other functions H is intended, other estimation methods are required. Such methods are available for the mixture model.

Many statisticians are willing to presume that the unknown prior distribution M belongs to a certain parametric family, usually the location and scale family of normal distributions. With M from a parametric family, we remain with the model $H^*_{(\delta_1,\dots,\delta_m),M}$ in the realm of parametric families, a case for which the usual asymptotic optimality results for ML estimators apply.

With modern computing facilities available, estimating $(\delta_1, \ldots, \delta_m)$ under less stringent assumptions on $\mathcal{M}$, the family of possible prior distributions, has become feasible. Among the estimation methods available for mixture models is "MML" and "minimum distance". However: If we admit a large family of mixing distributions $\mathcal{M}$ (say all distributions on $\mathbb{B}$ with positive and continuous Lebesgue density), a problem occurs which seems to have been overlooked hitherto: that of the identifiability of $(\delta_1, \ldots, \delta_m)$. This means that $(\delta'_1, \ldots, \delta'_m) \neq (\delta_1, \ldots, \delta_m)$ implies $H^*_{(\delta'_1,\ldots,\delta'_m),M'} \neq H^*_{(\delta_1,\ldots,\delta_m),M}$ for all $M', M \in \mathcal{M}$. To be sure, this condition does not necessarily imply that $(\delta_1, \ldots, \delta_m)$ is consistently estimable from an increasing number of observations from $H^*_{(\delta_1,\ldots,\delta_m),M}$. Identifiability is, however, an indispensable requirement for consistent estimability by any method.

The question whether $(\delta_1, \ldots, \delta_m)$ is identifiable from $H^*_{(\delta_1,\ldots,\delta_m),M}$, if M belongs to some large family $\mathcal{M}$, seems far from trivial.

Identifiability of $(\delta_1, \ldots, \delta_m)$ for some $m = m_0$ implies identifiability for all $m > m_0$. Hence, the question is: What is the smallest m for which $\delta \in \mathbb{R}^m$ are identifiable? If $H = \Psi$, identifiability starts with $m = 1$. Regrettably, this is the only case with this desirable property: If every $\delta \in \mathbb{R}$, $\delta \neq 0$, is identifiable from $H^*_{\delta,M}$, then $H = \Psi$ (see Pfanzagl, 1991a, Theorem 5). This result is not yet disastrous: Identifiability starting with some reasonably small m would suffice for practical purposes. A question like this cannot be answered simultaneously for all $H \neq \Psi$. Hence we turn our attention to those functions H which are considered in the literature as alternatives to $H = \Psi$.

This is primarily $H = \Phi$. Starting, perhaps, with Thurstone's (1927) "case V" for laws of comparative judgement, the use of Φ was suggested by several authors. Birnbaum (1968, p. 400) considers Ψ "as a close approximation to ... Φ, introduced to help solve or to avoid mathematical or theoretical problems that arrive with Φ." Most of these authors think of using for $\mathcal{M}$ a parametric family, but not all of them. Bock and Aitkin (1981, p. 444, p. 449) suggest "to free the method from arbitrary assumptions about the distribution of ability" and estimate $(\delta_1, \ldots, \delta_m)$ from $\Phi^*_{(\delta_1,\ldots,\delta_m),M}$ by using an EM-algorithm for the estimation of M. The bad news is that — except for a set of Lebesgue measure zero — the difficulty parameters $(\delta_1, \ldots, \delta_m)$ are not identifiable from $H^*_{(\delta_1,\ldots,\delta_m),M}$, however large m may be, if $H = \Phi$. (See Pfanzagl, 1991a, Theorem 2).

A comparable non-identifiability result holds for $H(u) = \exp[-\exp[-u]]$, a special extreme value–distribution suggested in Falmagne (1985, pp. 137ff.).

It is, of course, the size of the family $\mathcal{M}$ which is responsible for the non-identifiability of $(\delta_1, \ldots, \delta_m)$. If $\mathcal{M}$ is a location and scale family, then, even for $m = 1$, any $\delta \neq 0$ is identifiable from $H^*_{(\delta_1,\ldots,\delta_m),M}$, whatever the function H. (See Pfanzagl, 1991a, Theorem 3.)

The question remains whether MML estimators offer any advantage — as compared to CML estimators — in cases where both methods apply, i.e., if $H = \Psi$. This is certainly true if M is known to belong to a parametric family: In this case, $\Psi^*_{(\delta_1,\ldots,\delta_m),M}$ is a member of a parametric family; the MML estimator for $(\delta_1,\ldots,\delta_m)$ is a component of the ML estimator in a parametric family and therefore asymptotically efficient under the usual regularity conditions. The CML estimator, being asymptotically efficient if $\mathcal{M}$ is a nonparametric family, will be asymptotically inefficient, in general, if applied in a case where M is known to be a member of a certain parametric family.

If some authors (like Bock and Aitkin, 1981, p. 454, or Thissen, 1982, p. 184) observe in numerical examples a close agreement between CML estimators and MML estimators based on a normal prior distribution, this does not prove the contrary. To evaluate the difference between two types of estimators requires a more refined investigation with simulated data.

As stated in Theorem 3, the CML estimator is asymptotically efficient if $\mathcal{M}$ is nonparametric. Hence, in this case, the MML estimator cannot be better, asymptotically. In an important paper, de Leeuw and Verhelst (1986, pp. 190/2) establish the asymptotic equivalence of CML and MML estimators in the case that $\mathcal{M}$ is nonparametric. This does not just mean that the difference between the two estimator–sequences converges to 0. It means that the two estimators are identical if the sample size is sufficiently large, for all sequences $\boldsymbol{x}_j$, $j \in \mathbb{N}$, except for a set of measure zero under $(\Psi^*_{(\delta_1,\ldots,\delta_m),M})^{\mathbb{N}}$. (See Pfanzagl, 1991b, for a more elementary proof.)

What are the consequences of this theorem?

(i) Since CML estimators are strongly consistent according to Theorem 2 without any condition on the mixing distribution M, the same holds true for MML estimators.

Consistency of MML estimators was proved by Follman in his thesis (see Follman, 1988, p. 558) under the assumption that $|\delta_i| \le k$ for $i = 1,\ldots,m$, and m is odd.

(ii) Since CML estimators are asymptotically efficient according to Theorem 3, if the mixing distribution is an unknown distribution with positive and continuous density, the same holds true for MML estimators.

Strangely enough, de Leeuw and Verhelst use their beautiful result to argue (see p. 193) that, because of this asymptotic equivalence, CML estimators are asymptotically efficient. This argument obviously takes for granted that MML estimators *are* asymptotically efficient, a result which, to my knowledge, has never been established for nonparametric mixture models.

For the purpose of estimating $(\delta_1,\ldots,\delta_m)$ under the assumption of the Rasch model, one may, therefore, choose between CML and MML estimators according to convenience. This choice will, of course, also depend on other intended uses of the data. Moreover, this equivalence result is an asymptotic one. Even if one is confident that this "asymptotic" equivalence is an "approximate" equivalence for the usual sample sizes, it remains to be investigated whether it persists under small deviations from the Rasch model.

Computing the MML estimator means maximizing

$$((\delta_1,\ldots,\delta_m),M) \to \prod_{j=1}^{n} H^*_{(\delta_1,\ldots,\delta_m),M}(\boldsymbol{x}_j).$$

Maximizing probability measures M exist under weak conditions, but they are not unique if $\mathcal{M}$ is a nonparametric family. In other words, even if $(\delta_1,\ldots,\delta_m)$ is identifiable from $H^*_{(\delta_1,\ldots,\delta_m),M}$, the probability measure M, in general, is not. This is intuitively clear: With a small number of items, the distribution of abilities, M, expresses itself in $H^*_{(\delta_1,\ldots,\delta_m),M}$ only to some extent (the larger m is, the better).

If the sufficient statistic S in (19.4) assumes only a finite number of values, then, for any ϑ_0, M_0, there exists a probability measure M with finite support such that $P_{\vartheta_0,M} = P_{\vartheta_0,M_0}$. In the case $P_{\vartheta,M} = \Psi^*_{(\delta_1,\ldots,\delta_m),M}$, $(m+1)/2$ supporting points are enough. If the non-identifiability prevents us from estimating M, this non-identifiability is of advantage if we are interested in $(\delta_1,\ldots,\delta_m)$ only: Among the distributions M representing the same probability measure $\Psi^*_{(\delta_1,\ldots,\delta_m),M}$ we may choose the simplest one. For practical purposes, a distribution M with much less than $(m+1)/2$ points may suffice (see Lindsay et al., 1991).

19.4 References

Aczél, J. (1966). *Lectures on functional equations and their applications.* New York: Academic Press.

Andersen, E.B. (1973a). *Conditional inference and models for measuring.* Copenhagen: Mentalhygiejnisk Forskningsforlag.

Andersen, E.B. (1973b). A goodness of fit test for the Rasch model. *Psychometrika, 38,* 123–140.

Andersen, E.B. (1980). *Discrete statistical models with social science applications.* Amsterdam: North–Holland.

Balakrishnan, N. (Ed.) (1992). *Handbook of the logistic distribution.* New York: Marcel Dekker.

Birnbaum, A. (1968). Some latent trait models and their use in inferring an examinee's ability. In F.M. Lord & M.R. Novick, *Statistical theories of mental test scores* (pp. 397-479). Reading, MA: Addison–Wesley.

Bock, R.D., & Aitkin, M. (1981). Marginal maximum likelihood estimation of item parameters: Application of an EM algorithm. *Psychometrika, 46,* 443–459.

Chambers, E.A., & Cox, D.R. (1967). Discrimination between alternative binary response models. *Biometrika, 54,* 573–578.

de Leeuw, J., & Verhelst, N. (1986). Maximum likelihood estimation in generalized Rasch models. *Journal of Educational Statistics, 11,* 183–196.

Falmagne, J.C. (1985). *Elements of psychophysical theory.* Oxford Psychology Series Vol. 6. Oxford: Clarendon Press.

Feller, W. (1939/41). On the logistic law of growth and its empirical verifications in biology. *Acta Biotheoretica, A, 5,* 51–66.

Fischer, G.H. (1968). *Psychologische Testtheorie.* Bern: Huber.

Fischer, G.H. (1974). *Einführung in die Theorie psychologischer Tests.* Bern: Huber.

Fischer, G.H. (1981). On the existence and uniqueness of maximum likelihood estimates in the Rasch model. *Psychometrika, 46,* 59–77.

Fischer, G.H. (1988). Spezifische Objektivität: Eine wissenschaftstheoretische Grundlage des Rasch-Modells. In K.D. Kubinger (Ed.), *Moderne Testtheorie* (pp. 87-111). Weinheim und München: Psychologie Verlags Union.

Follman, D. (1988). Consistent estimation in the Rasch model based on nonparametric margins. *Psychometrika, 53,* 553–562.

Gigerenzer, G. (1981). *Messung und Modellbildung in der Psychologie.* München: Ernst Reinhardt.

Hambleton, R.K., & Swaminathan, H. (1985). *Item response theory.* Boston: Kluwer and Nijhoff.

Hillgruber, G. (1990). Schätzung von Parametern in psychologischen Testmodellen. Master's thesis. Cologne: Department of Mathematics, University of Cologne.

Lindsay, B., Clogg, C.C., & Grego, J. (1991). Semiparametric estimation in the Rasch model and related exponential response models, including a simple latent class model for item analysis. *Journal of the American Statistical Association, 86,* 96–107.

Pfanzagl, J. (1971). *Theory of measurement* (2nd ed.). New York: Wiley.

Pfanzagl, J. (1990a). *Estimation in semiparametric models. Some recent developments.* Lecture Notes in Statistics, Vol. 63. New York: Springer–Verlag.

Pfanzagl, J. (1990b). *Incidental versus random nuisance parameters.* Preprints in Statistics 128. Cologne: Department of Mathematics, University of Cologne. (To appear in: *Annals of Statistics.*)

Pfanzagl, J. (1991a). *On the identifiability of structural parameters in mixtures; applications to psychological tests.* Preprints in Statistics 130. Cologne: Department of Mathematics, University of Cologne. (To appear in: *Journal of Statistical Planning and Inference.*)

Pfanzagl, J. (1991b). *A case of asymptotic equivalence between conditional and marginal maximum likelihood estimators.* Preprints in Statistics 132. Cologne: Department of Mathematics, University of Cologne. (To appear in: *Journal of Statistical Planning and Inference.*)

Pfanzagl, J. (1992). *On the consistency of conditional maximum likelihood estimators.* Preprints in Statistics 134. Cologne: Department of Mathematics, University of Cologne. (To appear in: *Annals of the Institute of Statistical Mathematics.*)

Pfanzagl, J., & Wefelmeyer, W. (1982). *Contributions to a general asymptotic statistical theory.* Lecture Notes in Statistics, Vol. 13. New York: Springer–Verlag.

Roskam, E.E., & Jansen, P.G.W. (1984). A new derivation of the Rasch model. In E. Degreef & J. van Buggenhaut, (Eds.), *Trends in mathematical psychology* (pp. 293-307). Amsterdam: North–Holland.

Thissen, D. (1982). Marginal maximum likelihood estimation for the one–parameter logistic model. *Psychometrika, 47,* 175–186.

Thurstone, L.L. (1927). A law of comparative judgement. *Psychological Review, 34,* 273–286.

Wright, B.D., & Douglas, G.A. (1977). Best procedures for sample–free item analysis. *Applied Psychological Measurement, 1,* 281–294.

20

The Uniqueness Structure of Simple Latent Trait Models

Hans Irtel[1]

ABSTRACT A latent trait system is a set of subjects A, a set of items X, and a response function r, mapping $A \times X$ into the real numbers. Numerical representations of such a system map A and X into the reals, such that r is represented by a numerical operation. It is shown for an additive latent trait system that its internal structure may be characterized by its automorphism group and that homogeneity and uniqueness of this group make the system ratio scalable. A non-additive case is also considered. Here the two factors are combined in a non-additive way, but the system's internal structure induces an independent system on one of the two factors which is interval-scalable.

20.1 Uniqueness Problems in Latent Trait Systems

Latent trait models have been introduced by Rasch (1960) and Birnbaum (1968) in order to improve the theoretical foundations of psychometric measurement. A consequence is that these models may be analyzed by measurement-theoretic methods which rigorously specify a set of sufficient conditions for the existence of the parameters and their uniqueness properties. Unfortunately, there are three types of uniqueness problems involved which have been confused in the past. The first one is based on the fact that most measurement models do not define the homomorphisms which map the empirical structure they deal with into a given numerical structure uniquely. This is what is usually called the *uniqueness problem* in measurement theory and its solution is the set of admissible transformations of a scale (Suppes and Zinnes, 1963; Krantz, Luce, Suppes, and Tversky, 1971). The second type of uniqueness question is how different but equally suitable numerical models for one and the same empirical structure are related. This question is of less importance to measurement theory since the essential fact about any scale is that all suitable numerical representations are isomorphic, as also are the corresponding sets of admissible transformations (Krantz et al., 1971; Colonius, 1979). A measurement-theoretic analysis of latent trait models is usually based on response probabilities and this creates a third type of uniqueness problem which is related to the statistical problem of how to estimate the response probabilities and how the uniqueness of the estimates depends on properties of a finite set of response data. This problem has been dealt with by Rasch (1960), Andersen (1973), and Fischer (1981).

[1]University Regensburg, Department of Psychology, 8400 Regensburg, Germany

Statistical properties of a latent trait model, like the existence of sufficient statistics for its parameters, affect the statistical parameter estimation problem, but do not affect its scale type. The scale type of a measurement model can be thought of as the limit of over-all uniqueness that one can get with an ideal set of data, namely the response ***probabilities.***

This paper deals only with the first type of uniqueness question. Its main point is to analyze the uniqueness of latent trait systems by using structural properties of the response probabilities without referring to numerical representations. This avoids any confusion with the second type of uniqueness problem mentioned earlier. In fact, several authors have confused the lack of uniqueness of homomorphisms into a fixed numerical relational system with isomorphic mappings between different numerical relational systems (Colonius 1979; Wottawa 1980; Roskam 1983), although others have made rather clear that the second type of problem does not affect the scale type nor the set of meaningful statements which are possible with a given scale (Krantz et al. 1971).

Statistical estimation problems are not addressed here, which means that we start our analysis with a response function r which assigns a real valued number $r(a,x)$ to every pair (a,x) of subject and item. The response function is assumed to be fixed and derived from response probabilities, which do not have any nontrivial admissible transformations.

Definition 1. *Let A and X be nonempty sets and let r be a mapping from $A \times X$ into the set of real numbers $\mathbf{R}$. Then, $\mathcal{L} = \langle A \times X, r \rangle$ is called* a latent trait system.

We first consider only strictly additive latent trait systems. This means that the response function may be decomposed into independent components which are combined in an additive manner. These systems may be called *Rasch* type since they are essentially equivalent to what has become known as the *Rasch* model:

$$r(a,x) = \theta(a) - \epsilon(x). \tag{20.1}$$

The existence of such an additive decomposition imposes structural restrictions on the response function. These restrictions may be considered "empirical" properties of a latent trait system which, given the response function, are empirically testable (Hamerle and Tutz, 1980).

Definition 2. *A latent trait system $\mathcal{L} = \langle A \times X, r \rangle$ is called* strictly additive independent *iff, for all a, b in A and all x, y in X,*

$$r(a,x) - r(b,x) = r(a,y) - r(b,y). \tag{20.2}$$

A latent trait system is additive decomposable in the sense of (20.1) iff it is strictly additive independent (Hamerle and Tutz, 1980). Latent trait theory does not use strictly additive representations of response probabilities directly, but uses mapping functions which restrict their range to the interval $[0, 1]$. An example is the so called *logistic* function $L(x) = 1/(1 + e^{-x})$ which has been used by Rasch (1960). The representation for response probabilities then is

$$p(a, x) = \frac{1}{1 + e^{-(\theta(a)-\epsilon(x))}}. \tag{20.3}$$

Since here the mapping function is specified explicitly and is strictly monotone, one may invert it and get

$$\ln \frac{p(a, x)}{1 - p(a, x)} = \theta(a) - \epsilon(x).$$

Transforming the response probabilities p by the inverse of the logistic creates a response function which has an additive representation in the sense of (20.1). It is important to note that the transformation also affects the empirical condition (20.2). In terms of response probabilities condition (20.2) then becomes

$$\frac{p(a, x)}{1 - p(a, x)} \frac{1 - p(b, x)}{p(b, x)} = \frac{p(a, y)}{1 - p(a, y)} \frac{1 - p(b, y)}{p(b, y)}.$$

Pfanzagl (1971), Hamerle (1979), and Fischer (1988) look at a representation which is more general than (20.1):

$$r(a, x) = H[\theta(a) - \epsilon(x)], \tag{20.4}$$

where H is some strictly monotone increasing function with appropriate domain. Pfanzagl (1971) and Hamerle (1979) derive θ, ϵ, and H from purely ordinal constraints on r which essentially form an additive conjoint measurement structure (Luce and Tukey, 1964). This means that only ordinal information in r is used and that any strictly monotone transformation of r results in an equivalent representation. However, since the type of ordinal constraints used in additive conjoint measurement ("double cancellation" or "quadruple condition") is not suitable as a restriction in statistical parameter estimation, these models are hard to apply in situations where response probabilities cannot be estimated by relative frequencies.

Clearly, if H is invertible, then (20.4) implies the following generalization of (20.2),

$$G\{H^{-1}[r(a, x)], H^{-1}[r(b, x)]\} = G\{H^{-1}[r(a, y)], H^{-1}[r(b, y)]\}, \tag{20.5}$$

with $G(x, y) = x - y$. Fischer (1988) derives (20.4) from (20.5) for a more general class of functions G. He also shows that, if the response function values are stochastically independent response probabilities and if G is some function of the likelihood of a response event involving the subjects a, b and the item x, then

H must be the logistic. This also shows that, within the latent trait framework, (20.4) becomes empirically applicable only if the function H is fixed and known. The reason is simply that, for H unknown, (20.5) is not empirically testable, nor is it useful as a side condition in parameter estimation. An exception is $H(x) = cx + d$ (Suppes and Zinnes, 1963; Fischer, 1988) which makes (20.5) equivalent to (20.2), such that (20.4) and (20.1) are *data eqivalent* in the sense of Adams, Fagot, and Robinson (1965). The parameters c and d are not identifiable and thus may be set equal to 1 and 0, respectively, without loss of generality.

In the following we ignore all problems related to statistical parameter estimation and only look at latent trait systems with fixed and known response functions. These are treated as the empirical basis for latent trait models in the same way as fundamental measurement theory treats "empirical" relational systems as the empirical basis for measurement. With this restriction, the uniqueness properties that will be derived are properties that become effective only if an actual set of data comes sufficiently close to those properties of latent trait systems that will be required. In this sense, they describe the optimum degree of uniqueness one can get within a given latent trait system.

20.2 Intrinsic Uniqueness

The following presentation applies methods for characterizing the scale type of a measurement structure which have been developed by Narens (1985). These methods are not based on numerical representations but make it possible to characterize the scale type of a latent trait system by using only its intrinsic primitives.

Our analysis is based on a special type of *automorphisms* of a latent trait system $\mathcal{L} = \langle A \times X, r \rangle$, which are generally one-to-one mappings ζ from $A \times X$ onto itself such that $r[\zeta(a, x)] = r(a, x)$. This is similar to what Roberts and Rosenbaum (1988/89) call "tight value automorphisms" in the context of finite weak orderings with an additional binary relation ("valued digraphs"). This terminology is slightly different from the one used by Narens (1985). He looks at automorphisms which preserve relations like $r(a, x) \geq r(b, y)$ iff $r[\zeta(a, x)] \geq r[\zeta(b, y)]$. Since latent trait systems are based on real valued functions, we have to use value preserving automorphisms. We simply use the term *automorphism*, but keep in mind that these mappings not only preserve relations but values. We use the term *order automorphism* for denoting the conventional automorphisms which preserve relations.

The following concept has been introduced by Luce and Cohen (1983). They look at automorphisms of conjoint structures which strongly resemble latent trait systems. The difference is that conjoint structures have an empirical ordering relation as their basic primitive, while latent trait systems have a response function.

Definition 3. *Let $\mathcal{L} = \langle A\times X, r\rangle$ be a latent trait system. An automorphism ζ of $\mathcal{L}$ is* factorizable *iff there exist one-to-one functions α from A onto itself and ξ from X onto itself, such that, for all a in A and all x in X, $\zeta(a,x) = (\alpha(a), \xi(x))$. The functions α and ξ will be denoted* component transformations *induced by ζ.*

We use $\mathcal{M}$ to denote the set of all factorizable automorphisms of a latent trait system $\mathcal{L}$, and $\mathcal{M}_A$ and $\mathcal{M}_X$ for the induced component transformations on A and X, respectively.

Lemma 1. *Let $\mathcal{L} = \langle A\times X, r\rangle$ be a latent trait system, $\mathcal{M}$ its set of factorizable automorphisms, and $\mathcal{M}_A$ and $\mathcal{M}_X$ its respective sets of induced component transformations. Then, with function composition, $\mathcal{M}$, $\mathcal{M}_A$, and $\mathcal{M}_X$ are groups.*

Proof: The proof closely follows the proof of Lemma 1 of Luce and Cohen (1983). Let ι_A, ι_X, and $\iota = \langle \iota_A, \iota_X\rangle$ be the identity maps on A, X, and $A\times X$, respectively. Then, these are the identities of the three groups. Let $\langle\alpha,\xi\rangle$ be in $\mathcal{M}$, then $\langle\alpha,\xi\rangle^{-1} = \langle\alpha^{-1},\xi^{-1}\rangle$ is also in $\mathcal{M}$, since $r(\alpha^{-1}(a),\xi^{-1}(x)) \neq r(a,x)$ leads to the contradiction $r(\alpha^{-1}(\alpha(a)),\xi^{-1}(\xi(x))) \neq r(a,x)$. From the associativity of function composition, it follows that $\mathcal{M}$, $\mathcal{M}_A$, and $\mathcal{M}_X$ are groups.

In order to characterize the structural properties of a latent trait system, we use its automorphism group. Two properties have turned out to be especially useful: homogeneity and uniqueness.

Definition 4. *Let $\mathcal{L} = \langle A\times X, r\rangle$ be a latent trait system. Its group $\mathcal{M}$ of factorizable automorphisms is* 1/2-point homogenous on *A iff, for any a, b in A there exists a factorizable automorphism $\langle\alpha,\xi\rangle$ in $\mathcal{M}$ such that $\alpha(a) = b$. $\mathcal{M}$ is* 1/2-point unique on *A iff, for all a in A and for all factorizable automorphisms $\langle\alpha,\xi\rangle$, $\langle\alpha',\xi'\rangle$ in $\mathcal{M}$, the condition $\alpha(a) = \alpha'(a)$ implies that $\langle\alpha,\xi\rangle = \langle\alpha',\xi'\rangle$. Analogous definitions may be given for* 1/2-point homogenous on *X and* 1/2-point unique on *X. $\mathcal{L}$ will be called* 1/2-point homogenous *and* 1/2-point unique *iff the respective properties hold for both components simultanously.*

In order to use automorphisms to characterize the structural properties of latent trait systems, we have to make sure that they exist. This means that we have to assume infinite sets of subjects and items, since our systems are essentially ordered (Def. 7). One way to construct an automorphism on a latent trait system is by inverting the response function.

Definition 5. *A latent trait system $\mathcal{L} = \langle A\times X, r\rangle$ is called* invertible *iff, for every fixed x in X, the function $f_x(a) = r(a,x)$, and for every fixed a in A, the function $f_a(x) = r(a,x)$ are one-to-one functions from A and X, respectively, into $\mathbb{R}$ such that, for every s in $r(A\times X)$, the inverse $f_x^{-1}(s)$ is a unique element a in A, and the inverse $f_a^{-1}(s)$ is a unique element x in X.*

Theorem 1. *Let $\mathcal{L} = \langle A \times X, r \rangle$ be a latent trait system, and let $\mathcal{M}$ be its group of factorizable automorphisms. If $\mathcal{L}$ is invertible, then $\mathcal{M}$ is 1/2-point unique.*

Proof: Suppose both $\langle \alpha, \xi \rangle$ and $\langle \alpha', \xi' \rangle$ are in $\mathcal{M}$, and $\alpha(a) = \alpha'(a) = b$ for some a in A. Then, $r(b, \xi(x)) = r(b, \xi'(x))$, and it follows from invertibility that $\xi = \xi'$. By invertibility in A we then get $\alpha = \alpha'$, and thus $\langle \alpha, \xi \rangle = \langle \alpha', \xi' \rangle$. This shows 1/2-point uniqueness on A. 1/2-point uniqueness on X may be shown in the same way. □

This theorem shows that, if a latent trait system $\mathcal{L}$ is invertible, then its set of factorizable automorphisms is 1/2-point unique. This result is closely related to Theorem 12 of Luce and Cohen (1983). They show that the group of factorizable order automorphisms of conjoint structures which are order independent and unrestrictedly solvable either satisfies 1- or 2-point uniqueness (Def. 8).

Theorem 1 also shows that, for invertible systems, we have to look at the homogeneity properties of the automorphism groups. Note that the *degree of homogeneity (uniqueness)* of a group of automorphisms is the largest (smallest) n, such that it is n-point homogenous (unique), given such a n exists. If such a n does not exist, then the degree of homogeneity (uniqueness) is said to be infinite. Also, remember that for any group of automorphisms with a finite degree of uniqueness the degree of homogeneity is less than or equal to its degree of uniqueness (Narens, 1985).

Up to this point condition (20.2) has not been used. The study of additive systems means using real addition as an operation for combining the two factors of a latent trait system. A stronger version of invertibility is needed in order to define the corresponding automorphisms in the system.

Definition 6. *A latent trait system $\mathcal{L} = \langle A \times X, r \rangle$ is called* A-solvable *iff, for any fixed x in X and any real number s, there exists a unique a in A such that $r(a, x) = s$. X-solvability is defined in an analogous way. $\mathcal{L}$ is* solvable *iff it is A- and X-solvable.*

Theorem 2. *Let $\mathcal{L} = \langle A \times X, r \rangle$ be a solvable latent trait system which is strictly additive independent, and let $\mathcal{M}$ be its group of factorizable automorphisms. Then $\mathcal{M}$ is 1/2-point homogenous and 1/2-point unique.*

Proof: Let a' and b' be in A and define $\alpha(a') = b'$. Then, by strict additive independence,

$$k_\alpha = r(\alpha(a'), x) - r(a', x)$$

is independent of x. For all a in A define $\alpha(a)$ as the solution of

$$r(\alpha(a), x) - r(a, x) = k_\alpha,$$

which again is independent of x. Then define ξ on X as the solution of

$$r(a, \xi(x)) - r(a, x) = -k_\alpha,$$

for some a in A, and note that by strict additive independence ξ is independent of a. Then, $\langle \alpha, \xi \rangle$ is factorizable. It also is an automorphism of $\mathcal{L}$ because, by definition,

$$\begin{aligned} r(\alpha(a), \xi(x)) &= r(\alpha(a), x) - k_\alpha \\ &= (r(a, x) + k_\alpha) - k_\alpha \\ &= r(a, x). \end{aligned}$$

This shows that $\mathcal{M}$ is 1/2-point homogenous on A. 1/2-point homogeneity on X may be shown in the same way. The 1/2-point uniqueness of $\mathcal{L}$ follows from Theorem 1. □

The proof of Theorem 2 shows what the automorphisms of a strictly additive latent trait system look like. They have the form

$$\alpha(a) - f_x^{-1}(f_x(a) + k_\alpha).$$

An equivalent form has been derived by Colonius (1979) in showing that each member of the automorphism group of any *representation* of a strictly additive independent latent trait system of the form

$$r(a, x) = F[\theta(a), \xi(x)],$$

with F, strictly increasing in the first and strictly decreasing in the second argument, may be characterized by a single parameter. Theorem 2 leads to an equivalent conclusion, but does not refer to any representation at all. It thus shows that 1/2-point uniqueness and 1/2-point homogeneity are intrinsic properties of a strictly additive latent trait system and are not imposed on it by the selection of a special representation as has been claimed, among others, by Colonius (1979), Wottawa (1980), and Roskam (1983).

20.3 Component Structures

20.3.1 The Additive Case

Factorizable automorphisms are especially useful for looking at component structures of latent trait systems. Let $\mathcal{L} = \langle A \times X, r \rangle$ be a solvable latent trait system which satisfies strict additive independence, and let $\mathcal{M}$ be its group of factorizable automorphisms. Choose x from X and define a function δ on $A \times A$ by

$$\delta(a, b) = r(a, x) - r(b, x). \tag{20.6}$$

Then δ is independent of x by strict additive independence, and the set of component transformations, $\mathcal{M}_A$, is the automorphism group of the system $\langle A, \delta \rangle$:

$$\begin{aligned} \delta(\alpha(a), \alpha(b)) &= r(\alpha(a), x) - r(\alpha(b), x) \\ &= r(\alpha(a), \xi(x)) - r(\alpha(b), \xi(x)) \\ &= r(a, x) - r(b, x) \\ &= \delta(a, b). \end{aligned}$$

More definitions are needed for looking at the homogeneity and uniqueness properties of component structures.

Definition 7. *A latent trait system* $\mathcal{L} = \langle A \times X, r\rangle$ *is called* ordinal independent *in* A *iff, for all* a, b *in* A *and all* x, y *in* X,

$$r(a,x) \geq r(b,x) \text{ iff } r(a,y) \geq r(b,y). \tag{20.7}$$

Note that strict additive independence implies ordinal independence in A. Ordinal independence in A is sufficient to define a relation $\succeq_A$ on A by

$$a \succeq_A b \text{ iff } r(a,x) \geq r(b,x), \tag{20.8}$$

for some x in X, such that $\langle A, \succeq_A\rangle$ is a weak order.

Definition 8. *Let* $\mathcal{L} = \langle A, \succeq_A, r\rangle$ *be a system, where* A *is a set,* $\langle A, \succeq_A\rangle$ *is a weak order, and* r*, for some finite* k*, is a mapping from* A^k *into* $\mathbf{R}$*, and let* $\mathcal{M}$ *be a subgroup of the automorphisms of* $\mathcal{L}$*.* $\mathcal{M}$ *satisfies* n-point homogeneity *iff, for any* $a_1, \ldots, a_n$ *and* $b_1, \ldots, b_n$ *in* A *which satisfy* $a_1 \succ_A \cdots \succ_A a_n$ *and* $b_1 \succ_A \cdots \succ_A b_n$*, there exists an automorphism* α *in* $\mathcal{M}$ *such that* $\alpha(a_i) = b_i$ *for* $i = 1, \ldots, n$*.* $\mathcal{M}$ *satisfies* n-point uniqueness *iff, for all* α*,* β *in* $\mathcal{M}$ *and* $a_1, \ldots, a_n$ *in* A *with* $a_1 \succ_A \cdots \succ_A a_n$*, the condition* $\alpha(a_i) = \beta(a_i)$*, for* $i = 1, \ldots, n$*, implies* $\alpha = \beta$*.*

It is easy to show that the automorphism group of the system $\langle A, \delta\rangle$, which has been defined earlier is 1-point homogenous and 1-point unique, and thus $\langle A, \delta\rangle$, is ratio scalable. Since δ also is independent of X it allows specifically objective ratio-scale comparisons of subjects (Irtel, 1987).

20.3.2 A Nonadditive Case

In this section we look at a latent trait system which preserves many of the properties of strict additivity but does so only for one of the two components. Such a system has been introduced by Birnbaum (1968). Its main difference to a strictly additive system is that invariant quantitative statements about subjects involve three of them. This is a common situation with interval scales where ratios of intervals are invariant under admissible transformations.

Definition 9. *A latent trait system* $\mathcal{L} = \langle A \times X, r\rangle$ *is called* affine independent *in* A *iff, for all* a*,* b*,* c *in* A *and all* x*,* y *in* X*, whenever these expressions are defined,*

$$\frac{r(a,x) - r(b,x)}{r(c,x) - r(b,x)} = \frac{r(a,y) - r(b,y)}{r(c,y) - r(b,y)}. \tag{20.9}$$

Note that affine independence in A implies ordinal independence in A (20.7). Thus, it also allows the definition of a weak order $\langle A, \succeq_A \rangle$ which is independent of X. So Def. 8 may be applied also.

Theorem 3. *Let $\mathcal{L} = \langle A \times X, r \rangle$ be a latent trait system, affine independent in A and A-solvable. Let $\mathcal{M}_A$ be the set of component transformations α on A which satisfy*

$$\frac{r(\alpha(a), x) - r(\alpha(b), x)}{r(\alpha(c), x) - r(\alpha(b), x)} = \frac{r(a, x) - r(b, x)}{r(c, x) - r(b, x)} \tag{20.10}$$

for all a, b, c in A and x in X, whenever these expressions are defined. Then, under function composition, $\mathcal{M}_A$ is a group, is 2-point homogenous and 2-point unique.

Proof: Showing that $\mathcal{M}_A$ is a group is left to the reader. We show 2-point homogeneity.

Let a, b, a', and b' be four elements of A with $a \succ_A b$ and $a' \succ_A b'$. Define $\alpha(a) = a'$ and $\alpha(b) = b'$. Then, let k_α be defined by

$$k_\alpha = \frac{r(a', x) - r(b', x)}{r(a, x) - r(b, x)}.$$

Note that because of affine independence in A the constant k_α does not depend on x. Then, extend α on all of A by defining $\alpha(c)$ for all c in A as the solution of

$$r(\alpha(c), x) = k_\alpha[r(c, x) - r(b, x)] + r(b', x).$$

We have to show that α is an element of $\mathcal{M}_A$. By definition, we have

$$\frac{r(\alpha(c), x) - r(b', x)}{r(c, x) - r(b, x)} = k_\alpha = \frac{r(a', x) - r(b', x)}{r(a, x) - r(b, x)}.$$

This implies that α is in $\mathcal{M}_A$, since $a' = \alpha(a)$ and $b' = \alpha(b)$. Thus, $\mathcal{M}_A$ is 2-point homogenous since the choice of a, b, a', and b' was arbitrary. It remains to show that $\mathcal{M}_A$ is 2-point unique.

Suppose we have two transformations of $\mathcal{M}_A$, α and β, which coincide at two points a and b. It immediately follows from (20.10) that α and β also coincide at any other point c in A: From

$$\begin{aligned}
\frac{r(\alpha(a), x) - r(\alpha(b), x)}{r(\alpha(c), x) - r(\alpha(b) x)} &= \frac{r(a, x) - r(b, x)}{r(c, x) - r(b, x)} \\
&= \frac{r(\beta(a), x) - r(\beta(b), x)}{r(\beta(c), x) - r(\beta(b), x)} \\
&= \frac{r(\alpha(a), x) - r(\alpha(b), x)}{r(\beta(c), x) - r(\alpha(b), x)}
\end{aligned}$$

we get $r(\alpha(c), x) = r(\beta(c), x)$, which implies $\alpha(c) = \beta(c)$ by A-solvability. □

Affine independence corresponds to the Birnbaum model (Birnbaum, 1968). It induces a 3-argument function,

$$\gamma(a,b,c) = \frac{r(a,x) - r(b,x)}{r(c,x) - r(b,x)}, \tag{20.11}$$

on A which is independent of X. Theorem 3 shows that the automorphism group of the induced system $\langle A, \gamma \rangle$ is 2-point homogenous and 2-point unique. Thus, the system is interval scalable.

20.4 References

Adams, E.W., Fagot, R.F., & Robinson, R. (1970). On the empirical status of axioms in fundamental theories of measurement. *Journal of Mathathematical Psychology, 7,* 379–409.

Andersen, E.B. (1973). Conditional inference for multiple-choice questionnaires. *British Journal of Mathematical and Statistical Psychology, 26,* 31-44.

Birnbaum, A. (1968). Some latent trait models and their use in inferring an examinee's ability. In F.M. Lord, & M.R. Novick, *Statistical theories of mental test scores.* Reading/Mass.: Addison-Wesley.

Colonius, H. (1979). Zur Eindeutigkeit der Parameter im Rasch-Modell. *Psychologische Beiträge, 21,* 414–416.

Fischer, G.H. (1974). *Einführung in die Theorie psychologischer Tests.* Bern: Huber.

Fischer, G.H. (1981). On the existence and uniqueness of maximum-likelihood estimates in the Rasch model. *Psychometrika, 46,* 59–77.

Fischer, G.H. (1988). Spezifische Objektivität: Eine wissenschaftstheoretische Grundlage des Rasch-Modells. In K.D. Kubinger (Ed.), *Moderne Testtheorie – Ein Abriß samt neuesten Beiträgen.* Weinheim: Psychologie Verlags Union.

Hamerle, A. (1979). Über die meßtheoretischen Grundlagen von Latent-Trait-Modellen. *Archiv für Psychologie, 21,* 153–167.

Hamerle, A., & Tutz, G. (1980). Goodness of fit tests for probabilistic measurement models. *Journal of Mathematical Psychology, 21,* 153–167.

Irtel, H. (1987). On specific objectivity as a concept in measurement. In E.E. Roskam & R. Suck (Eds.), *Progress in mathematical psychology-I.* Amsterdam: North-Holland.

Krantz, D.H., Luce, R.D., Suppes, P., & Tversky, A. (1971). *Foundations of measurement* (Vol. 1). New York: Academic Press.

Luce, R.D., & Cohen, M. (1983). Factorizable automorphisms in solvable conjoint structures I. *Journal of Pure and Applied Algebra, 27,* 225–261.

Luce, R.D., & Tukey, J.W. (1964). Simultaneous conjoint measurement: A new type of fundamental measurement. *Journal of Mathematical Psychology, 1,* 1-27.

Narens, L. (1985). *Abstract measurement theory.* Cambridge, MA: MIT Press.

Pfanzagl, J. (1971). *Theory of measurement.* Würzburg: Physica-Verlag.

Rasch, G. (1960). *Probabilistic models for some intelligence and attainment tests.* Copenhagen: Paedagogiske Institut.

Roberts, F.S., & Rosenbaum, Z. (1988/89). Tight and loose value automorphisms. *Discrete Applied Mathematics, 22,* 169–179.

Roskam, E.E. (1983). Allgemeine Datentheorie. In H. Feger & J. Bredenkamp (Eds.), *Messen und Testen, Enzyklopädie der Psychologie, Forschungsmethoden, Vol. 3.* Göttingen: Hogrefe.

Suppes, P., & Zinnes, J. (1963). Basic measurement theory. In R.D. Luce, R.R. Bush, & E. Galanter (Eds.), *Handbook of mathematical psychology* (Vol. 1). New York: Wiley.

Wottawa, H. (1980). *Grundriß der Testtheorie.* München: Juventa Verlag.

21

Some LBTL and LLTM Relationships

Gerhard H. Fischer[1] and Norbert Tanzer[2]

ABSTRACT This paper investigates some relationships between the Bradley-Terry-Luce model, extended by imposing linear constraints on the parameters (LBTL), on the one hand, and the Rasch model, similarly extended by constraining the item parameters (linear logistic test model, LLTM), on the other. These relationships not only lead to a unified treatment and the formulation of analogous uniqueness results for the maximum likelihood estimators of the LBTL and conditional maximum likelihood estimators of the LLTM, but also yield a new uniqueness theorem for the joint maximum likelihood estimators for the LLTM. Possible applications of the results are indicated.

21.1 Introduction

This paper deals with extensions of two well-known models from mathematical psychology and psychometrics, the Bradley–Terry–Luce model (BTL) and the Rasch model (RM). In both cases, linear constraints are imposed on the parameters, allowing one to model the effect of hypothetical factors influencing the likelihood of the observations, to estimate such effects, and to test hypotheses about them. The present extension of the BTL will be called "linear BTL" (LBTL for short). The extended RM is called the "linear logistic test model" (LLTM; Fischer, 1973, 1983a). It will be shown that any LBTL can be interpreted formally as an LLTM, and *vice versa*, which implies, in particular, that the uniqueness conditions for maximum likelihood estimation in the two models are isomorphic. These conditions will be treated here in a uniform manner; this leads to some new results.

The BTL (Zermelo, 1929; Bradley and Terry, 1952; Ford, 1957; Luce, 1959) is one of the best-known and most discussed models in mathematical psychology and psychometrics. It models the probability of preferring an object O_i over an object O_j (denoted $O_i \succ O_j$) in the form

$$P(O_i \succ O_j) = \frac{\exp(\tau_i)}{\exp(\tau_i) + \exp(\tau_j)}, \tag{21.1}$$

where the τ_i are parameters characterizing preferences for (attractiveness of) the different objects. Alternatively, (21.1) can be interpreted as describing the probability that O_i dominates O_j, e.g., when the O_i are players in some compet-

[1]University of Vienna, Department of Psychology, Liebiggasse 5, A-1010 Wien, Austria. E-mail address A6212DAC at VM.UNIVIE.AC.AT

[2]University of Graz, Department of Psychology, Universitätsplatz 2, A-8010 Graz, Austria

itive game, in which case the τ_i are interpreted as measures of playing skill. The model is overparameterized, because (21.1) is invariant under transformations $\tau_i \to \tau_i + c$ for all τ_i, so that any one of the τ_i can be arbitrarily set to zero. The τ_i will therefore be taken to be "normalized" by the imposition of a condition such as $\tau_k = 0$ or $\sum_i \tau_i = 0$. Hereafter without further qualification, "uniqueness" relates to normalized parameter vectors $\boldsymbol{\tau}$ only.

Zermelo (1929) was the first to study this model (in connection with chess tournaments). He showed that unique maximum likelihood (ML) estimates of $\boldsymbol{\tau}$ exist iff, for any partition of the objects into two subsets, $\mathcal{J}_1$ and $\mathcal{J}_2$, at least one subject has preferred some object $O_i \in \mathcal{J}_1$ over some object $O_j \in \mathcal{J}_2$. This condition is easy to understand because, if it were violated, this would mean that all subjects have preferred all objects in $\mathcal{J}_2$ over all objects in $\mathcal{J}_1$, and the ML estimates $\hat{\tau}_j$ for $O_j \in \mathcal{J}_2$ would become infinitely large in comparison to the $\hat{\tau}_i$ for $O_i \in \mathcal{J}_1$. This necessary and sufficient (n. s.) uniqueness condition was rediscovered independently by Ford (1957) and will therefore be referred to as the "Zermelo-Ford condition".

The condition can be elegantly formulated in terms of graph theory: A directed graph (digraph) consists of vertices V_i and a set of directed lines (dilines) joining some of the V_i to some other V_j; such dilines will be denoted by $\overrightarrow{V_iV_j}$. A digraph is strongly connected if there is a directed path (a sequence of dilines) from any vertex V_i to any other vertex V_j, e.g., $\overrightarrow{V_iV_h}, \overrightarrow{V_hV_l}, \ldots, \overrightarrow{V_uV_j}$. Graph theory provides very efficient algorithms for determining whether a given digraph is strongly connected (cf. Christofides, 1975). If a digraph is not strongly connected, it has at least two "strong components", subsets of vertices each forming strongly connected subgraphs. The smallest possible strong component is a single vertex, which is considered strongly connected to itself.

Let a digraph G be introduced by assigning one–one vertices V_i to objects O_i and inserting a diline $\overrightarrow{V_iV_j}$ iff some subject has responded "$O_i \succ O_j$". The Zermelo-Ford condition is then equivalent to strong connectedness of G (Harary et al., 1965). By this means it is easy to determine whether a given set of data, be it complete or incomplete, satisfies the critical condition.

The Rasch model (RM) is defined by the equations

$$P(+ \mid S_v, I_i) = \frac{\exp(\theta_v - \beta_i)}{1 + \exp(\theta_v - \beta_i)}, \qquad \text{for } i = 1, \ldots, k;\ v = 1, \ldots, n, \tag{21.2}$$

where

θ_v is the ability parameter of subject S_v, and

β_i the difficulty parameter of item I_i.

Again, this model is overparameterized: Any transformation $\theta_v \to \theta_v + c$ can immediately be compensated by $\beta_i \to \beta_i + c$. Some normalization, such as $\beta_1 = 0$

TABLE 21.1. The structure of an ill-conditioned item–score matrix with two strong components. Asterisks denote missing observations. "0/1/*" signifies that elemets in the submatrix may be either "0" or "1" or "*".

$\mathcal{J}_1$: I_1	…				$I_{k'}$	$\mathcal{J}_2$: $I_{k'+1}$	…				I_k	
						1	*	1	1	1	*	
						1	1	1	1	1	*	
0/1/*						1	1	1	*	1	*	$\mathcal{S}_1$
						*	*	1	1	1	1	
						1	1	1	*	1	1	
0	0	0	0	*	0							
0	0	0	*	0	0							
0	0	*	0	0	0	0/1/*						$\mathcal{S}_2$
0	*	*	0	0	0							
0	0	0	0	0	0							

or $\sum_i \beta_i = 0$, is required. Throughout this paper uniqueness of a ML solution in a RM refers to normalized solutions only.

The formal similarity of (21.2) to (21.1) means that the RM can be viewed as an incomplete BTL for the comparison of persons (parameters θ_v) with items (parameters β_i): When a subject attempts to solve a problem (item), either the subject dominates (the problem is solved), or the item dominates (the problem resists solution). Important structural properties of the item–response matrix can again be characterized by a digraph, either of the kind defined above, or alternatively and more parsimoniously by a digraph C defined as follows: A vertex V_i is assigned to each item I_i, and a diline $\overrightarrow{V_iV_j}$ is inserted iff some subject S_v has given a correct response to item I_i and an incorrect response to item I_j. The existence of a unique joint maximum likelihood (JML) solution $(\hat{\boldsymbol{\theta}}, \hat{\boldsymbol{\beta}})$ of the person and item parameters (under suitable normalization) is then assured by the jointly n. s. conditions that (a) the digraph C is strongly connected, and (b) all raw scores r_v satisfy the restriction $0 < r_v < k_v$, where k_v is the number of items responded to by S_v (Fischer, 1981).

If, however, C decomposes into two (or more) strong components, a reordering of items and subjects produces a data matrix like in Table 21.1: There are two (or more) subsets of items, $\mathcal{J}_1 = \{I_1, \dots, I_{k'}\}$ and $\mathcal{J}_2 = \{I_{k'+1}, \dots, I_k\}$, and two subsets of subjects, $\mathcal{S}_1 = \{S_1, \dots, S_{n'}\}$ and $\mathcal{S}_2 = \{S_{n'+1}, \dots, S_n\}$, such that all subjects $S_v \in \mathcal{S}_1$ have responded correctly (except possibly for missing observations) to all items $I_j \in \mathcal{J}_2$, and all subjects $S_w \in \mathcal{S}_2$ have responded incorrectly (again except possibly for missing observations) to all items $I_i \in \mathcal{J}_1$.

Obviously, items $I_i \in \mathcal{J}_1$ appear infinitely more difficult than items $I_j \in \mathcal{J}_2$, entailing that the ML estimates of the parameters β_i for $I_i \in \mathcal{J}_1$ and of θ_v for $S_v \in \mathcal{S}_1$ become infinitely large (or undetermined) relative to those for $I_j \in \mathcal{J}_2$ and $S_w \in \mathcal{S}_2$. $\mathcal{J}_1$ and $\mathcal{J}_2$ correspond to different strong components of C, denoted by C_1 and C_2, respectively. Since items and vertices were assigned one–one to each other, we may safely use brief formulations like "I_i belongs to C_l".

Data structured as in Table 21.1 are said to be "ill-conditioned"; if, on the other hand, digraph C is strongly connected, the data are called "well-conditioned" (Fischer, 1981).

21.2 The LBTL

We now define the LBTL as a BTL with a set of linear constraints imposed on the object parameters τ_i,

$$\tau_i = \sum_{l=1}^{m} q_{il}\eta_l + c \qquad \text{for } i = 1, \ldots, k, \text{ with } m < k. \tag{21.3}$$

The matrix $(\boldsymbol{Q},\boldsymbol{1})$, that is, the matrix $\{q_{il}\}$ augmented with an additional column of 1's (henceforth denoted $\boldsymbol{Q}^+$), is assumed to have rank $m+1$. This condition ensures that the τ_i in (21.3) are uniquely expressible in terms of the η_l; it will be presumed to hold throughout this paper. The η_l are less in number than the τ_i $(m < k)$; as we shall see, this may impose uniqueness on the ML estimates, even when the Zermelo–Ford condition is not satisfied. The normalization of $\boldsymbol{\tau}$ remains the same as in the BTL.

Amazingly, this extension of the BTL seems not to have been considered in the psychometric literature, except for unpublished papers of Fischer (1983b) and Tanzer (1984). Fischer suggested its application as a tool for deriving diagnostic rules in traffic psychology from expert judgements. Possible applications to food–tasting experiments immediately come to mind where the η_l are effects, on the preference for brands, of certain ingredients.

Another suggested application belongs to the nature-nurture discussion (Fischer and Formann, 1981; Fischer, 1993): Suppose that monozygotic twins reared apart (MZRA) take an intelligence test. Let twin T_a of a given pair have been raised in environment E_j, and twin T_b of the same pair in environment E_l, and suppose that T_a solves some item I_i that is not solved by T_b. These two item responses taken together constitute a dominance relation $E_j \succ E_l$ that can be modeled by means of the BTL. (This approach to the measurement of environmental effects, however, requires a more thorough justification, given by Fischer, 1993.) Suppose that some twins have *partly* grown up in different environments, or that their development has been influenced by a combination of several environmental factors, the total effect of which can be described as a sum of main effects (and possibly interactions). This leads one to consider the linear parametric structure (21.3).

For deriving ML estimates, let a data matrix $\boldsymbol{S}$ be specified as follows: Suppose O_i has been compared, independently, to O_j n_{ij} times, $n_{ij} \geq 0$; in these comparisons the response "$O_i \succ O_j$" occurred s_{ij} times, and "$O_j \succ O_i$", $n_{ij} - s_{ij}$ times. (Note that this notation also comprises incomplete designs.) The likelihood of the data $\boldsymbol{S}$ is

$$L(\boldsymbol{S}) = \prod_{i<j} \frac{\exp[s_{ij}\tau_i + (n_{ij} - s_{ij})\tau_j]}{[\exp(\tau_i) + \exp(\tau_j)]^{n_{ij}}},$$

and its logarithm,

$$\ln(L) = \sum_{i<j} \left\{ s_{ij}\tau_i + (n_{ij} - s_{ij})\tau_j - n_{ij} \ln\left[\exp(\tau_i) + \exp(\tau_j)\right] \right\}.$$

Differentiating with respect to η_a for some $a \in \{1, \ldots, m\}$ and equating to zero yields

$$\begin{aligned} \frac{\partial \ln L}{\partial \eta_a} &= \sum_{i<j} \left\{ s_{ij} \frac{\partial \tau_i}{\partial \eta_a} + (n_{ij} - s_{ij}) \frac{\partial \tau_j}{\partial \eta_a} - \right. \\ &- \left. \frac{n_{ij}}{\exp(\tau_i) + \exp(\tau_j)} \left(\exp(\tau_i) \frac{\partial \tau_i}{\partial \eta_a} + \exp(\tau_j) \frac{\partial \tau_j}{\partial \eta_a} \right) \right\} = 0. \end{aligned}$$

Inserting $\partial\tau_i/\partial\eta_a = q_{ia}$,

$$\sum_{i<j} \left\{ s_{ij}q_{ia} + (n_{ij} - s_{ij})q_{ja} - n_{ij} \frac{q_{ia}\exp(\tau_i) + q_{ja}\exp(\tau_j)}{\exp(\tau_i) + \exp(\tau_j)} \right\} = 0,$$

and rearranging,

$$\sum_{i<j} s_{ij}(q_{ia} - q_{ja}) = \sum_{i<j} n_{ij}(q_{ia} - q_{ja}) \frac{\exp(\tau_i)}{\exp(\tau_i) + \exp(\tau_j)}, \tag{21.4}$$

for all $a = 1, \ldots, m$. In passing, note that the comparison of two objects O_i and O_j with the same dependence on all parameters τ_l (i.e., $q_{il} = q_{jl}$ for all l) tells us nothing about the parameter vector $\boldsymbol{\eta}$.

For solving the ML equations (21.4) for $a = 1, \ldots, m$ under normalization $\tau_k = 0$, the Newton-Raphson method is very effective. The required second-order partial deriatives are

$$\frac{\partial^2 \ln L}{\partial\eta_a \partial\eta_b} = -\sum_{i<j} n_{ij}(q_{ia} - q_{ja})(q_{ib} - q_{jb}) \frac{\exp(\tau_i + \tau_j)}{[\exp(\tau_i) + \exp(\tau_j)]^2} \tag{21.5}$$

for all a, $1 \leq a \leq m$, and b, $1 \leq b \leq m$. Solving (21.4) for $\hat{\boldsymbol{\eta}}$ poses no problem.

Despite an apparent neglect of the LBTL in psychometrics, the statisticians El-Helbawy and Bradley (1977) have studied BTL models with linear restrictions on the object parameters of the form

$$\boldsymbol{B\tau} = \mathbf{0}, \tag{21.6}$$

where $\boldsymbol{B}$ is a given matrix of m zero-sum orthonormal contrast vectors (cf. also Bradley, 1984). El-Helbawy and Bradley (1977) derived, by means of an involved proof, the following sufficient condition for the uniqueness of $\hat{\boldsymbol{\tau}}$: A unique normalized ML estimator $\hat{\boldsymbol{\tau}}$ under restrictions (21.6) exists if the data matrix $\boldsymbol{S}$ satisfies the Zermelo-Ford condition. Furthermore, these authors showed that nested hypotheses of the form of (21.6), $\boldsymbol{B}$ and $\boldsymbol{B}_0$, say, can be compared by means of an asymptotic likelihood–ratio test procedure.

The Bradley and El-Helbawy model, however, is a special case of the LBTL: The latter is more general because it does not require orthonormality of the weight matrix $\boldsymbol{Q}$. Moreover, we give a n. s. condition and not just a sufficient one. As a consequence, a weaker uniqueness condition will be formulated for the LBTL.

Theorem 1 *Let digraph G have strong components G_t, $t = 1, \ldots, u$. A unique ML solution $\hat{\boldsymbol{\eta}}$ for the LBTL defined by (21.1) and (21.3) exists iff the linear system of equations (21.7),*

$$\sum_{l=1}^{m} q_{il}\alpha_l = \delta_t \qquad \text{for all } V_i \in G_t,\ t = 1, \ldots, u, \tag{21.7}$$

with unknown δ_t sufficing the inequality constraints (21.8),

$$\delta_s \geq \delta_t \qquad \text{if a diline } \overrightarrow{V_a V_b} \text{ exists for some } V_a \in G_s \text{ and some } V_b \in G_t, \tag{21.8}$$

has only the trivial solution $\boldsymbol{\alpha} = \mathbf{0}$.

If the system (21.7)–(21.8) has some solution $\boldsymbol{\alpha} \neq \mathbf{0}$, then $c\boldsymbol{\alpha}$, for any constant $c > 0$, is also a solution, i.e., then there are infinitely many solutions. It will be shown that the vector $\boldsymbol{\alpha}$ has an interesting interpretation: Under the transformation $\boldsymbol{\eta} \to \boldsymbol{\eta} + c\boldsymbol{\alpha}$ the likelihood function is non–decreasing. If $\boldsymbol{\alpha} \neq \mathbf{0}$, this clearly means that either there exists no maximum of the likelihood function, or there exist infinitely many maxima (in the weak sense). So, only for $\boldsymbol{\alpha} = \mathbf{0}$ there can be a unique maximum.

This linear system can be called an "alternative system" (AS), because either the ML equations possess a unique solution and the AS has only the (trivial) null-solution, or the AS has some non-null solution, $\boldsymbol{\alpha} \neq \mathbf{0}$, and the ML equations have no unique solution. As far as uniqueness is concerned, it suffices to look at the linear AS. However, in practical applications the AS may still be burdensome; to what extent this is the case depends largely on the number of strong components G_t, $t = 1, \ldots, u$, because the effective number of inequalities in (21.8) is $u - 1$. Clearly, if $u = 1$, there are no inequality constraints, and we have a particularly simple special case:

If $u = 1$, i.e., if digraph G is strongly connected (the Zermelo-Ford condition satisfied), the AS reduces to the homogeneous system of linear equations (21.7) with $\delta_t = \delta$, possessing only the trivial null-solution $\boldsymbol{\alpha} = \mathbf{0}$ due to the rank condition $r(\boldsymbol{Q}^{+}) = m + 1$. Hence, the normalized ML estimate $\hat{\boldsymbol{\eta}}$ is unique. This sufficient result, corresponding to that of El-Helbawy and Bradley (1977), can be formulated as follows:

Corollary 1 *If the data $\boldsymbol{S}$ in the LBTL with $r(\boldsymbol{Q}^{+}) = m+1$ satisfy the Zermelo-Ford condition, the ML estimate $\hat{\boldsymbol{\eta}}$ is unique.*

Theorem 1 will be proved with the help of a recent result of Jacobsen (1989) on uniqueness in discrete exponential families. Jacobsen considers likelihood functions of the form

$$L(\boldsymbol{a}_v|\boldsymbol{\eta}) = \frac{\exp\{\langle \boldsymbol{t}(\boldsymbol{a}_v), \boldsymbol{\eta}\rangle + g(\boldsymbol{a}_v)\}}{\sum_{\boldsymbol{x} \in E_v} \exp\{\langle \boldsymbol{t}(\boldsymbol{x}), \boldsymbol{\eta}\rangle + g(\boldsymbol{x})\}}, \tag{21.9}$$

where $\boldsymbol{a}_v$ is a (possibly vector-valued) observation, $\boldsymbol{\eta}$ a (vector-valued) parameter, $\boldsymbol{t}(\boldsymbol{a}_v)$ a (vector-valued) minimal sufficient statistic, E_v the sample space corresponding to observation $\boldsymbol{a}_v$, $\boldsymbol{x}$ any point in E_v, $g: E_v \to \mathbf{R}$ some real-valued function on E_v, and $\langle \cdot, \cdot \rangle$ the inner vector product. As an example with discrete observations, imagine $\boldsymbol{a}_v$ to be the response of subject S_v when asked for a comparison of two objects O_i and O_j; $\boldsymbol{\eta}$ the vector of parameters under the LBTL; $\boldsymbol{t}(\boldsymbol{a}_v)$ a vector-valued sufficient statistic assuming the value $\boldsymbol{q}_i$ if the response was "$O_i \succ O_j$", and $\boldsymbol{q}_j$ for "$O_j \succ O_i$", where $\boldsymbol{q}_i$ denotes the vector $(q_{i1}, \ldots, q_{im})$ (see the proof below); E_v the two possible outcomes of this one pair comparison; $\boldsymbol{x}$ any response in E_v; and $g = 0$ for all $\boldsymbol{x}$.

If the observations are discrete (as in this example), the likelihood function (21.9) can be rewritten as

$$L(\boldsymbol{a}_v|\boldsymbol{\eta}) = \Big(1 + \sum_{l \in D_v} \exp\{\langle \boldsymbol{t}(\boldsymbol{x}_l) - \boldsymbol{t}(\boldsymbol{a}_v), \boldsymbol{\eta}\rangle + g(\boldsymbol{x}_l) - g(\boldsymbol{a}_v)\Big)^{-1}, \tag{21.10}$$

where D_v is a countable index set that may depend on index v.

Amongst other uniqueness conditions for models of the form (21.10), Jacobsen proves the following result (hereafter denoted Jacobsen's Theorem):

Jacobsen's Theorem *For a given set of observations $\boldsymbol{a}_v$, a unique maximum of the likelihood function $\prod_v L(\boldsymbol{a}_v|\boldsymbol{\eta})$ exists iff the linear system (21.11),*

$$\langle \boldsymbol{t}(\boldsymbol{x}_l) - \boldsymbol{t}(\boldsymbol{a}_v), \boldsymbol{\mu}\rangle \geq 0 \quad \textit{for all } l \in D_v \textit{ and all } v, \tag{21.11}$$

has only the null-solution $\boldsymbol{\mu} = \mathbf{0}$.

Proof of Theorem 1. For all S_v, the likelihood of the preference "$O_i \succ O_j$" is

$$P(O_i \succ O_j) = \frac{1}{1+\exp[\langle(\boldsymbol{q}_j - \boldsymbol{q}_i), \boldsymbol{\eta}\rangle]};$$

similarly, the likelihood of "$O_j \succ O_i$" is

$$P(O_j \succ O_i) = \frac{1}{1+\exp[\langle(\boldsymbol{q}_i - \boldsymbol{q}_j), \boldsymbol{\eta}\rangle]}.$$

Jacobsen's condition (21.11) reduces in this case to the equations

$$\sum_{l=1}^{m}(q_{jl} - q_{il})\mu_l \geq 0 \quad \text{for all "}O_i \succ O_j\text{"},$$

$$\sum_{l=1}^{m}(q_{il} - q_{jl})\mu_l \geq 0 \quad \text{for all "}O_j \succ O_i\text{"}.$$

For all O_i and O_j belonging to the same strong component G_t of the data matrix both equations hold, and hence

$$\sum_{l=1}^{m} q_{il}\mu_l = \sum_{l=1}^{m} q_{jl}\mu_l = -\delta_t \quad \text{for all } V_i, V_j \in G_t, \tag{21.12}$$

the common value of $\sum_l q_{il}\mu_l$ and $\sum_l q_{jl}\mu_l$ being denoted by $-\delta_t$ for convenience. For any two objects O_a and O_b with $O_a \in G_s$, $O_b \in G_t$, $s \neq t$, where at least one preference "$O_a \succ O_b$" has been recorded, the inequality

$$\sum_{l=1}^{m}(q_{bl} - q_{al})\mu_l \geq 0$$

must hold, implying

$$\delta_s \geq \delta_t \quad \text{for } V_a \in C_s,\ V_b \in C_t \text{ with "}O_a \succ O_b\text{"}. \tag{21.13}$$

Hence, (21.12), with $\boldsymbol{\alpha} = -\boldsymbol{\mu}$, together with (21.13) immediately yields the AS of Theorem 1.

Suppose there are at least two strong components, G_s and G_t, and (21.12)–(21.13) have a solution $\boldsymbol{\mu} \neq \mathbf{0}$; then it is seen from transforming $\boldsymbol{\eta}$ into $\boldsymbol{\eta} - c\boldsymbol{\mu} = \boldsymbol{\mu} + c\boldsymbol{\alpha}$ that the likelihood function is non-decreasing under that transformation. □

21.3 The LLTM

The LLTM is a RM (21.2) with linear constraints imposed on the item parameters,

$$\beta_i = \sum_{j=1}^{m} q_{ij}\eta_j + c, \qquad m < k, \tag{21.14}$$

where

η_j is the j-th "basic parameter" of the model, e.g., in typical applications, the difficulty of cognitive operation B_j required by some of the items, and

q_{ij} the given weight of B_j as a factor of item difficulty.

The $k \times m$ matrix $\boldsymbol{Q} = ((q_{ij}))$, which is assumed to be known, is called the "item structure" of the LLTM.

The item parameters β_i are normalized by setting, e.g., $\beta_1 = 0$ or $\sum_i \beta_i = 0$. Given such normalization, one would like the decomposition (21.14) of $\boldsymbol{\beta}$ into $\boldsymbol{\eta}$ to be unique. Then the matrix $\boldsymbol{Q}^+ = (\boldsymbol{Q}, \mathbf{1})$ must have full column rank; hereafter we assume $r(\boldsymbol{Q}^+) = m + 1$.

The LLTM as an extension of the RM emerged from a regression analysis of item parameter estimates $\hat{\beta}_i$ by Scheiblechner (1972) who attempted to explain the psychological complexity of logical propositions in terms of three cognitive operations. Fischer (1972, 1973) incorporated the linear constraints (21.14) into the Rasch model (21.2) and derived conditional maximum likelihood (CML) estimation equations for the LLTM. Fischer and Formann (1972) and Fischer (1974) presented algorithms and FORTRAN programs for CML estimation and for conditional likelihood ratio (CLR) tests of null-hypotheses of the form (21.14).

The initial motivation for the LLTM, and also its most obvious field of application, was the explanation of item difficulty in terms of cognitive operations, thereby testing psychological hypotheses about the cognitive structure of item material. A "cognitive structure" is thereby represented by the weight matrix $\boldsymbol{Q}$. Differently restrictive nested hypotheses, Q and Q_0 say, can be elegantly compared by means of CLR tests. A considerable number of such applications have been made in test analysis and test construction (e.g., Formann, 1973; Formann and Piswanger, 1979; Fischer and Pendl, 1980; Nährer, 1980; Embretson, 1985; Hornke and Habon, 1986; Hornke and Rettig, 1988; Gittler and Wild, 1988), in cognitive psychology (Heinrich, 1975; van Maanen et al., 1989), in didactics and education (Spada, 1976; Kubinger, 1979, 1980; Spada and May, 1982; Sijtsma, 1982; Spada and McGraw, 1985), and in cross-cultural psychology (Piswanger, 1975, see Fischer and Formann, 1982a; van de Vijver, 1988). Besides these kinds of applications, the LLTM has been seen to be very appropriate for the measurement of change and the assessment of training or treatment effects (e.g., Fischer, 1989, 1991, and older references therein), where the basic parameters are the effects of treatments, trend, and treatment interactions. Many applications have been made in this area too: to the therapy of children with impaired hearing (Heckl, 1976), effects of pre-school education (Rop, 1977; Zeman, 1980), psychotherapy (Zimprich, 1980; Widowitz, 1987; Mutschlechner, 1987); group dynamics (Witek, 1980); attitude change (Barisch, 1989), personnel training (Iby, 1987; Schmied, 1987). Finally, the LLTM can be understood as a model for cognitive development (Spada and Kluwe, 1980; Wilson, 1989; Embretson, 1991; Fischer, 1992) or as a tool for monitoring environmental effects on development (Fischer and Formann, 1982b; Fischer, 1993). (For a further discussion of the LLTM, see Traub and Lam, 1985; Hambleton, 1989; Wilson, 1992.)

Four methods of estimation of the "basic" parameter vector $\boldsymbol{\eta}$ have been proposed: Regressing the $\hat{\beta}_i$ of the RM (21.2) on the columns of $\boldsymbol{Q}$ (Scheiblechner, 1972), the conditional maximum likelihood approach (CML; Fischer, 1973, 1983a), the joint maximum likelihood method (JML; Fischer and Formann, 1982a), and the marginal maximum likelihood method (MML; Mislevy, 1988; Glas and Verhelst, 1989). But the present paper will be concerned only with CML and JML and their respective uniquenesses.

Some notation will be needed: Let $\boldsymbol{A}$ be the (possibly incomplete) $n \times k$ item–score matrix with elements

$$a_{vi} = \begin{cases} 1 & \text{if } S_v\text{'s response to } I_i \text{ was correct,} \\ 0 & \text{if } S_v\text{'s response to } I_i \text{ was incorrect,} \\ p, & 0 < p < 1, \text{ if no response of } S_v \text{ to } I_i \\ & \text{was observed,} \end{cases}$$

where p is an arbitrary dummy constant introduced simply for formal completeness. Let there also be a design matrix $\boldsymbol{B} = ((b_{vi}))$ with elements

$$b_{vi} = \begin{cases} 1 & \text{if } a_{vi} \in \{0,1\}, \text{ that is, if some response of } S_v \text{ to} \\ & \text{item } I_i \text{ was observed, and} \\ 0 & \text{if } a_{vi} = p, \text{ that is, if no response of } S_v \text{ to item } I_i \\ & \text{was observed.} \end{cases}$$

$\boldsymbol{B}$ contains no information beyond that in $\boldsymbol{A}$ and is introduced solely as a matter of convenience. In terms of $\boldsymbol{A}$ and $\boldsymbol{B}$, the item marginal sums are

$$s_i = \sum_v b_{vi} a_{vi},$$

the subjects' raw scores

$$r_v = \sum_i b_{vi} a_{vi},$$

and the number of items to which S_v responded

$$k_v = \sum_i b_{vi}.$$

We presume $k_v > 0$ for all S_v. Under this assumption the unconditional (or joint) likelihood of $\boldsymbol{A}$ can be written as

$$L^{(J)}(\boldsymbol{A}) = \prod_v \prod_i \frac{\exp[b_{vi}a_{vi}(\theta_v - \beta_i)]}{[1+\exp(\theta_v - \beta_i)]^{b_{vi}}}. \tag{21.15}$$

Maximizing (21.15) jointly with respect to the basic parameters and the person parameters yields the JML equations (see Fischer and Formann, 1982a),

$$\sum_i q_{ij} \left[s_i - \sum_v b_{vi} \frac{\exp(\theta_v - \beta_i)}{1+\exp(\theta_v - \beta_i)} \right] = 0 \quad \text{for } j = 1, \ldots, m, \tag{21.16}$$

and

$$r_v = \sum_i b_{vi} \frac{\exp(\theta_v - \beta_i)}{1+\exp(\theta_v - \beta_i)} \qquad \text{for } v = 1, \ldots, n, \tag{21.17}$$

where the β_i in both sets of equations are subject to the constraints (21.14). These equations are easy to solve iteratively for the estimates $\hat{\eta}_j$, $j = 1, \ldots, m$, and $\hat{\theta}_v$, $v = 1, \ldots, n$.

This method, however, has not been used in practice, because JML lacks consistency even in the RM (see Andersen, 1973a, 1980, p. 244), which is a special case of the LLTM. So JML cannot be consistent in the LLTM either. The reason is that each and every fresh subject introduces an additional parameter θ_v so that, as $n \to \infty$, the number of unknown parameters tends to infinity. However, Haberman (1977) has shown that consistency can be attained in the RM if both the numbers of subjects *and* of items tend to infinity. This implies that the bias of the JML method should be small when the test is long. So the JML method can be expected to yield an acceptable approximation to the true parameter values in sufficiently long tests. But since, in many interesting applications, the tests are short (even $k = 2$ in certain repeated measurement designs, see Fischer, 1989), JML is not a sufficiently general solution to the estimation problem. Moreover, no rigorous hypothesis tests are known within the JML framework.

A better estimation method is CML: In this method the "incidental" or "nuisance" parameters θ_v are conditioned out of the likelihood equations which then depend only on the "structural" parameters η_j. Hence, as $n \to \infty$, the number of unknown parameters no longer increases; this is an important prerequisite for reasonable asymptotic behavior of the estimators. The asymptotic properties of CML estimators for models of the Rasch-type have been studied, e. g., by Andersen (1973b), and consistency and asymptotic normality have been established (see also Pfanzagl, 1993, Chapter 19 of this volume).

The conditional likelihood of $\boldsymbol{A}$, given the design vectors (b_{vi}), for $i = 1, \ldots, k$, $v = 1, \ldots, n$, and the raw scores r_v of all subjects, is

$$L^{(C)}(\boldsymbol{A}) \;=\; \prod_v L_v^{(C)}(a_{v1}, \ldots, a_{vk}) = \prod_v \Big(\prod_i \frac{\epsilon_i^{a_{vi}b_{vi}}}{\gamma_{r_v}} \Big) =$$

$$= \Big(\prod_v \gamma_{r_v}\Big)^{-1} \exp\{-\sum_j \eta_j t_j\}, \tag{21.18}$$

where $L_v^{(C)}$ is the conditional likelihood of the responses of S_v, $\epsilon_i = \exp(-\beta_i)$, $\gamma_{r_v}(\epsilon_1 b_{v1}, \ldots, \epsilon_k b_{vk})$ is the elementary symmetric function of order r_v of the variables $\epsilon_i b_{vi}$, and the t_j are the minimal sufficient statistics for $\boldsymbol{\eta}$, namely,

$$t_j = \sum_{i=1}^{k-1} (s_i - s_k)(q_{ij} - \bar{q}_{.j}), \tag{21.19}$$

with $\bar{q}_{.j}$ for the average of $q_{1j}, \ldots, q_{kj}$ (see Fischer, 1983a). Maximizing (21.18) yields the CML equations

$$\sum_{i=1}^{k} q_{ij}\Big(s_i - \sum_v \epsilon_i b_{vi} \gamma_{r_v-1}^{(i)} \gamma_{r_v}^{-1}\Big) = 0 \quad \text{for} \quad j = 1, \ldots, m, \tag{21.20}$$

where $\gamma_{t-1}^{(i)}$ denotes the partial deviative of $\gamma_t(z_1, \ldots, z_k)$ with respect to z_i; this again is an elementary symmetric function, now of order $t-1$ of the variables $z_1, \ldots, z_{i-1}, z_{i+1}, \ldots, z_k$ (Fischer, 1983a). Since the conditional model (21.18) defines an exponential family in "minimal" or "canonical" form (on that term, cf. Andersen, 1980), the CML estimates $\hat{\boldsymbol{\eta}}$ are unique if they exist. Ascertaining the existence of a finite CML estimator $\hat{\boldsymbol{\eta}}$ for a given LLTM and given data $\boldsymbol{A}$ therefore is a prerequisite for any meaningful application of the LLTM.

The existence of a finite (and hence unique) CML estimator $\hat{\boldsymbol{\eta}}$ is guaranteed by the n. s. condition specified in the following theorem:

Theorem 2 *A unique finite CML solution $\hat{\boldsymbol{\eta}}$ for the LLTM defined by (21.2) and (21.14) exists iff the linear system of equations (21.21),*

$$\sum_{j=1}^{m} q_{ij}\alpha_j = \delta_l \quad \text{for all } V_i \in C_l, \; l = 1, \ldots, u, \tag{21.21}$$

with unknown constants δ_l satisfying the inequality constraints (21.22),

$$\delta_s \le \delta_t \quad \text{if a diline } \overrightarrow{V_a V_b} \text{ exists for some } V_a \in C_s \text{ and some } V_b \in C_t, \tag{21.22}$$

has only the trivial solution $\boldsymbol{\alpha} = \mathbf{0}$.

This result was given by Fischer (1983a). The system (21.21)–(21.22) is again called an "alternative system" (AS), and again, if the AS has some solution $\boldsymbol{\alpha} \neq \mathbf{0}$, there are infinitely many solutions. The vector $\boldsymbol{\alpha}$ has the same interesting interpretation as in the LBTL: Each term $L_v^{(C)}$ in the conditional likelihood function (21.18) is non-decreasing under the transformation $\boldsymbol{\eta} \rightarrow \boldsymbol{\eta} + c\boldsymbol{\alpha}$, for any $c > 0$ (Fischer, 1983a).

All that has been said about the AS of the LBTL applies here again *mutatis mutandis.* A special case of Theorem 2 arises if C is strongly connected (well-conditioned data): the AS reduces to a system of linear equations $\sum_j q_{ij}\alpha_j = \delta$ for all $i = 1, \ldots, k$, which has only the null-solution $\boldsymbol{\alpha} = \mathbf{0}$ iff $\boldsymbol{Q}^+$ is of full column rank (as we generally presumed). Hence, Theorem 2 implies the following result:

Corollary 2 *If digraph C is strongly connected, a unique CML solution $\hat{\boldsymbol{\eta}}$ for the LLTM (21.2) and (21.14) exists iff $r(\boldsymbol{Q}^+) = m + 1$.*

The content of Corollary 2 illustrates an interesting relation between the LLTM and the RM: Although the LLTM restricts the item parameter β_i of the RM, the formal structure of the LLTM is more general than that of the RM. The RM can be nested in the LLTM by letting, e. g., $q_{ii} = 1$ for $i = 2, \ldots, k$, and $q_{ij} = 0$ otherwise, such that $\beta_i = \eta_i$ for $i = 1, \ldots, k$. (Note that this implies the normalization $\beta_1 = \eta_1 = 0$, eliminating one of the basic parameters, such that there remain $k - 1$ independent basic parameters.) This moreover entails that $\boldsymbol{Q}^+$ has rank k, implying that the $m = k - 1$ normalized CML parameter estimates of the RM are unique iff C is strongly connected. Hence we obtain the well-known standard result about the uniqueness of CML estimates in the RM (Fischer, 1981) as a very special case of Theorem 2. The n. s. condition for the LLTM is clearly weaker than the one for the RM: any set of data having a unique RM solution $\hat{\boldsymbol{\beta}}$ must also have a solution $\hat{\boldsymbol{\eta}}$ under any LLTM, whatever the restrictions (21.14), as long as $r(\boldsymbol{Q}^+) = m + 1$; but data that do *not* admit a solution under the RM may still have a unique solution $\hat{\boldsymbol{\eta}}$ under a suitably chosen LLTM. The intuitive explanation of this is that linear restrictions of the form (21.14) tie the item parameters together and thereby may prevent divergence of the $\hat{\beta}_i$.

21.4 A proof of Theorem 2 based on the Jacobsen Theorem

This section presents a new proof of Theorem 2 (using the result of Jacobsen, 1989), shorter than that of Fischer (1983a).

Proof of Theorem 2. Apply (21.11) to the conditional likelihood function (21.18) of the LLTM, which, considering the negative sign of t_j in (21.18), yields the AS

$$\sum_j [t_j(\boldsymbol{x}_l) - t_j(\boldsymbol{a}_v)]\mu_j \leq 0, \tag{21.23}$$

for all response vectors $\boldsymbol{x}_l$ compatible with the given raw score $r_v = \sum_i a_{vi}b_{vi}$, and for all S_v. Recall that, when $b_{vi} = 0$, we have set a_{vi} equal to some dummy constant p, $0 < p < 1$; so, whenever $b_{vi} = 0$, we again put $x_{li} = p$ for all $l \in D_v$.

Since (21.23) depends on a single response vector $\boldsymbol{a}_v$, t_j in (21.19) now becomes

$$\begin{aligned} t_j(\boldsymbol{a}_v) &= \sum_{i=1}^{k-1}(a_{vi}b_{vi} - a_{vk}b_{vk})(q_{ij} - \bar{q}_{.j}) \\ &= \sum_{i=1}^{k} a_{vi}b_{vi}q_{ij} - r_v\bar{q}_{.j} - a_{vk}b_{vk}\underbrace{\sum_{i=1}^{k}(q_{ij} - \bar{q}_{.j})}_{=0} \\ &= \sum_{i=1}^{k} a_{vi}b_{vi}q_{ij} - r_v\bar{q}_{.j}. \end{aligned} \tag{21.24}$$

Similarly, for response vector $\boldsymbol{x}_l$, $l \in D_v$,

$$t_j(\boldsymbol{x}_l) = \sum_{i=1}^{k} x_{li}b_{vi}q_{ij} - r_v\bar{q}_{.j}. \tag{21.25}$$

Inserting (21.24) and (21.25) in the left hand side of (21.23), and denoting this expression M_v,

$$M_v(\boldsymbol{x}_l, \boldsymbol{\mu}) = \sum_{i=1}^{k}(x_{li} - a_{vi})b_{vi}\sum_j q_{ij}\mu_j. \tag{21.26}$$

Now distinguish two cases: Ill-conditioned and well-conditioned data. Suppose, first, that the data are ill-conditioned, such that digraph C has $u > 1$ strong components, $C_1, \ldots, C_u$, and suppose that $\boldsymbol{\alpha}$ is a solution of the AS (21.21)–(21.22). Let items and strong components be ordered, for convenience, such that $\delta_1 \geq \ldots \geq \delta_t \geq \ldots \geq \delta_u$, the order of the items within strong components being arbitrary. Suppose that items $I_1, \ldots, I_{k_1}$ belong to component C_1, items $I_{k_1+1}, \ldots, I_{k_2}$ to $C_2, \ldots$, items $I_{k_{t-1}+1}, \ldots, I_{k_t}$ to C_t, etc. Furthermore, let the subjects be ordered by their raw scores. Then the data matrix attains the characteristic structure exhibited by Table 21.2. Note that this ordering of items and subjects defines subject groups $\mathcal{S}_t$, such that subjects within the same group $\mathcal{S}_t$ have identical response vectors, except for those responses that were given to items $I_i \in C_t$, and except, possibly, for missing responses.

Inserting the solution $\boldsymbol{\alpha}$ of AS (21.21)–(21.22) in (21.26) gives, for all $S_v \in \mathcal{S}_t$ (cf. Table 21.2) and any $\boldsymbol{x}_l$, $l \in D_v$,

$$\begin{aligned} M_v(\boldsymbol{x}_l, \boldsymbol{\alpha}) &= \sum_{h=1}^{u}\sum_{i=k_{h-1}+1}^{k_h}(x_{li} - a_{vi})b_{vi}\delta_h \\ &= d_{v1}(\boldsymbol{x}_l)\delta_1 + \ldots + d_{vh}(\boldsymbol{x}_l)\delta_h + \ldots + d_{vu}(\boldsymbol{x}_l)\delta_u, \end{aligned} \tag{21.27}$$

where $k_0 = 0$, $k_u = k$ are defined for formal consistency, and $d_{vh}(\boldsymbol{x}_l)$ denotes the term $\sum_{i=k_{h-1}+1}^{k_h}(x_{li} - a_{vi})b_{vi}$. Obviously, $\sum_{h=1}^{u} d_{vh}(\boldsymbol{x}_l) = 0$ by definition of $\boldsymbol{x}_l$ (as

TABLE 21.2. The structure of ill-conditioned data with u strong components. Asterisks denote missing observations; the symbol "0/1/*" means that the respective submatrix may contain elements 0 or 1 or *.

C_1	....	C_t	C_{t+1}	....	C_u	
$1 \ldots\ldots k_1$	....	$k_{t-1}+1 \ldots k_t$	$k_t+1 \ldots k_{t+1}$	....	$k_{u-1}+1 \ldots k$	
	....	1 1 1 1 1 *	1 1 1 * 1 1	....	1 1 1 1 1 1	
0/1/*	....	....	....	....	....	$\mathcal{S}_1$
	....	1 * 1 1 1 1	1 1 1 1 1 *	....	* * 1 1 1 *	
0 0 0 0 0 0	....	* 1 1 1 1 1	1 1 1 1 1 *	....	1 1 1 1 1 1	
....	....	....	....	....	....	$\mathcal{S}_2$
0 0 0 0 * 0	....	1 1 1 1 1 *	1 1 1 * 1 1	....	1 1 1 1 1 1	
0 0 0 0 0 0	....		* 1 1 1 1 1	....	1 * 1 1 1 1	
....	....	0/1/*	....	....	....	$\mathcal{S}_t$
0 0 * 0 0 0	....		1 1 1 1 1 1	....	1 1 1 1 1 1	
....	....	....	....	....	....	⋮
....	....	....	....	....	....	⋮
....	....	....	....	....	....	⋮
....	....	....	....	....	....	⋮
....	....	....	....	....	....	⋮
....	....	....	....	....	....	
0 0 * 0 0 *	....	0 0 0 * 0 0	0 0 0 0 * 0	....		
....	....	0 0 0 0 0 0	....	....	0/1/*	$\mathcal{S}_u$
0 * 0 0 * 0	....	....	0 0 0 0 0 0	....		

compatible with r_v), such that

$$\sum_{h=1}^{t} d_{vh}(\boldsymbol{x}_l) = -\sum_{h=t+1}^{u} d_{vh}(\boldsymbol{x}_l), \tag{21.28}$$

for any fixed t, $1 \le t \le u$. Due to the particular structure of the data in Table 21.2, it happens for $S_v \in \mathcal{S}_t$ that $d_{vh}(\boldsymbol{x}_l) \ge 0$ for $h = 1, \ldots, t-1$, and $d_{vh}(\boldsymbol{x}_l) \le 0$ for $h = t+1, \ldots, u$, as far as the $d_{vh}(\boldsymbol{x}_l)$ are defined. (Undefined terms are taken equal zero.) Remembering that $\delta_1 \ge \ldots \ge \delta_t \ge \ldots \ge \delta_u$,

$$M_v(\boldsymbol{x}_l, \boldsymbol{\alpha}) = \sum_{h=1}^{t-1} d_{vh}(\boldsymbol{x}_l)\delta_h + d_{vt}(\boldsymbol{x}_l)\delta_t + \sum_{h=t+1}^{u} d_{vh}(\boldsymbol{x}_l)\delta_h$$

$$\ge \delta_t \sum_{h=1}^{t-1} d_{vh}(\boldsymbol{x}_l) + \delta_t d_{vt}(\boldsymbol{x}_l) + \delta_{t+1} \sum_{h=t+1}^{u} d_{vh}(\boldsymbol{x}_l). \tag{21.29}$$

Taking into account the fact that $d_{vh}(\boldsymbol{x}_l) \le 0$ for $h = t+1, \ldots, u$ implies $\sum_{h=t+1}^{u} d_{vh}(\boldsymbol{x}_l) \le 0$ and, by virtue of (21.28), $\sum_{h=1}^{t} d_{vh}(\boldsymbol{x}_l) \ge 0$, we finally conclude from (21.29) that

$$M_v(\boldsymbol{x}_l, \boldsymbol{\alpha}) \geq (\delta_t - \delta_{t+1}) \sum_{h=1}^{t} d_{vh}(\boldsymbol{x}_l) \geq 0. \tag{21.30}$$

Hence, vector $-\boldsymbol{\alpha}$ satisfies Jacobsen's condition (21.23); in other words, if there exists a solution $\boldsymbol{\alpha}$ of AS (21.21)–(21.22), then Jacobsen's AS (21.11) has the solution $\boldsymbol{\mu} = -\boldsymbol{\alpha}$.

Still assuming that the data $\boldsymbol{A}$ are ill-conditioned with strong components C_h, $h = 1, \ldots, u$, we next prove the reverse: Let there be a Jacobsen solution $\boldsymbol{\mu}$ of (21.23); we shall show that then $-\boldsymbol{\mu} = \boldsymbol{\alpha}$ is a solution of AS (21.21)–(21.22).

Consider any two items within the same strong component C_t, I_a and I_b, say. Belonging to the same component means that the respective vertices are connected by a path $\overrightarrow{V_a V_c}, \overrightarrow{V_c V_d}, \ldots, \overrightarrow{V_w V_b}$, whereby a diline $\overrightarrow{V_c V_d}$ corresponds to some subject S_v with $a_{vc} = 1$ and $a_{vd} = 0$. Within the sample space E_v of this subject S_v with raw score r_v, there exists a response vector $\boldsymbol{x}_l$ (compatible with r_v) which is identical to the observed response vector $\boldsymbol{a}_v$ except for the elements $x_{lc} = 0$ and $x_{ld} = 1$. Hence, $\overrightarrow{V_c V_d}$ together with (21.23) and (21.26) implies

$$\sum_j q_{dj}\mu_j - \sum_j q_{cj}\mu_j \leq 0,$$

such that the path $\overrightarrow{V_a V_c}, \overrightarrow{V_c V_d}, \ldots, \overrightarrow{V_w V_b}$ entails

$$\sum_j q_{bj}\mu_j - \sum_j q_{aj}\mu_j \leq 0. \tag{21.31}$$

Similarly, there exists a path $\overrightarrow{V_b V_f}, \overrightarrow{V_f V_g}, \ldots, \overrightarrow{V_y V_a}$, from which it follows that

$$\sum_j q_{aj}\mu_j - \sum_j q_{bj}\mu_j \leq 0. \tag{21.32}$$

So, (21.31) and (21.32) together imply

$$\sum_j q_{aj}\mu_j = \sum_j q_{bj}\mu_j = -\delta_t, \tag{21.33}$$

with a suitably chosen constant $-\delta_t$, for all pairs of items $I_a, I_b \in C_t$. Therefore, all equations (21.33) taken together yield (21.21) with $\boldsymbol{\alpha} = -\boldsymbol{\mu}$. Moreover, the inequalities (21.22) follow immediately from analogous arguments whenever, for some $V_a \in C_s$ and $V_b \in C_t$, there exists a diline $\overrightarrow{V_a V_b}$.

Finally, consider the case of well-conditioned data; then there exists just one strong component which is identical to C. By the same arguments that led to

(21.33) for any component C_t, Jacobsen's equations (21.23) together with (21.26) yield

$$\sum_j q_{ij}\mu_j = -\delta \tag{21.34}$$

for all items I_i. Since the rank $r(\boldsymbol{Q}^+)$ was assumed to be $m+1$, the linear system (21.34) can have only the trivial solution $\boldsymbol{\mu} = \mathbf{0}$. And because (21.34) is identical with the AS (21.21) in cases of well-conditioned data, both alternative systems have only the trivial null-solution. □

This result can also be summarized as follows:

Corollary 3 *The sets of solutions $\boldsymbol{\alpha}$ in (21.21)–(21.22) and of $\boldsymbol{\mu}$ in (21.23) are identical.*

21.5 A uniqueness condition for JML in the LLTM

Since the unconditional likelihood of data under the RM is at the same time the likelihood in an incomplete BTL, the unconditional likelihood of data under the LLTM must be interpretable as the likelihood in an incomplete LBTL. It must therefore be possible to derive a n. s. uniqueness condition for the JML estimators $(\hat{\boldsymbol{\eta}}, \hat{\boldsymbol{\theta}})$ for the LLTM from Theorem 1 (Tanzer, 1984). This result is completely parallel to the previous results.

Let digraph C be defined as before, and let C_t, $t = 1, \ldots, u$, be its strong components. Then:

Theorem 3 *A unique JML solution $(\hat{\boldsymbol{\eta}}, \hat{\boldsymbol{\theta}})$ for the LLTM defined by (21.2) and (21.14) exists iff*

(a) the linear system of equations (21.35),

$$\sum_j q_{ij}\alpha_j = \delta_l \quad \text{for all } V_i \in C_l,\ l = 1, \ldots, u, \tag{21.35}$$

with unknown constants δ_l sufficing the inequality constraints (21.36),

$$\delta_s \le \delta_t \quad \text{if a diline } \overrightarrow{V_a V_b} \text{ exists for some } V_a \in C_s \text{ and some } V_b \in C_t, \tag{21.36}$$

has only the trivial solution $\boldsymbol{\alpha} = \mathbf{0}$, and

(b) the raw scores of all S_v satisfy the restrictions $0 < r_v < k_v$.

Proof of Theorem 3. We give a short direct proof using Jacobsen's theorem. We first transform the Jacobsen AS such that it corresponds to condition (a) of

Theorem 3. The likelihood of $a_{vi} = 1$ is

$$P(A_{vi} = 1) = \Big\{1 + \exp\big(\sum_j q_{ij}\eta_j - \theta_v\big)\Big\}^{-1},$$

and that of $a_{vi} = 0$ is

$$P(A_{vi} = 0) = \Big\{1 + \exp\big(-\sum_j q_{ij}\eta_j + \theta_v\big)\Big\}^{-1},$$

where the θ_v are scaled such that c in (21.14) is zero. Denoting the vector of unknowns in Jacobsen's AS (21.11) by $(\mu_1, \ldots, \mu_m, \lambda_1, \ldots, \lambda_n)$, the equations of the AS become

$$\sum_j q_{aj}\mu_j - \lambda_v \geq 0 \quad \text{for all } a_{va} = 1, \tag{21.37}$$

$$\sum_j q_{bj}\mu_j - \lambda_v \leq 0 \quad \text{for all } a_{vb} = 0. \tag{21.38}$$

Hence, the presence of a diline $\overrightarrow{V_a V_b}$ implies $\sum_j q_{aj}\mu_j \geq \sum_j q_{bj}\mu_j$, and $\overrightarrow{V_b V_a}$ similarly implies $\sum_j q_{aj}\mu_j \leq \sum_j q_{bj}\mu_j$. For any V_a and V_b belonging to the same strong component C_t, there must exist some chains $\overrightarrow{V_a V_x}, \overrightarrow{V_x V_y}, \ldots, \overrightarrow{V_z V_b}$ and $\overrightarrow{V_b V_g}, \overrightarrow{V_g V_h}, \ldots, \overrightarrow{V_p V_a}$, implying

$$\sum_j q_{ij}\mu_j = -\delta_l \quad \text{for all } V_i \in C_l,\ l = 1, \ldots, u, \tag{21.39}$$

with suitably defined constants δ_l. Furthermore, for any two vertices $V_a \in C_s$ and $V_b \in C_t$ with $s \neq t$, we get the inequality constraints

$$\delta_s \leq \delta_t \quad \text{if there exists a diline } \overrightarrow{V_a V_b}. \tag{21.40}$$

So condition (a) of Theorem 3 is immediately obtained from (21.39) and (21.40) by replacing $\boldsymbol{\mu}$ by $-\boldsymbol{\alpha}$. It can thus be seen that condition (a) is a consequence of Jacobsen's AS, i.e., (a) is necessary for the existence of a unique JML solution $\hat{\boldsymbol{\eta}}$.

For proving the necessity of (b), two cases need to be distinguished:

(i) Suppose that (a) holds, such that the AS (21.35)–(21.36) has only the null-solution $\boldsymbol{\alpha} = \mathbf{0}$, but (b) fails to hold due to the presence of some subject S_w with $r_w = k_w$. Then all equations (21.37)–(21.38) can be satisfied with $\boldsymbol{\mu} = \mathbf{0}$, $\lambda_v = 0$ for all $v \neq w$, and arbitrary $\lambda_w < 0$. Hence, the Jacobsen AS (21.37)–(21.38) has a non-null-solution $(\boldsymbol{\mu}, \boldsymbol{\lambda})$, implying that no unique JML solution $(\hat{\boldsymbol{\eta}}, \hat{\boldsymbol{\theta}})$ exists.

(ii) Suppose as before that (a) holds, such that the AS (21.35)–(21.36) has only the null-solution $\boldsymbol{\alpha} = \mathbf{0}$, but that (b) fails to hold because some subject S_w has $r_v = 0$. Then all equations (21.37)–(21.38) can be satisfied with $\boldsymbol{\mu} = \mathbf{0}$, $\lambda_v = 0$ for all $v \neq w$, and $\lambda_w > 0$. The conclusion again is that no unique JML solution $(\hat{\boldsymbol{\eta}}, \hat{\boldsymbol{\theta}})$ exists.

This proves that (a) and (b) are necessary for the existence of a unique JML solution.

The joint sufficiency of conditions (a) and (b) can be seen as follows: Let graph C have the strong components C_l, $l = 1, \ldots, u$, and let conditions (a) and (b) be satisfied. We show that then Jacobsen's AS (21.37)–(21.38) has only the null-solution $(\boldsymbol{\mu}, \boldsymbol{\lambda}) = (\mathbf{0}, \mathbf{0})$, which implies that the JML estimate $\hat{\boldsymbol{\eta}}$ is unique. The proof is based on two observations: Firstly, if Jacobsen's AS (21.37)–(21.38) had a solution $(\boldsymbol{\mu}, \boldsymbol{\lambda})$ with $\boldsymbol{\mu} \neq \mathbf{0}$, (21.39)–(21.40) would have to have a solution $(\boldsymbol{\mu}, \boldsymbol{\delta}) \neq (\mathbf{0}, \mathbf{0})$ too, which would imply that (21.35)–(21.36) have the solution $(-\boldsymbol{\mu}, -\boldsymbol{\delta}) \neq (\mathbf{0}, \mathbf{0})$, contradictory to assumption (a). Second, if Jacobsen's AS (21.37)–(21.38) had a solution $(\boldsymbol{\mu}, \boldsymbol{\lambda})$ with $\boldsymbol{\mu} = \mathbf{0}$ and $\boldsymbol{\lambda} \neq \mathbf{0}$, there would have to exist at least one S_v with $\lambda_v < 0$ (or > 0, respectively), which by (21.37) and (21.38) implies $a_{vi} = 1$ for all I_i (or $a_{vi} = 0$ for all I_i respectively), so that $r_v = k_v$ (or $r_v = 0$, respectively); this would contradict assumption (b). Hence, $(\boldsymbol{\mu}, \boldsymbol{\lambda}) = (\mathbf{0}, \mathbf{0})$ must hold true, implying that the JML solution $\hat{\boldsymbol{\eta}}$ is unique. □

An immediate consequence of Theorem 3 is Corollary 4 which deals with the case of well–conditioned data: If C has only one strong component, the AS reduces to a system of linear equations $\sum_j q_{ij}\alpha_j = \delta$ for $i = 1, \ldots, k$, which has only the null-solution $\boldsymbol{\alpha} = \mathbf{0}$ iff $\boldsymbol{Q}^+$ is of full rank $m + 1$.

Corollary 4 *If digraph C is strongly connected, a unique JML solution* $(\hat{\boldsymbol{\eta}}, \hat{\boldsymbol{\theta}})$ *for the LLTM defined in (21.2) and(21.14) exists iff*

(a) $r(\boldsymbol{Q}^+) = m + 1$, *and*

(b) $0 < r_v < k_v$ *for all* S_v.

21.6 Uniqueness in the 2PL with fixed discriminations

A possible application of the above results is the uniqueness problem in the "Two–Parameter Logistic" (2PL) model with given *fixed* discrimination parameters (OPLM; Verhelst et al., 1993). This model has recently received attention as a practical compromise between the RM and the 2PL model. The 2PL is defined by

$$P(+|S_v, I_i) = \frac{\exp[\alpha_i(\theta_v - \beta_i)]}{1 + \exp[\alpha_i(\theta_v - \beta_i)]}, \tag{21.41}$$

where $\alpha_i > 0$ denotes the discrimination of item I_i, and all other parameters are defined as before. For removing the overparameterization, we assume that $\alpha_1 = 1$ and $\beta_1 = 0$ (normalization).

It is immediately seen that (21.41) can be rewritten as

$$P(+|S_v, I_i) = \frac{\exp(\theta_{vi} - \beta_i^*)}{1 + \exp(\theta_{vi} - \beta_i^*)} \tag{21.42}$$

with ability parameters

$$\theta_{vi} = q_i\theta_v \quad \text{for } i = 1, \ldots, k; v = 1, \ldots, n; \tag{21.43}$$

the $\beta_i^* = q_i\beta_i$, for $i = 1, \ldots, k$, are transformed difficulty parameters. We replace α_i by q_i, because we consider the discrimination parameters as given and fixed, so that they are constants, not estimable parameters. It is interesting to see what the conditions for uniqueness of a JML solution $(\hat{\boldsymbol{\beta}}, \hat{\boldsymbol{\theta}})$ in this model are. One easily applicable result, directly obtainable from Theorem 1, is the following:

Corollary 5 *Let the $q_i > 0$ be given. If (a) the data are well-conditioned, and (b) $0 < r_v < k_v$ for all S_v, then the normalized JML estimates $(\hat{\boldsymbol{\beta}}, \hat{\boldsymbol{\theta}})$ of model (21.42)–(21.43) are unique.*

Proof of Corollary 5: We first exclude all subjects with $r_v = 0$ or $r_v = k_v$, that is, subjects with none or all responses correct, for which no finite $\hat{\theta}_v$ can be obtained. Note that the exclusion of these subjects has no influence on digraph C of the data. Henceforth we assume (b) to be satisfied.

The model in (21.42)–(21.43) is an incomplete LBTL. Applying Jacobsen's (1989) theorem, we see that the AS, with obvious notation, is

$$\begin{aligned} q_i\mu_v - \lambda_i \geq 0 \quad &\text{for } a_{vi} = 1, \\ q_i\mu_w - \lambda_i \leq 0 \quad &\text{for } a_{wi} = 0. \end{aligned} \tag{21.44}$$

This implies that $q_i\mu_v \geq q_i\mu_w$, i.e., $\mu_v \geq \mu_w$ if $a_{vi} = 1$ and $a_{wi} = 0$.

Now let a new digraph C^* be introduced where vertices V_v^* are assigned one–one to subjects, with adjacency matrix elements $c_{vw}^* = 1$ if there exists some item I_i such that S_v has response $a_{vi} = 1$ and S_w has $a_{wi} = 0$, and $c_{vw}^* = 0$ otherwise. We shall show that the assumptions (a) and (b) imply that C^* is strongly connected: First observe that from (a) it follows that each column of $\boldsymbol{A}$ must contain at least one 1 and one 0 (see Table 21.1); and that (b) implies the same for each row of $\boldsymbol{A}$. Then it is clear that, if C^* comprised at least two strong components, $\boldsymbol{A}$ would be ill-conditioned (see Table 21.1), contrary to assumption (a). Therefore, C^* must be strongly connected too.

From the strong connectedness of C^*, however, it follows that the inequalities (21.44) imply $\mu_v \geq \mu_w$ and $\mu_v \leq \mu_w$ for all pairs (S_v, S_w), which yields $\mu_v = \mu$ for all S_v. Reinserting this in (21.44), we find $q_i\mu = \lambda_i$ for all I_i. Normalizing $q_1 = 1$ and $\beta_1 = \lambda_1 = 0$, we find that $q_1\mu = \lambda_1$ implies $\mu = 0$, which in turn implies $\lambda_i = 0$ for all I_i. Hence, the Jacobsen AS (21.44) has only the trivial null-solution, and hence the JML solution $(\hat{\boldsymbol{\beta}}, \hat{\boldsymbol{\theta}})$ is unique. □

21.7 Discussion and Summary

The LBTL is introduced as a BTL model with linear constraints on the object parameters. It is a generalization both of the BTL model and of an extended BTL model with restrictions (21.6) as studied by El-Helbawy and Bradley (1977). In spite of this greater generality of the LBTL, our approach to solving the estimation equations (21.4), outlined in Section 21.2 of the present paper, is simpler and more efficient than the method proposed by El-Helbawy and Bradley (1977). The sufficient uniqueness condition of El-Helbawy and Bradley is a special case of our Theorem 1.

The LLTM can be viewed as a RM with linear constraints imposed on the item parameters. From an empirical point of view, the LLTM is therefore more restrictive (it has fewer free parameters) than the RM and will in general fit the data less well. As a formal model, however, the LLTM is more general, because the RM can be nested in the LLTM. It is not, therefore, surprising that the uniqueness conditions for CML estimation in the LLTM (Theorem 2) are weaker than those for the RM, where uniqueness necessarily requires that the data $\boldsymbol{A}$ be well-conditioned. The data matrix $\boldsymbol{A}$ being well-conditioned, however, is sufficient for the uniqueness of the CML estimators in both models, in the LLTM under the trivial rank condition $r(\boldsymbol{Q}^+) = m + 1$ (Corollary 2). Since the likelihood function of an LBTL can be reinterpreted as a conditional likelihood of an LLTM, it is plausible that the uniqueness conditions for both models are completely analogous.

The question of the uniqueness of the JML estimators in the LLTM can be answered by virtue of the fact that the unconditional likelihood of any RM is also a BTL likelihood: So it turns out that the uniqueness conditions for JML estimation in the LLTM are the same as for CML estimation, except for the additional restriction that all raw-scores must satisfy $1 < r_v < k_v$; the latter requirement, by the way, is the same as in JML estimation in the RM (Fischer, 1981).

A nice side result is that the sufficient condition for uniqueness of a JML solution in the 2PL with given fixed discriminations are, in case of well-conditioned data, the same as in the LLTM. The well-known problems of JML estimation in the 2PL must therefore be wholly due to a trade-off between the α_i on the one hand and the β_i and θ_v on the other.

Regarding potential uses of the results in this paper, most "classical" LLTM applications to item analysis are based on samples of at least several hundred subjects, ensuring that the data are well-conditioned; in all such cases, ascertaining the uniqueness of the CML solution is no great problem since it depends only on the rank $r(\boldsymbol{Q}^+)$. In "non-standard" LLTM applications, such as to the measurement of change (Fischer, 1989) or to the investigation of environmental effects on development (Fischer, 1993), however, the data are structurally incomplete to a high degree, and this may result in an ill-conditioned data matrix. In this case, the uniqueness results above become an integral part of the formulation of a psychometric model.

The new result on JML estimates in the LLTM may be of lesser practical relevance because JML is hardly used in practice. Nevertheless, there are cases where the JML result may prove important: For example, a recent paper of Verhelst and Glas (1993) takes up the old problem of dynamic generalizations of the RM, where a response–contingent change of the person parameter is assumed during the taking of the test. Formally, this dynamic test model can be considered an LLTM for "virtual items". The authors show that CML is not feasible because the conditional likelihood of any response pattern is wholly determined if the design vector and the raw score are given, but they point out that MML estimates still do exist; they in fact argue that MML is the solution to the given problem. Our result on JML in the LLTM adds a new facet to this discussion:

Parametric MML is based on the unconditional likelihood weighted with the density of an assumed distribution of the person parameters. We have shown that the unconditional likelihood function of an LLTM can have a unique maximum only if the conditional likelihood also has one, which is not the case here. Hence, the unconditional likelihood of the dynamic RM would never have a unique maximum. So, if MML yields a unique result, this uniqueness must essentially be due to the distribution assumed for the person parameters. Since the researcher usually has no sound *a priori* knowledge of that distribution, the results based on such assumption appear questionable. More theoretical research seems to be needed for a definite answer to this problem.

In concluding we state that rigorous uniqueness results are important tools in the study of the psychometric models discussed in this chapter.

21.8 References

Andersen, E.B. (1973a). Conditional inference for multiple–choice questionnaires. *British Journal of Mathematical and Statistical Psychology, 26,* 31–44.

Andersen, E.B. (1973b). *Conditional inference and models for measuring.* Copenhagen: Mentalhygiejnisk Forlag.

Andersen, E.B. (1980). *Discrete statistical models with social science applications.* Amsterdam: North–Holland Publishing Company.

Barisch, S. (1989). *Einstellung zur Epilepsie und Einstellungsänderung durch Information.* [Attitudes towards epilepsy and attitude change via information.] Master's thesis. Vienna: University of Vienna.

Bradley, R.A. (1984). Paired comparisons: Some basic procedures and examples. In P.R. Krishnaiah & P.K. Sen (Eds.), *Handbook of statistics, Vol. 4* (pp. 299–326). New York: Elsevier Science Publishers.

Bradley, R.A., & Terry, M.E. (1952). Rank analysis of incomplete block designs. I. The method of paired comparisons. *Biometrika, 39,* 324–345.

Christofides, N. (1975). *Graph theory. An algorithmic approach.* London: Academic Press.

El-Helbawy, A.T., & Bradley, R.A. (1977). Treatment contrasts in paired comparisons: Convergence of a basic iterative scheme for estimation. *Communications in Statistics - Theory and Methods, 6,* 197–207.

Embretson, S.E. (1985). Multicomponent latent trait models for test design. In S.E. Embretson (Ed.), *Test design: Developments in psychology and psychometrics* (pp. 195–218). Orlando: Academic Press.

Embretson, S.E. (1991). A multidimensional latent trait model for measuring learning and change. *Psychometrika, 56,* 495–514.

Fischer, G.H. (1972). A measurement model for the effect of mass-media. *Acta Psychologica, 36,* 207–220.

Fischer, G.H. (1973). The linear logistic test model as an instrument in educational research. *Acta Psychologica, 37,* 359–374.

Fischer, G.H. (1974). *Einführung in die Theorie psychologischer Tests.* [Introduction to the theory of psychological tests.] Bern: Huber.

Fischer, G.H. (1981). On the existence and uniqueness of maximum-likelihood estimates in the Rasch model. *Psychometrika, 46,* 59–77.

Fischer, G.H. (1983a). Logistic latent trait models with linear constraints. *Psychometrika, 48,* 3–26.

Fischer, G.H. (1983b). *Zum Problem der Validierung diagnostischer Entscheidungen in der Verkehrspsychologie.* [On the problem of validating diagnostic decisions in traffic psychology.] Unpublished manuscript. Vienna: Department of Psychology, University of Vienna.

Fischer, G.H. (1989). An IRT-based model for dichotomous longitudinal data. *Psychometrika, 54,* 599–624.

Fischer, G.H. (1991). A new methodology for the assessment of treatment effects. *Evaluación Psicológica - Psychological Assessment, 7,* 117–147.

Fischer, G.H. (1992). The Saltus Model revisited. *Methodika, 6,* 87–98.

Fischer, G.H. (1993). The measurement of environmental effects: An alternative to the estimation of heritability in twin data. *Methodika* (in press).

Fischer, G.H., & Formann, A.K. (1972). Conditional maximum-likelihood estimation of item parameters for a linear logistic test model. *Research Bulletin, No. 9.* Vienna: Department of Psychology, University of Vienna.

Fischer, G.H., & Formann, A.K. (1981). Zur Schätzung der Erblichkeit quantitativer Merkmale. [On the estimation of the heritability of quantitative traits.] *Zeitschrift für Differentielle und Diagnostische Psychologie, 2,* 189–197.

Fischer, G.H., & Formann, A.K. (1982a). Some applications of logistic latent trait models with linear constraints on the parameters. *Applied Psychological Measurement, 4,* 397–416.

Fischer, G.H., & Formann, A.K. (1982b). Veränderungsmessung mittels linear-logistischer Modelle. [Measuring change by means of linear logistic models.] *Zeitschrift für Differentielle und Diagnostische Psychologie, 3,* 75–99.

Fischer, G.H., & Pendl, P. (1980). Individualized testing on the basis of the dichotomous Rasch model. In L.J.Th. van der Kamp, W.F. Langerak, & D.N.M. de Gruijter (Eds.), *Psychometrics for educational debates* (pp. 171–188). New York: Wiley.

Ford, L.R.Jr. (1957). Solution of a ranking problem from binary comparisons. *American Mathematical Monthly, 64,* 28–33.

Formann, A.K. (1973). *Die Konstruktion eines neuen Matrizentests und die Untersuchung des Lösungsverhaltens mit Hilfe des linear logistischen Testmodells.* [The construction of a new matrices test and the investigation of test behavior by means of the linear logistic test model.] Doctoral dissertation. Vienna: University of Vienna.

Formann, A.K., & Piswanger, K. (1979). *Wiener Matrizen-Test. Ein Rasch-skalierter sprachfreier Intelligenztest.* [The Viennese Matrices Test: A Rasch-scaled nonverbal intelligence test.] Weinheim: Beltz Test.

Gittler, G., & Wild, B. (1988). Der Einsatz des LLTM bei der Konstruktion eines Itempools für das adaptive Testen (pp. 115–139). [The use of the LLTM in constructing an item pool for adaptive testing.] In K.D. Kubinger (Ed.), *Moderne Testtheorie.* Weinheim: Beltz.

Glas, C.A.W., & Verhelst, N.D. (1989). Extensions of the partial credit model. *Psychometrika, 54,* 635–659.

Haberman, S.J. (1977). Maximum likelihood estimates in exponential response models. *The Annals of Statistics, 5,* 815–841.

Hambleton, R.K. (1989). Principles and selected applications of item response theory. In R.L. Linn, (Ed.), *Educational measurement* (pp. 147–200). New York: Macmillan and London: Collier Macmillan.

Harary, F., Norman, R.Z., & Cartwright, D. (1965). *Structural models: An introduction to the theory of directed graphs.* New York: Wiley.

Heckl, U. (1976). *Therapieerfolge bei der Behandlung sprachgestörter Kinder.* [Effects of therapy in speech-handicapped children.] Doctoral dissertation. Vienna: University of Vienna.

Heinrich, I. (1975). Beeinflussung des logischen Schlußprozesses durch semantische Variation seiner Elemente. [The influence of semantic variation of the elements of syllogisms on the reasoning process.] *Psychologische Beiträge, 17,* 497–518.

Hornke, L.F., & Habon, M.W. (1986). Rule-based item bank construction and evaluation within the linear logistic framework. *Applied Psychological Measurement, 10,* 369–380.

Hornke, L.F., & Rettig, K. (1988). Regelgeleitete Itemkonstruktion unter Zuhilfenahme kognitionspsychologischer Überlegungen. [Rule-based item construction using concepts from cognitive psychology.] In K.D. Kubinger (Ed.), *Moderne Testtheorie* (pp. 140–162). Weinheim: Beltz.

Iby, M. (1987). *Die Effektivität von Kommunikationsseminaren in der innerbetrieblichen Ausbildung der Zentralsparkasse.* [The efficacy of communication seminars for employees of the "Zentralsparkasse".] Unpublished doctoral dissertation. Vienna: University of Vienna.

Jacobsen, M. (1989). Existence and uniqueness of MLEs in discrete exponential family distributions. *Scandinavian Journal of Statistics, 16,* 335–349.

Kubinger, K.D. (1979). Das Problemlöseverhalten bei der statistischen Auswertung psychologischer Experimente. Ein Beispiel hochschuldidaktischer Forschung. [The task solving behavior in the statistical analysis of psychological experiments. An example of research in didactics.] *Zeitschrift für Experimentelle und Angewandte Psychologie, 26,* 467–495.

Kubinger, K.D. (1980). Die Bestimmung der Effektivität universitärer Lehre unter Verwendung des linear-logistischen Testmodells von Fischer - Neue Ergebnisse. [Determining the effects of teaching at universitiy by means of the linear logistic test model of Fischer - New results.] *Archiv für Psychologie, 133,* 69–70.

Luce, R.D. (1959). *Individual Choice Behavior.* New York: Wiley.

Mislevy, R.J. (1988). Exploiting auxiliary information about items in the estimation of Rasch item difficulty parameters. *Applied Psychological Measurement, 12,* 281–296.

Mutschlechner, R. (1987). *Der Patient im Krankenhaus. - Ein Versuch, die Wirksamkeit einer psychologischen Betreuung nachzuweisen.* [The patient in hospital. - An attempt to assess the effects of psychological treatments.] Doctoral dissertation. Vienna: University of Vienna.

Nährer, W. (1980). Zur Analyse von Matrizenaufgaben mit dem linear logistischen Testmodell. [On the analysis of matrices items by means of the linear logistic test model.] *Zeitschrift für experimentelle und angewandte Psychologie, 27,* 553–564.

Pfanzagl, J. (1993). On item parameter estimation in certain latent trait models. In G.H. Fischer & D. Laming (Eds.), *Contributions to mathematical psychology, psychometrics, and methodology* (pp. 249–263). New York: Springer-Verlag.

Piswanger, K. (1975). *Interkulturelle Vergleiche mit dem Matrizentest von Formann.* [Cross–cultural comparisons with Formann's matrices test.] Doctoral dissertation. Vienna: University of Vienna.

Rop, I. (1977). The application of a linear logistic model describing the effects of pre–school curricula on cognitive growth. In H. Spada & W. F. Kempf (Eds.), *Structural models of thinking and learning* (pp. 281–293). Bern: Huber.

Scheiblechner, H. (1972). Das Lernen und Lösen komplexer Denkaufgaben. [The learning and solving of complex reasoning items.] *Zeitschrift für experimentelle und angewandte Psychologie, 19,* 476–506.

Schmied, Ch. (1987). *Die Effektivität von Managementseminaren, durchgeführt bei den Österreichischen Bundesbahnen.* [The efficacy of management seminars for employees of the Austrian Federal Railways.] Doctoral dissertation. Vienna: University of Vienna.

Sijtsma, K. (1982). Een lineair logistisch model ter verklaring van de moeilijkheidsparameters van breukrekenitems. [A linear logistic model for the explanation of difficulty parameters of fractional arithmetics items.] In J.G.L.C. Lodewijks & P.R.J. Simons (Eds.), *Strategieën in leren en ontwikkeling* (pp. 69–79). Lisse: Swets & Zeitlinger.

Spada, H. (1976). *Modelle des Denkens und Lernens.* [Models of thinking and learning.] Bern: Huber.

Spada, H., & Kluwe, R. (1980). Two models of intellectual development and their reference to the theory of Piaget. In R. Kluwe & H. Spada (Eds.), *Developmental models of thinking* (pp. 1–30). New York: Academic Press.

Spada, H., & May, R. (1982). The linear logistic test model and its application in educational research. In D. Spearritt (Ed.), *The improvement of measurement in education and psychology* (pp. 67–84). Hawthorne, Victoria: The Australian Council for Educational Research.

Spada, H., & McGraw, B. (1985). The assessment of learning effects by means of linear logistic test models. In S.E. Embretson (Ed.), *Test design: Developments in psychology and psychometrics* (pp. 169–194). Orlando: Academic Press.

Tanzer, N. (1984). On the existence of unique joint maximum-likelihood estimates in linear logistic latent trait models for incomplete dichotomous data. *Research Bulletin, No. 25.* Vienna: Department of Psychology, University of Vienna.

Traub, R.E., & Lam, R. (1985). Latent structure and item sampling models for testing. *Annual Review of Psychology, 36,* 19–48.

Van de Vijver, F.J.R. (1988). Systematizing the item content in test design. In R. Langeheine & J. Rost (Eds.), *Latent trait and latent class models* (pp. 291–307). New York: Plenum.

Van Maanen, L., Been, P., & Sijtsma, K. (1989). The linear logistic test model and heterogeneity of cognitive strategies. In E.E. Roskam (Ed.), *Mathematical psychology in progress* (pp. 267–287). New York: Springer.

Verhelst, N.D., & Glas, C.A.W. (1993). A dynamic generalization of the Rasch model. *Psychometrika, 58* (in press).

Verhelst, N.D., Glas, C.A.W., & Verstralen, H.H.F.M. (1993). *OPLM: a computer program and manual.* Arnhem: CITO.

Widowitz, E. (1987). *Der Effekt autogenen Trainings bei funktionellen Erkrankungen.* [The effect of "autogenous training" on the functional syndrome.] Master's thesis. Vienna: University of Vienna.

Wilson, M. (1989). Saltus: A psychometric model of discontinuity in cognitve development. *Psychological Bulletin, 105 (2),* 276–289.

Wilson, M. (Ed.) (1992). *Objective measurement. Theory into practice.* Vol. 1. Norwood, NJ: Ablex Publishing Company.

Witek, J. (1980). Die Effektivität gruppendynamischen Sensitivity Trainings. *Zeitschrift für Experimentelle und Angewandte Psychologie, 27,* 335–345.

Zeman, M. (1980). *Kursmäßige Legasthenikertherapie an Wiens Schulen.* [Treatment programs for dislexic children in Viennese schools.] Vienna: Ketterl.

Zermelo, E. (1929). Die Berechnung der Turnierergebnisse als ein Maximumproblem der Wahrscheinlichkeitsrechnung. [The computation of tournament results as a maximum-probability problem.] *Mathematische Zeitschrift, 29,* 436–460.

Zimprich, H. (1980). Behandlungskonzepte und -resultate bei psychosomatischen Erkrankungen im Kindesalter. [Treatment principles and results in psychosomatic diseases of children.] *Pädiatrie und Pädologie, Supplementum 6,* 131–198.

22

Optimum Design in Item Response Theory: Test Assembly and Item Calibration

W. J. van der Linden[1]

ABSTRACT The idea of optimizing experimental design to give estimators maximal efficiency has been around in the statistical literature for several decades, but its applicability to sampling problems in item response theory (IRT) has not been widely noticed. It is the purpose of this paper to show how optimum design principles can be used to improve item and examinee sampling in IRT-based test assembly and item calibration. For both applications a result based on the maximin principle is given. The maximin principle fits these applications naturally, because IRT models are nonlinear and involve criteria of optimality that are dependent on the unknown parameters.

22.1 Optimum Design in IRT: Applications to Test Assembly and Item Calibration

The main topics addressed in statistics are parameter estimation and hypothesis testing. For both topics the literature has produced fruitful methods of finding estimators and test statistics as well as important criteria to evaluate their performances. Most statistical theory is based on the assumption of simple random sampling, and it seems safe to assert that the majority of the statisticians have hardly any interest in sampling beyond this assumption. Two exceptions to this practice are known, though. One is the interest in more complicated sampling procedures than simple random sampling, notably in the domain of survey research (Kalton, 1983; Särndal, Swensson, and Wretman, 1992). Another exception began with the pioneering work on optimum design of statistical experiments by R. A. Fisher, who, ironically, can also be considered the founder of mainstream statistics. Fisher's work originated in the domain of linear models. Linear models have the important aspect that they focus on the outcomes of statistical experiments in which the statistician may have control over some of the variables, and hence is able to design an experiment in which sampling with respect to estimation of the parameters is optimal. An example is the experiment underlying the estimation of the parameters in a bivariate linear regression model in which the predictor is a fixed variable and the statistician can select the levels of this predictor. Likewise, if the predictor is a random variable, the statistician may have control of the probabilities with which the levels of the

[1]University of Twente, Enschede, The Netherlands

predictor are present in the experiment. All such experiments share an hierarchical aspect which is reflected in the sampling procedures involved. First, the levels of the independent variables are sampled. Once the levels have been selected, the conditional distributions of the dependent variable given the levels are sampled. The first stage of sampling may or may not be random, but the second stage is always assumed to be (simply) random. Though the majority of Fisher's work immediately stimulated others to pursue his lines of research, the interest in optimum experimentation was modest at first. An exception was the Russian literature, which was unknown to most statisticians in the West until Fedorov's book was translated into English (Fedorov, 1972). In the late 1950s the scene changed, however, mainly due to a series of papers published by Kiefer (1959) and Kiefer and Wolfowitz (1960). The paper by Kiefer was a leading paper with discussion published in the *Journal of the Royal Statistical Society*. Nowadays, optimum design is a respected field of statistical research with numerous theoretical results and applications. Reviews of the literature on optimum design can be found in Atkinson (1982), Silvey (1980), and Steinburg and Hunter (1984). Though the original interest was in linear models, remarkable results have been derived for nonlinear models too. In spite of these advances, the literature on Item Response Theory (IRT) has hardly noticed their potential for its own models. Nevertheless, in two important experiments, IRT shares the hierarchical aspect with the linear models alluded to above. One experiment is test assembly where the psychometrician has to sample items from a pool to estimate the values of the ability parameters of examinees. The other experiment is item calibration, where examinees have to be sampled to estimate the values of the parameters of the items in the pool ("item calibration"). The main part of this chapter is devoted to an illustration of the fact that optimum design principles can be applied to optimize these two experiments. Since in this field the word "experiment" is a somewhat uncommon term to denote such activities as test assembly and item calibration, but it is quite common to view these activities as instances of item and examinee sampling, these applications will be addressed as problems of "optimum sampling design", or briefly "optimum design". For both test assembly and item calibration a result will be presented that is based on the maximin principle. Since for IRT models the quality of a sampling design is a function of unknown parameter values, the application of this principle is a natural way to prevent the psychometrician from incurring unexpectedly unfavorable results. Before continuing this discussion, however, some basic concepts will be defined, using the example of linear models, and a few technical difficulties inherent in the application of optimum design procedures to nonlinear models will be clarified.

22.2 Optimum Design in Linear Models

The following well-known statistical problem is used to introduce the problem of optimum experimental design and define a few basic concepts. Let x be a (fixed) predictor variable and Y a criterion variable the value of which is to be predicted by x. Further, let the interest be in estimating the parameters in the equation of Y on x that is used to make the prediction. It is assumed that the equation is linear. The problem of optimum design for this statistical problem can now be formulated as follows: For which distribution of values of x should we sample Y to get estimators for the regression parameters with optimal properties? In principle, different criteria of optimality can be chosen, but if the interest is in efficient estimators, the usual choice is (a function of) the (asymptotic) variance of the estimators or of its reciprocal known as Fisher's information measure. Throughout this chapter, it will be assumed that this choice has been made.

The *design point* of the experiment is the vector of values of x for which $Y|x$ is sampled. A separate vector can be defined to denote how often $Y|x$ is sampled for each design point. Together these two vectors define the *design* of the experiment. To deal with the case in which the fixed predictor in this example is replaced by a random variable, the frequencies in the latter vector can be replaced by probabilities. This problem will not be solved formally here. For our purposes it suffices to observe that a design with extreme values of x (both high and low) will be more informative than one with values in the middle. For the estimator of the slope of the regression line, this observation follows immediately form the fact that the slope of the sample regression line will be less sensitive to the variability in Y, the farther away the x values for which the observations on Y are obtained. The same holds for the estimator of the intercept. Of course, these relations only apply if the criterion of optimality is (a monotone) function of the variance of the estimator. If the interest is, for instance, in minimizing possible bias in the estimator, other sampling designs may be better. Also, if the interest is not in optimizing the estimators, but in maximizing the power of a statistical test against specific violations of the model, then a design with extreme values for x is unlikely to be best for most alternative hypotheses. Finally, it should be noted that the quality of the design depends on the type of model considered. If the interest is in nonlinear regression models, then, for instance, we will generally not be able to estimate the location of the peaks and valleys of the curves accurately if the values of x are concentrated at the extreme ends of the scale. Before going more deeply into problems related to optimum design for nonlinear models, let us first consider the multivariate generalization of the linear regression problem above. In the multivariate case, the problem is to estimate the vector of k parameters, $\boldsymbol{\beta}$, in the linear model

$$E(Y|\boldsymbol{x}) = \boldsymbol{x}'\boldsymbol{\beta},$$

where $\boldsymbol{x}$ is a vector with values for the k predictor variables. If the unknown parameters are estimated by least squares, the covariance matrix of the estimators of $\boldsymbol{\beta}$ is asymptotically equal to

$$\mathrm{Var}(\hat{\boldsymbol{\beta}}) = (\boldsymbol{X}'\boldsymbol{X})^{-1}\sigma^2,$$

where the $k \times k$ matrix $\boldsymbol{X}'\boldsymbol{X}$ is Fisher's information matrix (assumed to be of full rank here).

As the problem is now one of estimating multiple slope parameters, the choice of a criterion of optimality is not as straightforward as in the case of bivariate regression. Several criteria have been studied for this purpose, the ones best known being the following three:

D-optimality: Maximize $|\boldsymbol{X}'\boldsymbol{X}|$,

A-optimality: Minimize $\mathrm{tr}(\boldsymbol{X}'\boldsymbol{X})^{-1}$,

E-optimality: Minimize the maximum eigenvalue of $(\boldsymbol{X}'\boldsymbol{X})^{-1}$.

The criterion of D-optimality maximizes the determinant of Fisher's information matrix. As the inverse of the matrix is an (asymptotic) generalized variance measure, differing from $\mathrm{Var}(\hat{\boldsymbol{\beta}})$ only by the constant σ, the criterion minimizes $\mathrm{Var}(\hat{\boldsymbol{\beta}})$ in this sense. The criterion of A-optimality minimizes the trace of the matrix $(\boldsymbol{X}'\boldsymbol{X})^{-1}$ over the space of all possible designs, and hence does the same for the covariance matrix of the estimator of $\boldsymbol{\beta}$. Instead of the trace, the criterion of E-optimality minimizes the maximum eigenvalue of the covariance matrix.

Obviously, each of these criteria has different properties. The criterion of D-optimality has the important advantage that it uses the full information in Fisher's information matrix. In addition, being a determinant, it is invariant across linear transformation of the scale of the parameters. Important theorems have been proven relating results for these and other criteria to one another. It is not our intention to review these results, but merely to show that multiple-parameter problems involve covariance or information matrices and that different criteria may be formulated each optimizing a different aspect of such matrices. The reader interested in background information is referred to Fedorov (1972) or Kiefer (1959).

22.3 Optimum Design in IRT

Consider the case of an IRT model for random responses U_{ij} of examinee j on item i, and assume we have a pool of items and a population of examinees indexed by $i = 1, \ldots, I$ and $j = 1, \ldots, J$, respectively. A possible model could be the logistic model for binary response variables introduced by Rasch (1960) and given here below, but the treatment equally applies to other models and response formats. For an introduction to a large variety of models, see Hambleton and Swaminathan (1985). Further, let $\boldsymbol{\theta}$ be the vector of the abilities of the examinees in the population and $\boldsymbol{\pi}$ the vector of the parameters of the items in the pool. Depending on the model, $\boldsymbol{\pi}$ may consist of a series of subvectors of parameters denoting different properties of the items.

Now, as two distinct sets of parameters are present, the values of both of which have to be estimated to use the model in practice, two different optimum design problems can be identified:

1. For what values of $\boldsymbol{\pi}$ should we sample the distributions of $(U_{ij})_{I\times J}$ to estimate $\boldsymbol{\theta}$?

2. For what values of θ should we sample the distributions of $(U_{ij})_{I\times J}$ to estimate $\boldsymbol{\pi}$?

The former problem arises when a test has to be assembled that is optimal for measuring examinees with abilities given by the vector $\boldsymbol{\theta}$. In the latter problem, examinees are selected from a population to calibrate the items in the pool.

Observe that both problems have an hierarchical aspect that is identical to the one for the case of simple bivariate regression discussed earlier. The design points are the possible values for one set of the parameters. Once the points and their frequencies are selected, the response variables associated with these design points are sampled and the other parameters are estimated. Unlike the case of bivariate regression, however, it is not immediately clear how a solution to the above two problems could be implemented; in practice it is not known what values $\boldsymbol{\theta}$ and $\boldsymbol{\pi}$ have for the examinees and items at hand. This point will be taken up below, but first a simplified example will be given to show that the selection of design points does determine the information in the sample about the parameters.

Let

$$p_i(\theta) \equiv P_i(X=1) = [1+\exp(\delta_i - \theta)]^{-1}$$

represent the Rasch model for a fixed item i, with parameter δ_i to be estimated. The model can only be identified if at least two items are addressed, but this point does not bother us at the moment. Now assume that we have an infinite pool of examinees with known abilities and that the purpose is to estimate the unknown parameter δ_i using the method of maximum likelihood. It follows immediately from the model that, if J examinees are to be sampled, Fisher's information in the sample,

$$I(\delta_i) = \sum_{j=1}^{J} p_i(\theta_j)[1 - p_i(\theta_j)],$$

is maximal for a vector with design points $\theta_j = \delta_i$, $j = 1, \ldots, J$. In other words, the sample of examinees will be optimal if all examinees have abilities equal to the difficulty of the item.

As already observed, the example is not yet realistic; we do not know (exactly) the abilities of the examinees, nor do we know the (exact) value of the item difficulty parameter. As for the unknown abilities, in this respect IRT differs from a field like bio-assay, where models of the same logistic form as the Rasch model are used, for instance, to study the regression of mortality in a sample on the dosage level of some drug. In such applications, θ is a known dosage level and can be manipulated by the experimenter. However, bio-assay and IRT share the problem of the optimum design being a function of the (unknown) parameter to be estimated. As the remainder of this paper will show, the two problems of unknown parameter values can be circumvented in various ways and applications of IRT can still profit from applying optimum design principles. But first we will delve somewhat deeper into these problems.

Two Technical Problems and Their Solutions

The paradox of an optimum design dependent on the parameter to be estimated is typical of all nonlinear models. In the linear regression model, the information on the slope parameter in the data is the same for all intervals of equal length on the x-axis for which Y is sampled. However, if the model is replaced by a nonlinear regression model, and a parameter has to be estimated which, for instance, is related to a certain peak in the regression function, different intervals will yield different information, and generally the best interval is the one centered on the location of the (unknown) peak. More formally, Fisher's information matrix, and hence the asymptotic covariance matrix of maximum-likelihood estimators, is a function of the derivatives of the likelihood with respect to the unknown parameter(s). If the likelihood is nonlinear in the parameters, the derivatives are also functions of the unknown parameter(s). Hence, in nonlinear models all criteria of optimality based on Fisher's information are dependent on the parameter(s) to be estimated. Only for a linear model does this dependency vanish.

The problem is not hopeless, though. Possible strategies to deal with the dependency are:

1. Using a *local approach,* that is, maximizing (a function of) Fisher's information for plausible values of the parameter. This approach is not unknown in educational and psychological testing. For example, if the interest is in assembling a test for a certain population of examinees, it is mostly known in what interval the abilities will be and a test is usually designed to match this interval. The practice is formalized in IRT-based test assembly, where tests are assembled to meet a target for the test information *function* and the information function is taken to be Fisher's information as a function of the ability parameter. In the model for test assembly to be presented below, a finite number of values for the ability parameter are selected, but it is known that the intended result will hold approximately for small intervals covering these values.

2. Applying a *maximin criterion,* that is, maximizing the minimum of (a function of) Fisher's information over a realistic interval of possible values for the parameter. The maximin criterion guarantees a known minimum amount of information, wherever the true value of the parameter is. In this chapter, the maximin criterion is used both in the model for optimal test assembly and in a result for optimal item calibration.

The example in the previous section showed not only that the usual criteria of optimality depend on the values of the ability parameters, but also that the design points are values of an unknown parameter. This problem is typical of all models that, in addition to the parameters to be estimated, have incidental or nuisance parameters. Though it seems that optimum design solutions, therefore, cannot be experimentally realized, this conclusion is too premature. Possible strategies to circumvent problems due to the presence of incidental parameters are:

1. Using *initial estimates* for the incidental parameters. This solution is possible if information from previous experiments is available that provides estimates. An example is the use of average ability scores of schools from a previous assessment of educational progress to calibrate items for the next study. If previous information is poor, sensitivity analysis can be used to explore the effects of small changes in the initial estimates on the efficiency of the estimators.

2. Using *Bayes methods,* that is, assigning prior distributions to the parameters which are then eliminated by taking expectations. These prior distributions may be subjective or empirical. If previous information is present which is of a global nature and cannot be translated into an empirical point estimate, it may be translated into a subjective prior. If, in the example of assessment of educational progress above, estimates of the full distributions of ability scores are available, the distributions can be used as empirical priors.

3. Using *sequential methods* by repeating one of the above strategies and updating previous estimates at each stage. Examples of sequential designs applied to the problem of item calibration can be found in Berger (1992). In empirical Bayes methods, the second strategy is repeated using data to update an empirical prior. An empirical Bayes method for calibrating items under the Rasch model in a computerized testing environment is given in van der Linden and Eggen (1986).

Different selections from the above strategies are used in the treatment of the optimum test assembly and item calibration problems in the next two sections.

22.4 A Maximin Model for Test Assembly

There has recently been a series of papers using the methodology of $0-1$ Linear Programming (LP) to optimally sample items from a pool. The majority of these papers use objective functions based on Fisher's information in the test about the unknown ability of the examinees and can be classified as optimum design methods (Adema, 1990a, 1990b; Amstrong, Jones and Wu, 1992; Boekkooi-Timminga, 1987, 1989, 1990a, 1990b; Theunissen, 1985). Other papers deal with different problems, such as the optimization of classical test reliability (Adema and van der Linden, 1989), matching tests (van der Linden and Boekkooi-Timminga, 1989), or maximizing the generalizability of tests (Sanders, Theunissen and Baas, 1989, 1991).

As an illustration of the methodology, an optimization model for test assembly is presented which belongs to the class of Mixed Integer Programming (MIP) models for test assembly. A full treatment of the model can be found in van der Linden and Boekkooi-Timminga (1989). The model applies when test items are to be sampled from an item pool to produce a test that measures the unknown ability of an examinee with maximal information. The following selection of strategies is made:

1. The problem of the unknown ability parameters is solved by taking a local approach, that is, the information about the ability parameter is maximized for a set of previously chosen values θ_k, $k = 1, \ldots, K$.

2. Since this choice involves the problem of optimizing K different objective functions, one for each value θ_k, a maximin approach over these K functions is followed.

3. It is assumed that initial estimates for the values of the item parameters are available which may be taken to be equal to their true values.

It should be noted that, though the choice of the values θ_k is arbitrary, the consequences of the choice are minor. For the current IRT models, Fisher's information is a smooth function of θ, and its behavior at a selection of points will not differ dramatically from that at adjacent points. Moreover, the choice of the number of points as well as their spacing is at our disposal. Extensive simulation studies have shown that a selection of $3-5$ points leads to satisfactory results (Timminga, 1985).

The last assumption is not unusual in test item banking, where items are first calibrated using response data from a large number of examinees to guarantee sufficient accuracy and then the estimates of the values of the item parameters are treated as if they are the true parameter values. Surprisingly, however, not much is known about the seriousness of the assumption.

Let $I_i(\theta_k)$ denote the information on the value of θ_k, $k = 1, \ldots, K$, in responses on item i, for $i = 1, \ldots, I$. Further, the design of the sample is defined by the set of decision variables $x_{ij} \in \{0,1\}$, for $i = 1,\ldots,I$ and $j = 1,\ldots,J$, where $x_{ij} = 1$ if examinee j is given item i, and $x_{ij} = 0$ otherwise.

Now, using the fact that, for a fixed value of θ, the response variables are independent,

$$\sum_{i=1}^{I} I_i(\theta_k)x_{ij}$$

is the information in test $(x_{ij})_{I\times j}$ for examinee j for ability θ_k.

Suppose we constrain the test by giving examinee j exactly n_j items:

$$\sum_{i=1}^{I} x_{ij} = n_j.$$

The maximin model for designing a test for examinee j over all values $\theta_1,\ldots,\theta_k$ is given by:

$$\text{maximize } y$$

subject to

$$\sum_{i=1}^{I} I_i(\theta_k)x_{ij} \leq y, \qquad k = 1,\ldots,K,$$

$$\sum_{i=1}^{I} x_{ij} = n_j,$$

$$x_{ij} \in \{0,1\}, \qquad i = 1,\ldots,I,$$

$$y \leq 0.$$

Though the model was formulated for a single examinee, this was just for notational convenience; its generalization to the complete case of J examinees is straightforward.

The model implies a uniform target for Fisher's information function for the test. The generalization to nonuniform target functions can be realized by introducing weights r_k for each point θ_k, inserting these as coefficients of y in the first constraint of the model.

Also, linear constraints can be added to the model to deal with various specifications the test constructor may have in mind. Some examples are specifications with respect to test composition, administration time, item parameter values, inclusion or exclusion of certain items, or inter-item dependencies. A full review of the possibilities is given in van der Linden and Boekkooi-Timminga (1989).

Computational Aspects and Examples

As already observed, the above model is fully linear in the decision variables, and as the variables x_{ij} are 0/1 and y is real, the model belongs to the class of MIP models. The other models in the above references belong to the same class or pure Integer Programming (IP) models. For small problems, an implementation of the branch-and-bound algorithm can be used to calculate a solution to the problem. However, as (M)IP problems can not generally be solved in polynomial time (Nemhauser and Wolsey, 1988, chap. I.5), for larger problems, with more than, say, 50 items, the use of a heuristic is recommended.

Favorable results have been obtained for a heuristic proposed by Adema (1988; Adema, Boekkooi-Timminga and van der Linden, 1992). The heuristic uses the reduced costs as well as the value of the objective function in a solution to a relaxed version of the problem to reduce the search tree. For a variety of models, the heuristic generally has produced solutions to problems with 500-1.000 decision variables which were less than 1% from the true solution and needed no more than 1-2 mins. of CPU time on various PCs. Also, the heuristic always produces results that meet all of the constraints. Examples can be found in the papers by Adema, Boekkooi-Timminga and van der Linden referred to earlier.

If the constraints are only taxonomic, a so-called greedy algorithm can be used to quickly approximate the solution (Nemhauser and Wolsey, 1988, sect. I.3.3). A greedy algorithm has been implemented in the computer program *Optimal Test Design* (Verschoor, 1991).

Usually, the same test assembly problem can be modeled in different ways, and a considerable amount of CPU-time may be saved if the model can be given a special structure. A powerful example is given in Amstrong, Jones and Wu (1992), who partially modeled the problem of assembling several test forms to match a seed test as a transportation problem. Transportation problems can generally be solved in polynomial time. For a large problem, consisting of 510 items and six test forms, solutions were obtained within 1 min. and 54 secs. of CPU time on a PC with 286 processor.

22.5 A Maximin Result on Item Calibration

The literature about optimum design for item calibration is not abundant. To the author's knowledge, the problem has only been addressed in recent papers by Berger (1992) and Berger and van der Linden (1992). A related problem is the problem of optimal linking in test equating studies. The idea underlying test equating is that two sets of test items that are calibrated separately can be brought together on the same scale, using response data on a selection of items from the two sets administered to both samples of examinees. Obviously, in this type of equating an optimality problem is involved, as there is a trade-off between the errors in the parameter estimates for the two item sets and in the estimate of the linking equation. Several studies of "blocking", "spiraling" , and "interlacing" designs have been conducted to empirically assess the quality of these types of designs (see, e.g., Davey, 1992; Vale, 1986; Wingersky and Lord, 1984), but straightforward optimization has not yet been attempted.

In this section, equating will be viewed as an item calibration problem for an incomplete design. It will then be shown that the problem addressed in the papers above no longer exists and there is no need to use sampling designs with a special internal structure.

The following strategies are assumed to deal with the two problems of unknown parameters:

1. The problem of unknown ability parameters is circumvented by accepting a Bayes approach. The examinees are assumed to be sampled from a population with a certain ability distribution. A distribution function representing the ability distribution need not be specified.

2. The problem of unknown item difficulties is solved using the maximin principle. Again, this means that a conservative approach is taken and that the minimum value of the criterion over an interval of realistic values for the difficulty parameter is minimized.

Again, let the design be defined by decision variables $x_{ij} \in \{0,1\}$, for $i = 1,\ldots,I$ and $j = 1,\ldots,J$. The ability distribution from which the examinees are sampled is denoted by c.d.f. $F(\theta)$. Further, suppose that all item parameters δ_i are contained in the interval $[\delta_l, \delta_u]$. Finally, it is assumed that J examinees are sampled and that each examinee gets a test of length N.

Now let $C(\delta,\theta)$ be the criterion of optimality for the Rasch model; for example, the criterion of D-optimality mentioned earlier. The case of other models will be addressed below. We can define the expected value of the criterion for a random examinee as

$$E[C(\delta)] \equiv E_\theta[C(\delta,\theta)] = \int C(\delta,\theta)dF(\theta).$$

The main result is summarized in the following theorem:

Theorem 1 *For a Bayes approach, the maximin principle, and any criterion of optimality, the optimum design of a calibration sample is determined only by the marginal sums of values for the decision variables in the design and not by its internal structure.*

Proof: As we sample examinees randomly, the expected value of the criterion for parameter value δ is given by $E[C(\delta)]\sum_{j=1}^J x_{ij}$, where $\sum_{j=1}^J x_{ij}$ is the number of examinees who get item i. Let $\mu = \min_\delta E[C(\delta)]$ for $\delta \in [\delta_l, \delta_u]$. Then the optimization problem can be defined as

$$\text{maximize } \lambda$$

subject to

$$\mu\sum_{j=1}^J x_{ij} \geq \lambda, \qquad i = 1,\ldots,I,$$

$$\sum_{i=1}^{I} x_{ij} = N, \qquad j = 1, \ldots, I,$$

$$x_{ij} \in \{0, 1\}, \qquad i = 1, \ldots, J,$$

$$\lambda \leq 0.$$

Now the first set of constraints requires that $\sum_{j=1}^{J} x_{ij}$ shall not be smaller than a (common) constraint λ/μ for all items. The second constraint requires that $\sum_{j=1}^{J} x_{ij}$ be equal to N for all examinees, the maximum number of ones in the design being thereby equal to NJ. But since $\sum_{j=1}^{J} x_{ij} \leq \lambda/\mu$ for all i, with λ maximal, it follows that $\sum_{j=1}^{J} x_{ij} = NJ/I$. Hence we may conclude that the solution is determined by the marginal sums $\sum_{j=1}^{J} x_{ij}$ and $\sum_{i=1}^{I} x_{ij}$ and not by the values of the individual variables x_{ij}. □

Note that the result is not dependent on the form of $F(\theta)$. Further, though μ was defined as a function of the interval $[\delta_1, \delta_u]$, the same result is obtained for any pair of values for the endpoints of the interval. Though the result was formulated for the difficulty parameter in the Rasch model, it also holds for the parameters in the other current models. Therefore, the generality of the theorem is in no way restricted. A keen reader could have conjectured the result by observing that all of the examinees were assumed to be sampled randomly and that the items were "exchangeable".

The practical meaning of the result for designing a calibration sample is that – unless the psychometrician knows more about the items and examinees – the only thing that counts is the number of items administered to each examinee and the number of examinees assigned to each item. There is no need whatsoever to have a sampling design with a special internal structure, as, for instance, is the case if the item parameters are estimated in two separate sets and then linked together on a common scale.

It should be realized that estimates do not always exist for a given design. Fischer (1981) gives the necessary and sufficient conditions a data matrix has to satisfy to have unique maximum-likelihood estimates in the Rasch model.

22.6 Conclusion

The relevance of optimum design principles for sampling problems in IRT was demonstrated using the problems of test assembly and item calibration. However, other sampling problems exist for which the optimum design approach still has to be explored. Examples include the optimum design of samples to test the goodness of fit of an IRT model or to estimate norm distributions in studies of test norming.

It is anticipated that for these problems the application of optimum design principles will lead to further fruitful results.

22.7 References

Adema, J.J. (1988). *A note on solving large-scale zero-one programming problems* (Research Report 88-4). Enschede, The Netherlands: Department of Education, University of Twente.

Adema, J.J. (1990a). The construction of customized two-stage tests. *Journal of Educational Measurement, 27,* 241-253.

Adema, J.J. (1990b). *Models and algorithms for the construction of achievement tests.* Doctoral thesis. Enschede, The Netherlands: University of Twente.

Adema, J.J., Boekkooi-Timminga, E., & van der Linden, W.J. (1992). Achievement test construction using 0-1 linear programming. *European Journal of Operational Research, 55,* 103-111.

Adema, J.J., & van der Linden, W.J. (1989). Algorithms for computerized test construction using classical item parameters. *Journal of Educational Statistics, 14,* 279-290.

Amstrong, R.D., Jones, D.H., & Wu, I.-L. (1992). An automated test development of parallel tests from a seed test. *Psychometrika, 57,* 271-288.

Atkinson, A.C. (1982). Developments in the design of experiments. *International Statistical Review, 50,* 161-177.

Berger, M.P.F. (1992). Sequential sampling designs for the two-parameter item response theory model. *Psychometrika, 57,* 521-538.

Berger, M.P.F., & van der Linden, W.J. (1992). Optimality of sampling designs in item response theory models. In M. Wilson (Ed.), *Objective measurement: Theory into practice* (Vol. 1). Norwood, N.J.: Ablex.

Boekkooi-Timminga, E. (1987). Simultaneous test construction by zero-one programming. *Methodika, 1,* 101-112.

Boekkooi-Timminga, E. (1989). *Models for computerized test construction.* Doctoral thesis, University of Twente. De Lier, The Netherlands: Academisch Boeken Centrum.

Boekkooi-Timminga, E. (1990a). The construction of parallel tests from IRT-based item banks. *Journal of Educational Statistics, 15,* 129-145.

Boekkooi-Timminga, E. (1990b). A cluster-based method for test construction. *Applied Psychological Measurement, 14,* 341-354.

Davey, T.C. (1992, April). *Optimal common-item anchors for ability metric linking.* Paper presented at the annual meeting of the American Educational Research Association, San Francisco, California.

Fedorov, V.V. (1972). *Theory of optimal experiments.* New York: Academic Press.

Fischer, G.H. (1981). On the existence and uniqueness of maximum likelihood estimates in the Rasch model. *Psychometrika, 46,* 59-77.

Hambleton, R.K., & Swaminathan, H. (1985). *Item response theory: Principles and applications.* Boston: Kluwer-Nijhoff.

Kalton, G. (1983). Introduction to survey sampling *(Quantitative Applications in the Social Sciences,* Series Nr. 35). Newbury Park, CA: Sage.

Kiefer, J., & Wolfowitz, J. (1960). The equivalence of two extremum problems. *Canadian Journal of Mathematics, 12,* 363–366.

Nemhauser, G.L., & Wolsey, L.A. (1988). *Integer and combinatorial optimization.* New York: Wiley.

Rasch, G. (1960). *Probabilistic models for some intelligence and attainment tests.* Copenhagen: Paedogogiske Institut.

Sanders, P.F., Theunissen, T.J.J.M., & Baas, S.M. (1989). Minimizing the number of observations: A generalization of the Spearman-Brown formula. *Psychometrika, 54,* 587–598.

Sanders, P.F., Theunissen, T.J.J.M., & Baas, S.M. (1991). Maximizing the coefficient of generalizability under the constraint of limited resources. *Psychometrika, 56,* 87–96.

Särndal, C.-E., Swensson, B., & Wretman, J. (1992). *Model-assisted survey sampling.* New York: Springer-Verlag.

Silvey, S.D. (1980). *Optimal design.* London: Chapman and Hall.

Steinburg, D.M., & Hunter, W.G. (1984). Experimental design: Review and comment. *Technometrics, 26,* 71–130.

Theunissen, T.J.J.M. (1985). Binary programming and test design. *Psychometrika, 50,* 411–420.

Timminga, E. (1985). *Geautomatiseerd toetsontwerp: Itemselectie met behulp van binair programmeren* [Automated test design: Item selection using binary programming]. Master's thesis. Enschede, The Netherlands: University of Twente.

Vale, C.D. (1986). Linking item parameters onto a common scale. *Applied Psychological Measurement, 10,* 333–344.

van der Linden, W.J., & Boekkooi-Timminga, E. (1988). A zero-one programming approach to Gulliksen's matched random subsets method. *Applied Psychological Measurement, 12,* 201–209.

van der Linden, W.J., & Boekkooi-Timminga, E. (1989). A maximin model for test design with practical constraints. *Psychometrika, 53,* 237–247.

van der Linden, W.J., & Eggen, T.J.J.M. (1986). An empirical Bayes approach to item banking. *Applied Psychological Measurement, 10,* 345–354.

Verschoor, A. (1991). *Optimal test design (computer program).* Arnhem, The Netherlands: Cito.

Wingersky, M.S., & Lord, F.M. (1984). An investigation of methods for reducing sampling error in some IRT procedures. *Applied Psychological Measurement, 8,* 347–364.

23

Parameters of the Latent Distribution in Rasch's Poisson Counts Model

Margo G. H. Jansen[1]

ABSTRACT This paper concerns the situation in which k tests are given to n subjects and the test scores are Poisson distributed random variables. The Poisson parameter is assumed to be a product of a test difficulty and a subject ability parameter. The test parameters are supposed to be fixed, and the ability parameters random. A number of different latent distributions are discussed, and we show how the parameters of these distributions can be estimated and the fit to the data checked.
KEY WORDS. Rasch model, ability distribution.

23.1 Introduction

The Rasch Poisson Counts Model (RPCM) is a unidimensional latent trait model for tests rather than items. It can be used in situations where the test consists of multiple attempts on a single item within a given time-limit. This is often the case in the assessment of psychomotor skills (Spray, 1990), where the test score is the number of successful attempts. Another situation where the model might be appropriate arises, as a limiting case of a Binomial Trials Model (Rasch, 1960; Masters and Wright, 1984; Jansen and Van Duijn, 1991); there the test score is the number of incorrect responses out of n items with low, but not necessarily equal, error probabilities. In his monograph 'Probabilistic models for some intelligence and attainment tests' Rasch (1960) described a number of such applications to language and arithmetic tests. It is useful to think of the observed test scores y_{ij} as organized in a $n \times k$ matrix, with rows for subjects and columns for tests or trials. Let y_{i+} and y_{+j}, the row and column sums, denote the subject and test scores. The model assumes that the test scores y_{ij} are independent random Poisson distributed variates with means μ_{ij} which are themselves the products of two parameters, the ability of the subject and the difficulty of the test. In Rasch's original formulation both sets of parameters were taken to be fixed, so standard methods for analyzing two-way contingency tables without interaction could be used.

In the present approach, Rasch's model is elaborated by assuming a latent distribution with fixed but unknown parameters for the ability parameters. So, suppose there are k tests with parameters $\boldsymbol{\beta} = (\beta_1, \ldots, \beta_k)$, which are taken to

[1]Department of Education, University of Groningen, Grote Rozenstraat 38, 9712 TJ Groningen, The Netherlands

be fixed but unknown, given to a sample of n subjects with ability parameters $\boldsymbol{\theta} = (\theta_1, \ldots, \theta_n)$, selected from some prior distribution with fixed unknown (hyper)parameters $\boldsymbol{\Phi}$; the (joint) posterior density function, denoted by $p(\boldsymbol{Y} \mid \boldsymbol{\Phi})$, can be derived, and also the marginal likelihood function, $L(\boldsymbol{Y} \mid \boldsymbol{\Phi})$. Following the usual empirical Bayes approach, estimates for the fixed parameters can be obtained by maximizing the marginal likelihood.

23.2 The marginal likelihood

In dealing with the ability parameters much depends on the choice of the probability distributions for these variables. A number of different families of distributions will be considered. Since the subject parameters are non-negative quantities, the two-parameter gamma distribution is an obvious choice. Moreover, the gamma distribution, being conjugate to the Poisson, turns out to produce very tractable results. Although the gamma distribution is fairly flexible, it can hardly be expected to be applicable to all the kinds of empirical data for which the RPC model might be used. Several other considerably less tractable possibilities are mentioned in the literature on heterogeneous Poisson sampling (Anscombe, 1950; Johnson and Kotz, 1969), among them the uniform, the truncated normal, and the lognormal distribution. None of these, however, are used frequently. Here, attention will be restricted to the gamma and the log-normal distributions which, in my opinion, are the most promising candidates.

The probability of observing the response y_{ij} of subject i on test j is given by

$$P(\boldsymbol{Y}_{ij} = y_{ij}) = \frac{\exp(-\mu_{ij})\mu_{ij}^{y_{ij}}}{y_{ij}!}, \tag{23.1}$$

with

$$\mu_{ij} = \theta_i \beta_j. \tag{23.2}$$

In view of (23.1) and (23.2), the likelihood of $\boldsymbol{Y}$, the matrix of responses of n subjects on k tests considered as a function of $\boldsymbol{\theta}$ and $\boldsymbol{\beta}$, can be written in the form

$$L(\boldsymbol{Y} \mid \boldsymbol{\theta}, \boldsymbol{\beta}) = \frac{\exp(-\sum_i \theta_i \sum_j \beta_j) \prod_i \theta_i^{y_{i+}} \prod_j \beta_j^{y_{+j}}}{\prod_{ij} y_{ij}!}. \tag{23.3}$$

Let $p(\boldsymbol{\theta}|\boldsymbol{\Phi})$ denote the (multivariate) prior density of the ability parameters θ_i; then the posterior density function is

$$p(\boldsymbol{\theta} \mid \boldsymbol{\beta}, \boldsymbol{\Phi}, \boldsymbol{Y}) = H^{-1} L(\boldsymbol{Y} \mid \boldsymbol{\theta}, \boldsymbol{\beta}) p(\boldsymbol{\theta} \mid \boldsymbol{\Phi}), \tag{23.4}$$

where the constant of proportionality H is the marginal likelihood

$$L_m(\boldsymbol{Y} \mid \boldsymbol{\beta}, \boldsymbol{\Phi}) = \int L(\boldsymbol{Y} \mid \boldsymbol{\theta}, \boldsymbol{\beta}) p(\boldsymbol{\theta} \mid \boldsymbol{\Phi}) d(\boldsymbol{\theta}). \tag{23.5}$$

In the first place, let the subject parameters be sampled from a gamma distribution with shape parameter c and scale parameter $1/s$; it can then easily be shown that the posterior distribution $p(\boldsymbol{\theta}|c, s, \boldsymbol{Y})$ of $\boldsymbol{\theta} = (\theta_1, \ldots, \theta_n)$ is the product of independent gamma distributions, with parameter $(s + \sum \beta_j)$ and $(c + y_{i+})$. Furthermore, $\sum \beta_j$ can be set equal to 1; this does no more than make the β_j and the θ_i identifiable; then it can be shown that the marginal likelihood (23.5) is the product of three separate components, namely,

$$L_m(\boldsymbol{Y} \mid \boldsymbol{\beta}, \boldsymbol{\Phi}) = L(y_{*+} \mid \boldsymbol{\Phi}) \times L(y_{+*} \mid y_{++}, \boldsymbol{\beta}) \times L(\boldsymbol{Y} \mid y_{+*}, y_{*+}). \tag{23.6}$$

In (23.6) the first component is a negative binomial likelihood for the subject scores (row sums) contingent on the parameters of the latent distribution; the second, contingent on the sum of the row totals y_{++}, is a multinomial likelihood for the test scores (column sums) involving only the test parameters, and the third part, a multivariate hypergeometric, is a function of the data only.

The lognormal distribution is also a positively skewed distribution with two parameters, by μ and σ. If the ability parameters are sampled from a common lognormal distribution, the analysis is more complicated, since there is now no closed-form expression for the component in the marginal likelihood for the subject scores, or for the posterior distribution of $\boldsymbol{\theta}, p(\boldsymbol{\theta} \mid \boldsymbol{\beta}, \mu, \sigma^2, \boldsymbol{Y})$. The posterior distribution of θ_i, for given $(\boldsymbol{\beta}, \mu, \sigma^2)$, is proportional to

$$\exp\Big(- \theta_i \sum_j \beta_j + (y_{i+} - 1) \log \theta_i - \frac{(\log \theta_i - \mu)^2}{2\sigma^2} \Big). \tag{23.7}$$

It can be shown that the marginal distribution of the subject scores is a so-called "discrete lognormal distribution." The discrete lognormal distribution suffers from the disadvantage that its density function involves an integral (Anscombe, 1950), which might explain why applications are rare. For the marginal log-likelihood the following expression is obtained:

$$l_m(\boldsymbol{Y} \mid \boldsymbol{\beta}, \mu, \sigma^2) = \sum_i \log \int \exp\Big(- \theta_i \sum_j \beta_j + (y_{i+} - 1) \log \theta_i$$

$$- \frac{(\log \theta_i - \mu)^2}{2\sigma^2} \Big) d\theta_i + \sum_j y_{+j} \log \beta_j - \sum_{i,j} \log(y_{ij}!). \tag{23.8}$$

This can be rewritten as the sum of a discrete lognormal log-likelihood for the subject scores, a multinomial log-likelihood for the test scores, and a multivariate hypergeometric log-likelihood for the scores y_{ij}, given the row and column sums.

23.3 Parameter estimation

Estimates for the (hyper)parameters c and s and the test parameters $\boldsymbol{\beta}$ can be derived by maximizing the marginal likelihood (Jansen and Van Duijn, 1991). Note that due to the particularly pleasant mathematical properties of this model the MML and CML estimates for the test parameters are identical. As a matter of convenience, the gamma distribution will be reparameterized in terms of c and m, where $m = c/s$ is the mean. This leads to the following estimation equations:

$$m = \frac{1}{n}\sum_i y_{i+}, \tag{23.9}$$

and

$$\log(nc) - \log(y_{++} + nc) + \sum_i \psi(y_{i+} + c)/n - \psi(c) = 0. \tag{23.10}$$

Various approximate methods for estimating the parameters of a negative binomial distribution and calculating the standard errors can be found in the literature (Anscombe, 1950).

Estimation procedures in the lognormal case are less straightforward. As before, the β_j's are estimated by maximizing the multinomial component of expression (23.7). The parameters μ and σ are estimated by a procedure which closely resembles the version of the EM-algorithm used by Sanathanan and Blumenthal (1978) for estimating the parameters of the latent distribution in the context of the one-parameter logistic Rasch model. This approach treats the subject parameters $\theta_1, \ldots, \theta_n$ as missing data. Sanathanan and Blumenthal based their procedure on first obtaining CML estimates for the item parameters and then applying the EM-algorithm to $\boldsymbol{\theta}$ while holding the item parameters fixed at their estimated value. In our case, the parameters of the latent distribution can be estimated without any further reference to the test parameters, since the test parameters enter the equations only through their sum, which was set equal to 1.

Starting with some initial estimates for μ and σ^2, the following two steps are repeated until a convergence criterion is met:

1. estimate $\log\theta_i$ and $(\log\theta_i)^2$ by their posterior expectations given the current values of μ and σ^2, and

2. compute revised values for μ and σ^2 using the equations

$$\begin{aligned} \mu &= \sum \log\theta_i/n, \\ \sigma^2 &= \sum(\log\theta_i)^2/n - \mu^2. \end{aligned}$$

TABLE 23.1. Number of errors per occasion after grouping according to total scores

score group	grade 3	grade 4	score group	grade 3	grade 4	score group	grade 3	grade 4
1	8	2	7	11	3	17	10	7
2	7	5	8	12	4	18	13	5
3	19	8	9	6	3	19	14	5
4	6	6	11	12	10	20	14	6
5	16	9	13	9	4	22	11	11
6	5	1	14	9	5	24	13	11
						Σ	195	105
						$\hat{\beta}$	0.65	0.35

Finding the posterior expectations mentioned in step 1 involves performing as many one-dimensional numerical integrations as there are different subject scores. Moment estimators, which are easy to derive, can be used for the calculation of initial values (Johnson and Kotz, 1969). The asymptotic covariance matrix of (μ, σ) can be derived using the delta method as was shown by Sanathanan and Blumenthal (1978).

23.4 An example

The use of the two models will be illustrated with data from a study on spelling errors. A dictated spelling test was administered to a group of 49 primary school pupils on two occasions, first when they were in the third grade, and a second time about twelve months later when they were in the fourth grade. The scores consisted of the number of errors made by subject i on occasion j.

A Pearson chi-square test for the subject $\times$ test interactions, calculated after grouping the subjects according to their total numbers of errors, was not significant ($\chi^2 = 11.6$, $df = 17$). Some of the relevant data concerning the marginal distribution of the subject scores are summarized in Table 23.2, while Table 23.3 contains estimates of the so-called hyper-parameters and their standard errors.

It is difficult to compare the two models by inspecting the shape and scale parameters directly (for the lognormal distribution, $\xi = \exp(\mu)$ is a scale parameter, and σ determines the shape); a comparison of the predicted marginal distributions is more informative. Table 23.4 sets out the results of fitting a negative binomial, respectively discrete lognormal distribution, to the marginal distribution of the subject scores, and Figure 23.1 presents plots of the fitted and observed frequency distributions.

TABLE 23.2. Marginal distribution of the subjects' total scores

score	freq	score	freq	score	freq
0	0				
1	10	9	1	17	1
2	6	10	0	18	1
3	9	11	2	19	1
4	3	12	0	20	1
5	5	13	1	21	0
6	1	14	1	22	1
7	2	15	0	23	0
8	2	16	0	24	1
mean = 6.1		std. dev. = 6.12		$n = 49$	

TABLE 23.3. Estimates of the parameters of the prior distribution

	distribution			
	gamma		lognormal	
parameter	m	c	μ	σ
estimates	6.12	1.48	1.45	0.88
	(0.799)	(0.481)	(0.14)	(0.09)

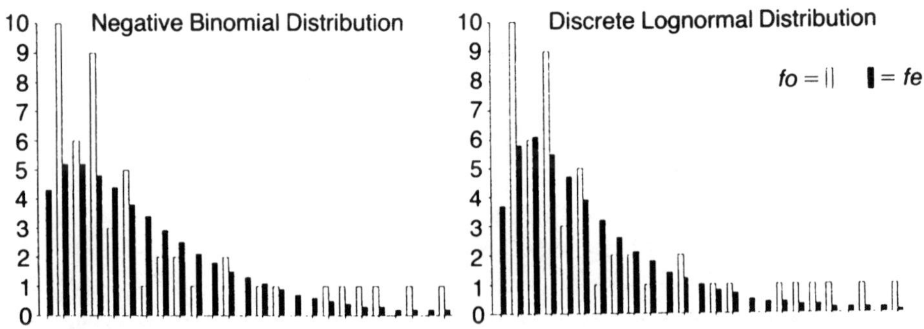

FIGURE 23.1. Comparisons of estimated and observed marginal distributions for the two models

For this particular case the discrete lognormal gives a reasonably good fit to the data, while the fit of the negative binomial is rather poor. Another case in which the discrete lognormal fitted much better than the negative binomial was discussed by Jansen (1991); this latter example might be seen as somewhat disappointing whith respect to computational and tractability considerations. But these two examples given should be looked on mainly as illustrations of the feasability of the approach, without drawing general conclusions.

23.5 Discussion

In its original formulation (in which both the subjects and the tests are treated as fixed factors) Rasch's multiplicative Poisson model can be shown to be equivalent to the independence model for a two-way contingency table, with subjects as rows and tests as columns. In the present approach, which I consider to be more realistic, a mixed model is used, that is, the column parameters are unknown fixed constants and the row parameters are treated as random variables. Furthermore a parametric distribution is specified for the row (subject) parameters. By doing this new problems are obviously introduced since there is now the risk of misspecifying the ability distribution.

As has already been mentioned, MML estimates for the test parameters are identical with those obtained from CML. It follows that violations of the assumptions concerning the ability distribution do not affect the test parameter estimates. If only the relative difficulties of the tests are of interest, the assumed distribution of ability parameters can be disregarded. In many applications, however, estimating the level and variability of ability in the population under study might be considered just as important. Moreover, one may wish to compare the ability distributions of a number of (sub) populations. Then, as has been shown, the MML framework enables these extensions of the RPCM to be modeled in a relatively straightforward way. By choosing a gamma distribution for the random parameters, estimates for the parameters of the (prior) distribution and measures of fit for the model can be easily derived. Choosing the lognormal leads to considerably less tractable results. Another interesting aspect is the link with multilevel methodology (Goldstein, 1991). The model can be viewed as a multilevel model for balanced data with a discrete outcome variable.

In the example an empirical data set was analyzed using the two types of latent distributions and the results of the two analyses compared. The discrete lognormal distribution gave a much better fit to the marginal distribution of the subject scores than the negative binomial. Further research is needed to discover how typical these results are. Special attention should also be given to the power of the statistical tests used for checking the assumed distributions.

TABLE 23.4. Goodness-of-fit measures for the two types of marginal distributions

distribution	χ^2	df
neg. binom.	20.75	10
discr. lognorm.	12.98	10

23.6 References

Anscombe, F.J. (1950). Sampling theory of the negative binomial and logarithmic series distributions. *Biometrika, 37*, 358–382.

Goldstein, H. (1990). Nonlinear multilevel models, with an application to discrete response data. *Biometrika, 78*, 45–51.

Jansen, M.G.H. (1992). Empirical Bayes estimation of the parameters in Rasch's multiplicative Poisson model. In R. Steyer, K.F. Wender, & K.F. Widaman (Eds.), *Proceedings of the 7th European Meeting of the Psychometric Society* (4 pag.). Stuttgart & New York: Gustav Fischer Verlag (in press).

Jansen, M.G.H., & Snijders, T.A.B. (1991). Comparisons of Bayesian procedures for estimating the parameters in a two-way contingency table. *Statistica Neerlandica, 45*, 51–65.

Jansen, M.G.H., & Van Duijn, M.A.J. (1992). Extensions of Rasch's multiplicative Poisson model. *Psychometrika, 57*, 405–414.

Johnson, N.L., & Kotz, S. (1969). *Distributions in statistics. Discrete distributions.* Boston: Houghton-Mifflin.

Masters, G.N., & Wright, B.D. (1984). The essential process in a family of measurement models. *Psychometrika, 49*, 529–544.

Rasch, G. (1980). *Probabilistic models for some intelligence and attainment tests.* Chicago: The University of Chicago Press. (Original work published in 1960).

Sanathanan, L., & Blumenthal, S. (1978). The logistic model and estimation of latent structure. *Journal of the American Statistical Association, 73*, 794–799.

Spray, J.A. (1990). One-parameter item response theory models for psychomotor tests involving repeated independent attempts. *Research Quarterly for Exercise and Sport, 61*, 162–168.

Part V
Methodology

24

On Connecting Network Analysis and Cluster Analysis

Anuška Ferligoj[1], Vladimir Batagelj[2], and Patrick Doreian[3]

ABSTRACT Actor equivalence is a fundamental concept for the analysis of network representations of social structure. In this paper it is shown that the partitioning of a network in terms of some kind of equivalence (e.g., structural or regular equivalence) is essentially a clustering problem. When searching for the best partition with standard clustering algorithms, a criterion function has to be defined, compatible with the chosen kind of equivalence. Such a criterion function can be constructed *indirectly* as a function of a compatible (dis)similarity measure between pairs of actors or *directly* as a function measuring the concordance of a given partition with an ideal partition for the chosen equivalence.

24.1 Introduction

Social network analysis is an integrated scientific specialty with all the characteristics of a normal science in the sense of Kuhn (1970). There is a professional association (the International Association of Social Network Analysts, INSNA) and a specialty journal (*Social Networks*) that was created specifically for the emerging field. Members of INSNA also receive *Connections* that serves as a professional newsletter linking the invisible college of network analysts. Within this invisible college there have been many research initiatives both with respect to the substance and to the development of network analytic methods. Efforts to cluster network actors and social relations emerged following the seminal paper by Lorrain and White (1971) on structural equivalence. The nodes of the networks have variously been individuals or nations or states since the methods developed on the basis of equivalence are completely portable. Implicitly, the claim is made by many network analysts that, through the use of these "equivalence methods", all networks can be analyzed in terms of equivalence. While there have been many inappropriate uses of these methods, they are nevertheless both general and powerful. By linking network analysis with cluster analysis through the specification of criterion functions based directly on network concepts, we provide a coherent foundation for clustering nodes in social networks.

Doreian (1988) noted that equivalence emerged as a fundamental concept for the analysis of network representations of social structure. Such claims have an

[1]University of Ljubljana, Faculty of Social Science, Kardeljeva pl. 5, 61 109 Ljubljana, Slovenia
[2]University of Ljubljana, Department of Mathematics, Jadranska 19, 61 111 Ljubljana, Slovenia
[3]University of Pittsburgh, Department of Sociology, Pittsburgh, PA 15260, USA

anecdotal flavor as there are many articles dealing with equivalence. Hummon and Carley (1992) provide a more systematic assessment. Their data comprised all articles appearing in the first twelve volumes of *Social Networks*, from 1978 to 1990. Those 227 articles cite a total of 5573 other articles, of which 3580 are unique. Using methods of main path analysis (Hummon and Doreian, 1989, 1990) that have been modified to deal with the presence of multiple main paths through a citation network of articles, Hummon and Carley discerned six main paths through this literature. One of these paths is primary and features 24 articles concerned with role analysis where the idea of equivalence dominates. Each main path is a collection of papers that are organized around a coherent research problem and represents a stream of research activity. It must be emphasized that the equivalence 'main path' (research front) is only one of the six identified. (Parenthetically, we note also that the main path methodology pioneered by Hummon and Doreian (1989, 1990) is not a portable methodology as it is geared to a particular form of scientific citation network.)

Hummon and Carley (1992) counted the citations research productions published both in *Social Networks* and elsewhere. The most frequently cited article published elsewhere is the seminal paper of Lorrain and White (1971). Seeking a representation of positions and roles, and using objects and morphisms within category theory, Lorrain and White provided the first definition of structural equivalence. Actors are structurally equivalent if they are connected to the rest of the network in identical ways. Another frequently cited paper, also published elsewhere, by Breiger, Boorman, and Arabie (1975) introduced the algorithm CONCOR as a method for partitioning nodes into subsets where the nodes within a subset are approximately structurally equivalent. In essence, the procedure is a divisive clustering procedure. Burt (1976) made the clustering aspect even more explicit with the program STRUCTURE. Using Euclidean distance as a measure of dissimilarity, he utilized procedures drawn from cluster analysis as a method of grouping the nodes in the network into non-equivalent positions. Within each of these positions, the nodes are approximately equivalent and the nodes of the network are thus partitioned.

The most frequently cited paper from *Social Networks* is the one by Sailer (1978) on the substitutability of nodes in a network, marking the beginning of a sustained effort to generalize the concept of structural equivalence. This paper is the starting point of the primary main path discerned by Hummon and Carley. An early paper on this main path, by White and Reitz (1983), provides a definition of regular equivalence. Intuitively, two actors are regularly equivalent if they are equivalently connected to equivalent others. This work laid the foundation for efforts to construct measures that capture the extent to which two actors are regularly equivalent in a network of ties. In most analyses, a matrix of these similarities is computed within the algorithm REGE, and then submitted to a clustering algorithm. Over the years here, as in STRUCTURE, early clustering algorithms were replaced by later clustering algorithms. The commitment scrutiny, it seems, of the properties of the clustering algorithms or the measures of (dis)similarities.

The dense middle part of the primary main path located by Hummon and Carley is occupied by a variety of papers. Some explore the various concepts of equivalence, some propose new methods for computing similarities or dissimilarities based on equivalence ideas, and some propose additional ways in which nodes can be grouped by taking into account the ties that link them. A helpful overview of some of these efforts is provided by Pattison (1988). There were also examinations of the different clusterings produced in the course of data analyses, but little was said about the clustering procedures themselves nor their role beyond producing partitions.

Within the papers featured at and near the current terminus of the primary main path is a collection of ideas that revolve around the ideas of automorphisms, orbits in a graph, and orbit equivalence; see Everett and Borgatti (1988) and Borgatti and Everett (1989). This work lays the foundations for a sustained consideration of equivalence in social networks. It seems likely that this work will continue primarily in terms of defining and measuring equivalence. A complementary line of inquiry is to examine the clustering aspects of partitioning nodes in networks with a view to establishing criterion functions explicitly designed to reflect the ideas of equivalence. In so doing, the clustering can be integral to the substantive concerns and not just a method for clustering similarities or dissimilarities obtained from some other source.

In this paper we show that the problem of establishing a partition of a network in terms of equivalence is a special case of a classical clustering problem. When searching for the best clustering (e.g., partition) with known clustering algorithms (e.g., a hierarchical agglomerative algorithm, or a relocation algorithm) a criterion function compatible with the equivalence considered has to be defined and used. Such a criterion function can be defined indirectly as a function of a compatible (dis)similarity measure between pairs of actors or directly with a function measuring the fit of a given partition to an ideal partition for the considered equivalence. Therefore two main approaches can be distinguished (Batagelj, Ferligoj, and Doreian, 1992):

- *indirect approach*: a reduction to a standard data analysis problem (cluster analysis, multidimensional scaling) by computing a dissimilarity matrix between units which is compatible with the selected type of equivalence;

- *direct approach*: the construction of a criterion function $P(\mathcal{C})$ which measures the fit of the clustering $\mathcal{C}$ to the data and solving the corresponding optimization problem. A relocation clustering procedure can be adapted for that purpose.

Both approaches will be discussed for structural and regular equivalences.

24.2 Basic definitions

Let $E = \{X_1, X_2, ..., X_n\}$ be a finite set of *units or actors*. The units are related by binary *relations,*

$$R_t \subseteq E \times E, \quad t = 1, \ldots, r,$$

which determine a *network*

$$\mathcal{N} = (E, R_1, R_2, ..., R_r).$$

In what follows we restrict our discussion to a single relation R described by a corresponding binary matrix $\mathbf{R} = [r_{ij}]_{n \times n}$, where

$$r_{ij} = \begin{cases} 1 & \text{if } X_i R X_j, \\ 0 & \text{otherwise.} \end{cases}$$

In some applications, r_{ij} can be a nonnegative real number expressing the strength of the relation R between units X_i and X_j.

One of the main procedural goals of network analysis is to identify, in a given network, *clusters* (classes) of units that share structural characteristics defined in terms of R. The units within a cluster have the same or similar patterns of connection to other units. They form a *clustering*

$$\mathcal{C} = \{C_1, C_2, ...C_k\},$$

which is a partition of the set E:

$$\bigcup_i C_i = E,$$

$$i \neq j \Rightarrow C_i \cap C_j = \emptyset,$$

Each partition determines an equivalence relation (and vice versa). Let us denote by $\sim$ the relation determined by partition $\mathcal{C}$.

A clustering $\mathcal{C}$ partitions also the relation R into *blocks*

$$R(C_i, C_j) = R \cap C_i \times C_j.$$

Each such block consists of actors belonging to clusters C_i and C_j and all arcs leading from cluster C_i to cluster C_j. If $i = j$, a block $R(C_i, C_i)$ is called a *diagonal* block. A *blockmodel* consists of structures obtained by identifying all units from the same cluster of the clustering $\mathcal{C}$. For an exact definition of a blockmodel, we have to be precise also about which blocks produce an arc in the reduced graph and which do not. The partition (ideally) is constructed by using structural information contained in R, and actors in the same cluster are equivalent to each other in terms of R alone. Such actors share a common structural position within the network. (In general, the statement can be extended to include $\{R_t\}$.)

Blockmodeling, as an empirical procedure, is based on the idea that actors in a network can be grouped according to the extent to which they are equivalent, according to some *meaningful* definition of equivalence. In general, and not surprisingly, different definitions of equivalence lead to distinct partitions. Regardless of the definition there are two basic approaches to the equivalence of units in a given network (compare Faust, 1988):

- the equivalent units have the same connection pattern to the same neighbors;
- the equivalent units have the same or similar connection pattern to (possibly) different neighbors.

The first type of equivalence is formalized by the notion of structural equivalence; the second by the notion of regular equivalence.

24.3 Definitions of some equivalences

A permutation $\varphi : E \to E$ is an *automorphism* of the relation R iff

$$\forall X, Y \in E : (XRY \Rightarrow \varphi(X)R\varphi(Y)).$$

The units X and Y are *structurally equivalent* (we write $X \equiv Y$) iff the permutation (transposition) $\pi = (XY)$ is an automorphism of the relation R (Borgatti and Everett, 1989).

In other words: X and Y are structurally equivalent iff

s1. $XRY \Leftrightarrow YRX$,
s2. $XRX \Leftrightarrow YRY$,
s3. $\forall Z \in E \setminus \{X, Y\} : (XRZ \Leftrightarrow YRZ)$,
s4. $\forall Z \in E \setminus \{X, Y\} : (ZRX \Leftrightarrow ZRY)$,

or in matrix form: $X_i \equiv X_j$ iff

s1'. $r_{ij} = r_{ji}$,
s2'. $r_{ii} = r_{jj}$,
s3'. $\forall k \neq i, j : r_{ik} = r_{jk}$,
s4'. $\forall k \neq i, j : r_{ki} = r_{kj}$.

The matrix form of the definition of structural equivalence can be extended also to the case where the r_{ij} are real numbers.

The units X and Y are *automorphically equivalent*, $X \cong Y$, iff there exists an automorphism $\varphi : E \to E$ of the relation R, such that

$$\varphi(X) = Y.$$

The equivalence relation $\approx$ on E is a *regular equivalence* on network $\mathcal{N} = (E, R)$ iff, for all $X, Y, Z \in E$, $X \approx Y$ implies both

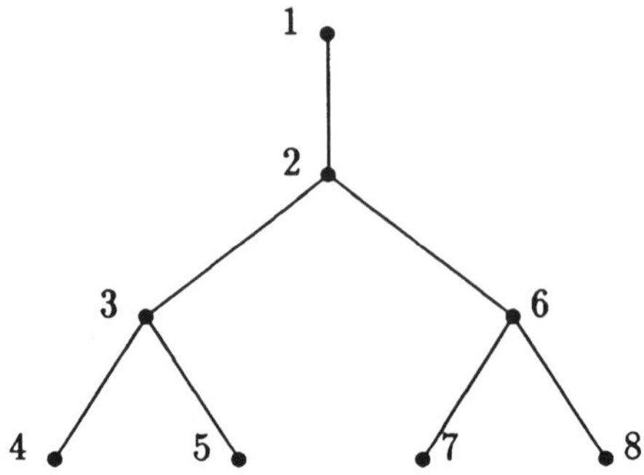

FIGURE 24.1. An Example

R1. $XRZ \Rightarrow \exists W \in E : (YRW \wedge W \approx Z)$, and
R2. $ZRX \Rightarrow \exists W \in E : (WRY \wedge W \approx Z)$.

The following relations can be proved for these three equivalences:

$$\equiv \subset \cong \subset \approx$$

Structural equivalence is a very stringent requirement in a network. Automorphic and regular equivalences have weaker requirements and are more often found in a given network. For a network there is, in general, a non-empty set of regular partitions which form a lattice (Borgatti and Everett, 1989).

For the relation R given with the graph in Figure 24.1, the three equivalences mentioned are:

$$\equiv = \{\{4,5\},\{7,8\},\{1\},\{2\},\{3\},\{6\}\},$$
$$\cong = \{\{4,5,7,8\},\{3,6\},\{1\},\{2\}\},$$
$$\approx = \{\{1,4,5,7,8\},\{2,3,6\}\}.$$

24.4 The indirect clustering approach

The problem of establishing a partition of a network in terms of a considered equivalence is a special case of clustering problem that can be formulated as an optimization problem: i. e., determine the clustering C^* for which

$$P(C^*) = \min_{C \in \Phi} P(C),$$

where C is a clustering of a given *set of units or actors* E, Φ is the set of all possible clusterings, and $P : \Phi \to \mathbb{R}$ the *criterion function.* The criterion function must reflect the considered equivalence.

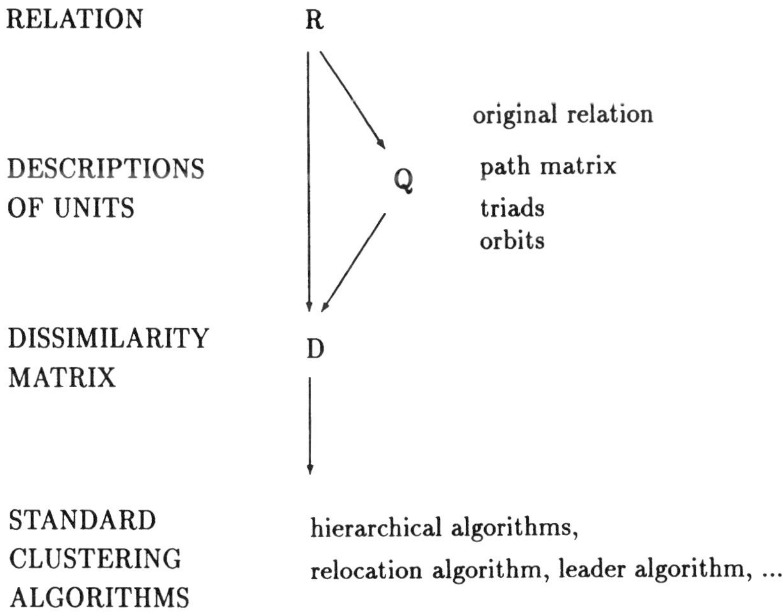

FIGURE 24.2. Indirect Clustering Approach

The indirect clustering approach solves the clustering problem by defining a (dis)similarity measure between pairs of units first, and then on this basis an appropriate clustering criterion function (e.g., Ward's, 1963, criterion function) is selected. Some typical steps as shown in Figure 24.2 have to be considered in this indirect approach.

The first step in this process is to specify a description of the units. A description might consist of the row and column in the original relational matrix of the selected unit or of the triadic spectrum of the selected unit (Hummell and Sodeur, 1987). Batagelj, Ferligoj and Doreian (1992) discuss these different descriptions. The next step in the indirect approach is the definition of a dissimilarity measure based on the descriptions of the units. This is a crucial step since not all dissimilarities are consistent with the chosen kind of equivalence. The most important requirement is that the selected dissimilarity measure be *compatible* with the chosen equivalence (Batagelj, Ferligoj and Doreian, 1992). The dissimilarity measure d is compatible with a chosen equivalence $\sim$ if, for each pair of units,

$$X_i \sim X_j \Leftrightarrow d(X_i, X_j) = 0.$$

Not all the dissimilarity measures commonly used are compatible with structural equivalence. However, the corrected Euclidean-like dissimilarity (Burt and

Minor, 1983):

$$d(X_i, X_j) = \sqrt{(r_{ii} - r_{jj})^2 + (r_{ij} - r_{ji})^2 + \sum_{\substack{s=1 \\ s \neq i,j}}^{n} ((r_{is} - r_{js})^2 + (r_{si} - r_{sj})^2)},$$

is compatible. From this formula it can be seen that the dissimilarity is calculated simultaneously between rows and columns in the relation matrix.

The indirect clustering approach is not suitable for establishing clusterings of a network in terms of regular or automorphic equivalence, since there appears not to be any way to construct a compatible (dis)similarity measure for those kinds of equivalence.

On the basis of the selected dissimilarity measure, different efficient clustering algorithms can be used, e.g., hierarchical algorithms, relocation algorithm, leader algorithm (Hartigan, 1975), etc.

24.5 Direct clustering approach

The second approach to solving the clustering problem is to construct an appropriate criterion function directly and then use a local optimization algorithm to obtain a 'good' clustering solution.

Suppose there is a single relation network, $\mathcal{N} = (E, R)$. Let Θ denote the set of all equivalence relations of a selected type (for example, regular or structural equivalences) over $\mathcal{N}$. Every equivalence relation $\sim$ on E determines a partition $\mathcal{C}$ of E, and vice versa. Let Φ denote the set of all partitions corresponding to the relations in Θ.

If a criterion function $P(\mathcal{C})$ can be constructed with the properties

P1. $P(\mathcal{C}) \geq 0$,

P2. $P(\mathcal{C}) = 0 \Leftrightarrow \sim \in \Theta$,

then the problem of establishing a partition of a network in terms of a considered equivalence can be expressed as a clustering problem. If there are exact equivalences, then (by P2) the minimal value of $P(\mathcal{C})$ is 0. In the case when Θ is empty, the optimization approach gives the solution(s) which differ(s) least from some ideal case.

One of the possible ways of constructing a criterion function that directly reflects the chosen equivalence is to measure the fit of a clustering to an ideal one with perfect relations both within clusters and between clusters according to the chosen equivalence.

Given a clustering $\mathcal{C} = \{C_1, C_2, \ldots, C_k\}$, let $\mathcal{B}(C_u, C_v)$ denote the set of all ideal blocks corresponding to block $R(C_u, C_v)$. Then the global error of a clustering $\mathcal{C}$ can be expressed as

$$P(\mathcal{C}) = \sum_{C_u, C_v \in \mathcal{C}} \min_{B \in \mathcal{B}(C_u, C_v)} d(R(C_u, C_v), B),$$

where the term $d(R(C_u, C_v), B)$ measures the difference (error) between the block $R(C_u, C_v)$ and the ideal block B. The function d has to be compatible with the chosen kind of equivalence.

24.5.1 A CRITERION FOR STRUCTURAL EQUIVALENCE

From the definition of structural equivalence it follows that only four possible ideal diagonal blocks can appear (Batagelj, Ferligoj and Doreian, 1992):

Type 0. $b_{ij} = 0$,
Type 1. $b_{ij} = \delta_{ij}$,
Type 2. $b_{ij} = 1 - \delta_{ij}$,
Type 3. $b_{ij} = 1$,

where δ_{ij} is the Kronecker delta, and $i, j \in C$.

For the nondiagonal blocks $R(C_u, C_v), u \neq v$, only blocks of type 0 and type 3 are admissible.

Given a clustering $\mathcal{C} = \{C_1, C_2, \ldots, C_k\}$, let $\mathcal{B}(C_u, C_v)$ denote the set of all ideal blocks corresponding to block $R(C_u, C_v)$. Then the term $d(R(C_u, C_v), B)$ can be expressed as

$$d(R(C_u, C_v), B) = \sum_{X \in C_u, Y \in C_v} |r_{xy} - b_{xy}|.$$

It is easy to verify that a criterion function $P(\mathcal{C})$ so defined is sensitive to structural equivalence:

$$P(\mathcal{C}) = 0 \Leftrightarrow \mathcal{C} \text{ defines structural equivalence.}$$

In addition, it is invariant under the transformation of complementing the relation $R \to E \times E \setminus R$. Note that empirically speaking this case is not infrequent.

24.5.2 A CRITERION FOR REGULAR EQUIVALENCE

The construction of a criterion function for regular equivalence is based on the following result (Batagelj, Doreian and Ferligoj, 1992): Let $\mathcal{C} = \{C_i\}$ be a clustering corresponding to a regular equivalence $\approx$ on the network $\mathcal{N} = (E, R)$. Then each block $R(C_u, C_v)$ is either null or it has the property that there is at least one '1' in each of its rows and in each of its columns. Conversely, if for a given clustering $\mathcal{C}$ each block has this property, then the corresponding equivalence relation is a regular equivalence.

From this proposition it follows that regular equivalence produces two types of blocks:

- *null blocks* in which all entries are '0'; and
- *1-covered blocks*, which have at least one '1' in each row and in each column.

Therefore, as a measure of regularity of a block the following quantity,

$$d(R(C_u, C_v), B) = \begin{cases} \text{\# of 1-covered rows/columns if } B \text{ is a null block,} \\ \text{\# of 0 rows/columns if } B \text{ is a 1-covered block,} \end{cases}$$

can be used. From the proposition it follows that a criterion function $P(\mathcal{C})$ so defined is sensitive to regular equivalence:

$$P(\mathcal{C}) = 0 \Leftrightarrow \mathcal{C} \text{ defines regular equivalence.}$$

24.5.3 Clustering algorithm

In the direct clustering approach in which an appropriate criterion function is constructed, one of the local optimization procedures can be used for solving the resulting blocking problem (Batagelj, Doreian and Ferligoj, 1992).

Often, for a given optimization problem (Φ, P), there are rules which relate each element of the set of feasible solutions Φ to certain elements of Φ. Such rules are called *local transformations.* They form the basis of the local optimization procedure which, starting with an element of Φ, repeatedly moves to another element, determined by a local transformation, to reach a lower value of the criterion function.

The elements which can be obtained by local transformations from a given element are called neighbors – local transformations determine the *neighborhood relation* $S \subseteq \Phi \times \Phi$ in the set Φ. The *neighborhood* of element $X \in \Phi$ is the set $S(X) = \{Y : XSY\}$. The element $X \in \Phi$ is a *local minimum* for the neighborhood structure (Φ, S) iff

$$\forall Y \in S(X) : P(X) \leq P(Y).$$

The basic scheme of the local optimization procedure is therefore very simple (see Foulds, 1984):

> determine the initial element $X_0 \in \Phi$, $X := X_0$;
> **while** $\exists Y \in S(X) : P(Y) < P(X)$, **repeat** $X := Y$.

Usually, the neighborhood relation in local optimization procedures over clusterings (partitions) is determined by the following two transformations:

- clustering $\mathcal{C}'$ is obtained from the clustering $\mathcal{C}$ by *moving* a unit X_k from cluster C_p to cluster C_q (*transition*);
- clustering $\mathcal{C}'$ is obtained from the clustering $\mathcal{C}$ by *interchanging* units X_u and X_v from different clusters C_p and C_q (*transposition*).

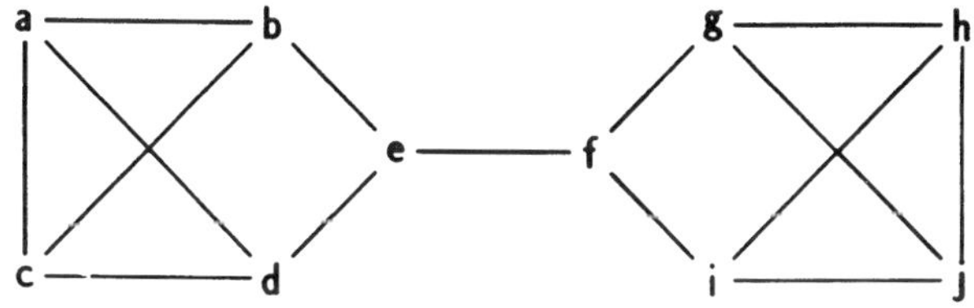

FIGURE 24.3. Everett Network

TABLE 24.1. Everett Network

	a	b	c	d	e	f	g	h	i	j
a	0	1	1	1	0	0	0	0	0	0
b	1	0	1	0	1	0	0	0	0	0
c	1	1	0	1	0	0	0	0	0	0
d	1	0	1	0	1	0	0	0	0	0
e	0	1	0	1	0	1	0	0	0	0
f	0	0	0	0	1	0	1	0	1	0
g	0	0	0	0	0	1	0	1	0	1
h	0	0	0	0	0	0	1	0	1	1
i	0	0	0	0	0	1	0	1	0	1
j	0	0	0	0	0	0	1	1	1	0

This local optimization procedure is also called the relocation algorithm in the cluster analysis literature (e.g., Gordon, 1981).

For obtaining a 'good' solution and such impression of its quality, repeat this procedure with different (random) initial partitions $\mathcal{C}_0$. If the procedure is repeated several times (several hundred times), all or most of the partitions of the selected type of equivalence (structural or regular) in a given network can be found.

24.6 Example

Figure 24.3 contains a directed graph on 10 nodes, also known as an Everett network (Borgatti and Everett, 1989), and Table 24.1 gives its relation matrix.

First, consider the structural equivalences in the Everett network. From an examination of the graph it is possible to establish the following pairs of structurally equivalent units: $\{a, c\}$, $\{b, d\}$, $\{g, i\}$, and $\{h, j\}$. These are also the only clusters with structurally equivalent units. The result obtained by the indirect approach is presented in Figure 24.4. The corrected Euclidean-like dissimilarity was calculated on the original relation. The corresponding dendrogram was ob-

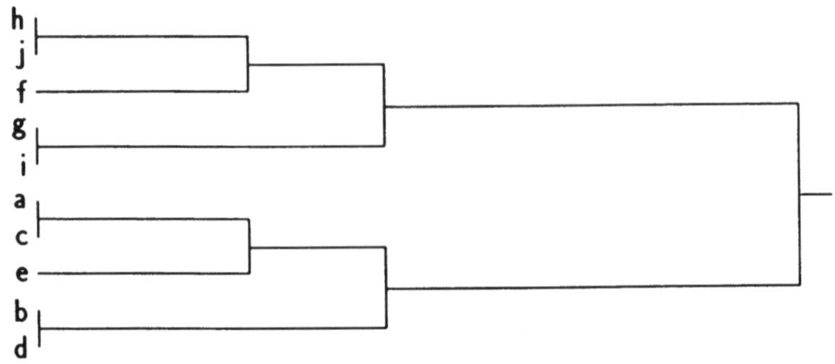

FIGURE 24.4. Everett Network - Indirect Approach, Ward

tained by the agglomerative algorithm with the Ward criterion function (e.g., Gordon, 1981, p. 41). The dendrogram nicely shows the four structurally equivalent pairs of units previously mentioned.

We searched for structural partitions into 1 to 10 clusters by using the direct approach based on the proposed criterion function for structural equivalence. The relocation procedure was repeated 300 times for each number of clusters. The results are presented in Table 24.2. They are congruent with the indirect solution presented by Ward dendrogram. As the analyzed network is rather simple, the exact solutions after 300 repetitions of the algorithm were obtained for each number of clusters.

We also searched for regular equivalences in the Everett network. By means of an exhaustive search, all exact regular equivalences in the network were found. Table 24.3 shows the distribution of the number of clusters in a partition and the number of different regular partitions obtained. The network is small enough for such an exhaustive search. If this relocation procedure were repeated often enough with the criterion function for regular equivalence above, all regular partitions would also be obtained. The table shows that for each number of clusters there is at least one regular partition. There are 94 different regular partitions into 2 clusters, and rather fewer for larger numbers of clusters. The regular partitions into 6 or more clusters are the structural equivalence partitions which are to be expected in this symmetrical network. The regular partitions obtained are quite interesting. For example, consider all the regular partitions into 3 clusters:

$$\mathcal{C}_1 = \{\{ a, c, h, j \}, \{ b, d, g, i \}, \{ e, f \}\},$$

$$\mathcal{C}_2 = \{\{ a, e, h \}, \{ b, d, g, i \}, \{ c, f, j \}\},$$

$$\mathcal{C}_3 = \{\{ a, e, j \}, \{ b, d, g, i \}, \{ c, f, h \}\},$$

TABLE 24.2. Number of Clusters and Structural Equivalence, Direct Approach

Number of Clusters	Minimum Value Criterion Function	Number of Different Obtained Partitions
1	30	1
2	14	1
3	10	2
4	6	1
5	4	2
6	0	1
7	0	4
8	0	6
9	0	4
10	0	1

TABLE 24.3. All Regular Partitions of Everett Network

Number of Clusters	Number of All Partitions	Number of Regular Partitions
1	1	1
2	511	94
3	9330	7
4	34105	12
5	42525	4
6	22827	1
7	5880	4
8	750	6
9	45	4
10	1	1

TABLE 24.4. One Regular Partition into Three Clusters

	a	c	h	j	b	d	g	i	e	f
a	0	1	0	0	1	1	0	0	0	0
c	1	0	0	0	1	1	0	0	0	0
h	0	0	0	1	0	0	1	1	0	0
j	0	0	1	0	0	0	1	1	0	0
b	1	1	0	0	0	0	0	0	1	0
d	1	1	0	0	0	0	0	0	1	0
g	0	0	1	1	0	0	0	0	0	1
i	0	0	1	1	0	0	0	0	0	1
e	0	0	0	0	1	1	0	0	0	1
f	0	0	0	0	0	0	1	1	1	0

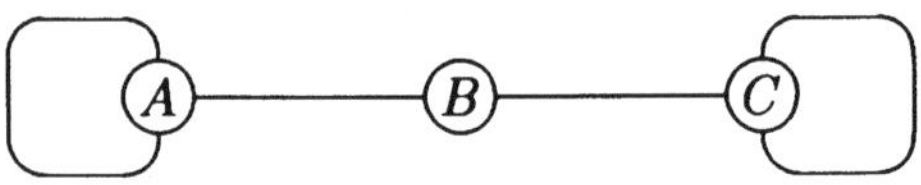

FIGURE 24.5. Blockmodel

$$\mathcal{C}_4 = \{\{ \text{a, f, h} \}, \{ \text{b, d, g, i} \}, \{ \text{c, e, j} \}\},$$

$$\mathcal{C}_5 = \{\{ \text{a, f, j} \}, \{ \text{b, d, g, i} \}, \{ \text{c, e, h} \}\},$$

$$\mathcal{C}_6 = \{\{ \text{a, c, e, f, h, j} \}, \{ \text{b, g} \}, \{ \text{d, i} \}\},$$

$$\mathcal{C}_7 = \{\{ \text{a, c, e, f, h, j} \}, \{ \text{b, i} \}, \{ \text{d, g} \}\}.$$

The first five regular partitions have the same second cluster, and the last two partitions have the same first cluster. Table 24.4 displays the relation matrix permuted into a form compatible with the blockmodel according to the first partition $\mathcal{C}_1$. As stated earlier, regular equivalence produces two types of blocks: null blocks and 1-covered blocks. All of the 9 blocks obtained are of these two types. The blockmodel is also presented as a three-node graph in Figure 24.5 where the first cluster { a, c, h, j } is presented by the node A, the second cluster { b, d, g, i } by the node B, and the third cluster { e, f } by the node C. This simple graph shows an easily comprehensible structure for the analyzed network. For example, each unit from the cluster C is connected to at least one unit of the same cluster and to at least one unit of cluster B. All computations were carried out with the package for structure analysis STRAN (Batagelj, 1991).

24.7 Conclusion

To establish a partition of networks in terms of a chosen equivalence, appropriate clustering algorithms can be successfully applied. First choose between direct and indirect approaches; then, for indirect approaches, there is a choice of a dissimilarity measure. In making this choice, it is necessary to use a measure that is compatible with the chosen kind of equivalence. For example, corrected Euclidean distances on relational data, or the shortest paths matrix, are both appropriate for measuring structural equivalence. However, some of the popular measures of dissimilarity, notably Manhattan distance and Euclidean distance, are not compatible with structural equivalence.

The direct approach uses a special goodness-of-fit measure as a criterion function to obtain partitions based on a chosen kind of equivalence. A local optimization procedure was used to minimize the appropriately constructed criterion function.

Regular equivalence is only one attempt to extend structural equivalence ideas. We conjecture that for other types of equivalence, defined with sufficient precision, a criterion function can be constructed, local optimization methods can be used to locate partitions, and partitions can be evaluated by use of the criterion function.

24.8 References

Batagelj, V. (1991). *STRAN - STRucture ANalysis.* Manual. Ljubljana: Department of Mathematics, University of Ljubljana.

Batagelj, V., Doreian, P., & Ferligoj, A. (1992). An optimizational approach to regular equivalence. *Social Networks, 14,* 121–135.

Batagelj, V., Ferligoj, A., & Doreian, P. (1992). Direct and indirect methods for structural equivalence. *Social Networks, 14,* 63–90.

Borgatti, S.P., & Everett, M.G. (1989). The class of all regular equivalences: Algebraic structure and computation. *Social Networks, 11,* 65–88.

Burt, R.S. (1976). Positions in networks. *Social Forces, 55,* 93–122.

Burt, R.S., & Minor, M.J. (1983). *Applied network analysis.* Beverly Hills: Sage.

Doreian, P. (1988). Equivalence in a social network. *Journal of Mathematical Sociology, 13,* 243–282.

Everett, M.G., & Borgatti, S.P. (1988). Calculating role similarities: An algorithm that helps determine the orbits of a graph. *Social Networks, 10,* 71–91.

Faust, K. (1988). Comparison of methods for positional analysis: Structural and general equivalences. *Social Networks, 10,* 313–341.

Foulds, L.R. (1984). *Combinatorial optimization for undergraduates.* New York: Springer-Verlag.

Gordon, A.D. (1981). *Classification.* London: Chapman and Hall.

Hartigan, J.A. (1975). *Cluster algorithms*. New York: Wiley.

Hummon, N.P., & Carley, K. (1992). Social networks as normal science. Presented at the Annual Sunbelt Social Network Conference, San Diego, February 13-16 1992.

Hummon, N.P., & Doreian, P. (1989). Connectivity in a citation network: The development of dna theory. *Social Networks, 11,* 39–63.

Hummon, N.P., & Doreian, P. (1990). Computational methods for social network analysis. *Social Networks, 12,* 273–288.

Hummell, H., & Sodeur, W. (1987). Strukturbeschreibung von Positionen in Sozialen Beziehungsnetzen. In F.U. Pappi (Ed.), *Methoden der Netzwerkanalyse* (pp. 177–202). München: Oldenbourg.

Kuhn, T. (1970). *The structure of scientific revolutions*. Chicago: Chicago University Press.

Lorrain, F., & White, H.C. (1971). Structural equivalence of individuals in social networks. *Journal of Mathematical Sociology, 1,* 49–80.

Pattison, P.E. (1988). Network models; some comments on papers in this special issue. *Social Networks, 10,* 383–411.

Sailer, L.D. (1978). Structural equivalence: Meaning and definition, computation and application. *Social Networks, 1,* 73–90.

Ward, J.H. (1963). Hierarchical grouping to optimize an objective function. *JASA, 58,* 236–244.

White, D.R., & Reitz, K.P. (1983). Graph and semigroup homomorphisms on networks of relations. *Social Networks, 5,* 193–234.

25

Single Case Studies: The Time Series vs. the Smoothing Regression Approach

Michael G. Schimek[1]

ABSTRACT Single case studies are usually evaluated by time series intervention analysis, a well developed parametric approach. This paper discusses the most advanced technique, theoretically speaking, of Box and Tiao (1975), pointing out conceptual problems and limitations. As an alternative, a new non-parametric approach is introduced for trend estimation, an approach which makes it possible to accommodate linear stochastic noise processes from the Box and Jenkins (1976) methodology. It is called Dependent Error Regression Smoothing (DERS; Schimek, 1992, 1994). Conceptual differences from parametric time series intervention analysis are explored. An example is cited from clinical psychology in which parametric approaches fail to provide any evidence for an intervention effect on an observational series. The example data set is reanalysed in the non-parametric framework of DERS, and the intervention effect under consideration can then be identified.
Keywords: *AR*, *ARMA*, intervention analysis, intervention function, *MA*, non-parametric regression, penalized least squares, prefiltering, pseudo-variance, single case study, smoothing parameter, smoothing spline, time series, trend estimation.

25.1 Introduction

A very common experimental paradigm, originally introduced to evaluate behaviour modification schemes, is the single case study (for reference see Glass, Willson and Gottman, 1975, and Kratochwill, 1978). It has been adopted in many fields of psychological research, although it has never gained much popularity.

The basic outline of a single case study is as follows:
"1. A baseline period in which behavior is observed and recorded, in some of its quantifiable aspects, without attempting to change the environment parameters in any way that would be contingent upon behavior.
2. A period of intervention in which something is done according to a theory that predicts at least one of the aspects of behavior quantified in item 1 will, as a consequence, change in a consistent and interpretable manner.
3. A return to baseline, in order to see if the patterns of behavior examplified in

[1]Medical Biometrics Group, University of Graz Medical Schools, Auenbruggerplatz 30/IV, A-8036 Graz, Austria
Many helpful discussions with C. Chatfield and B.W. Silverman from the University of Bath (UK) are acknowledged. The data were kindly provided by my Ph.D. student Brigitte Pusch. I thank K.G. Schmaranz for technical assistance.

item 1 spontaneously reinstate themselves when the intervention procedures of item 2 are discontinued." (Gregson, 1983, p.199).

This outline is often enriched by a period in which the reverse intervention of item 2 is applied, to cause behaviour changes in the opposite direction for the purpose of additional experimental control. A number of applications published so far show relaxations or specializations of this design but still follow the basic concept.

Problems arise more from inadequate statistical methods for data analysis than from inadequate design. Typical data are equidistant time series observations. The time-dependence (i.e., lack of stochastic independence) of these data is often ignored (see, e.g., the discussion in Gottman and Glass, 1978). Special parametric techniques have been developed to deal with univariate time series from single case studies. There are two popular approaches for such data subject to interventions: One uses the context of the General Linear Model (Glass, Willson and Gottman, 1975), the other the framework of linear stochastic processes (Box and Tiao, 1975). Both are conditionally equivalent, although the latter offers conceptual and practical advantages (see Schimek, 1988, for a discussion). Whichever method is applied, a reasonable number of observations are required within the baseline period. Some psychological studies fail because of a lack of baseline information. The total length of time series sufficient for statistical analysis depends on the design, the type of intervention effects, and – to a lesser extent – on the chosen method.

Analysing time series with embedded episodes due to interventions, even when these interventions follow an experimental or at least a quasi-experimental (*ex post facto*) design, is not trivial methodologically. The main problems include:

1. Only noisy observations are available (there is no information on the signal-to-noise ratio).

2. Mean stationarity is a prerequisite for identification and estimation of parametric models for the data. When data transformations (e.g., differencing) are required, only the linear stochastic structure is preserved. The trend structure to be identified – this is the structure mostly influenced by interventions – undergoes substantial changes (and may be filtered out).

3. Covariance stationarity is required as well for the reasons given above, not only for that part of the series which is free from any intervention (the pre-intervention baseline period of the series), but also for the part with embedded episodes (the covariance structure could also be affected by interventions).

4. The precise onset in time and end point of an embedded episode need to be known (the exact intervention period).

5. Changes in the intensity of the intervention should also be known.

6. Only a relatively limited class of models for describing episodes (intervention effects) are available (e.g., linear stochastic filters with deterministic input).

7. There is uncertainty about the connexion between the basic process (the time series free from intervention) and the embedded episode. The connexion could be additive (the usual assumption) or non-additive causing a change in the basic process.

Parametric approaches are in general too rigid to overcome the problems mentioned above. They always emphasize one feature of the data, e.g., the noise structure of the time series, and the estimation is carried out solely with respect to a goodness-of-fit criterion. As will appear, a non-parametric regression approach can be introduced by means of a so-called "roughness penalty", and estimation carried out under the additional criterion of the smoothness of the regression curve. As a direct consequence, some of the problems listed can now be treated, namely, items 1, 2, 4, 5, and 6 above.

25.2 The regression model, the trend, and the error component

Suppose observations y_i are taken of a continuous random variable Y (the response variable) of at least interval-scale level at equidistant time points t_i (the explanatory variable) satisfying $t_1 < t_2 < \ldots < t_n$. The dependence of the response variable Y on the explanatory variable t is formulated as the univariate regression model

$$y_i = g(t_i) + e_i, \tag{25.1}$$

for $i = 1, 2, \ldots, n$ (n observations). The function g is unknown, even the class of the functional form; it is referred to as a trend curve which is assumed to be smooth to some degree. The errors e_i are taken to follow a linear stochastic process in the time domain. Appropriate error models are autoregressive of order p, $AR(p)$, moving average of order q, $MA(q)$, and autoregressive moving average, $ARMA(p, q)$. For the random shocks ϵ_i (input of the error process) a zero-mean and a common variance are assumed. White noise is included as a special case.

In evaluating a single case study, the basic problem is that the measurements of the response variable Y are noisy. The primary goal is definitely not the identification and estimation of error models. It is instead, the trend g of equation (25.1). Under the assumption of asymptotically unbiased estimation the function g is the best indicator of mean changes in Y due to interventions. Most, though not all, of the embedded episodes usually studied are related to changes in mean level (i.e., trend). Those which modify the covariance or the noise structure of the time sample cannot be evaluated within the framework of linear stochastic processes. There is, therefore, no point in stressing error estimation because the noise model cannot be analysed with respect to intervention effects.

It is clear from the regression model (25.1) that the trend component g and the error component e are taken to be additive. This, of course, is a simplification; but, the estimation problem is still delicate because the ratio between these two components is unknown. All classical parametric approaches obscure this critical aspect by automatically emphasizing one component – usually the error component – for methodological reasons. In standard time series analysis, this is always the error (noise) component.

This paper considers a most widespread and useful methodology made popular by Box and Jenkins (1976) for modelling time-dependent errors. This methodology allocates most of the variation in the response variable Y to the error component as represented by an $ARMA$ process. There is no procedure for handling arbitrary trend functions; on the contrary, the method models linear stochastic processes which are solely defined for weak (covariance) stationary time samples. Hence, the first requirement is to remove any prevailing mean trend. Then, under the assumption of a common variance, the noise process can be identified and estimated. Following the Box and Jenkins methodology, trend is removed by successive differencing (prefiltering) until stationarity is achieved. The number of differencing operations is said to be the order of differencing, d. Hence, any linear stochastic error model $ARIMA(p,d,q)$ can be written as (B denoting the backshift operator)

$$e_i = \frac{\Theta_q(B)}{\Phi_p(B)(1-B)^d}\epsilon_i, \tag{25.2}$$

with the autoregressive operator of order p,

$$\Phi_p(B) = (1 - \phi_1 B - \phi_2 B^2 - \ldots - \phi_p B^p)$$

(for stationarity the roots of $\Phi_p(B) = 0$ must lie outside the unit circle), and the moving average operator of order q,

$$\Theta_q(B) = (1 - \theta_1 B - \theta_2 B^2 - \ldots - \theta_q B^q)$$

(for invertibility the roots of $\Theta_q(B) = 0$ lie must outside the unit circle).

What are the consequences of prefiltering the series? First-order differencing eliminates linear trends (or sequences of linear trends), second-order differencing eliminates quadratic trends (or sequences of linear and quadratic trends), and so on. But this is the exact opposite of what is needed in terms of trend recovery. Information relevant for the estimation of g is lost before the stage of trend estimation is reached . The standard motivation is that, for baseline estimation, trend information should be irrelevant. In clinical psychology, for instance, it is common practice to consider only mean stationary pre-intervention series. But there is no logical argument why the mathematical expectation (i.e., the mean level) of a series, either before or during interventions, should always be constant.

Anyway, what remains, uninfluenced by differencing as long as that operation is not carried to excess, is the noise structure under the assumption that it is adequately described by a linear stochastic error process. It will be seen that the conflict mentioned above between the goal of trend recovery and of the estimation of the noise can only be solved in the non-parametric regression context.

The identification and estimation of an $ARIMA$ model in the presence of a trend is far from trivial (see the example in Chatfield and Schimek, 1987, pp. 359–361). This stage of single case study evaluation certainly requires reasonable experience and has to be mastered for both time series intervention analysis (discussed next) and dependent error regression smoothing (introduced thereafter).

25.3 The time series intervention analysis approach

Time series approaches are usually based on the regression model

$$y_i = c + e_i, \tag{25.3}$$

where c is a constant, and e_i is defined as in (25.1). After fitting an adequate $ARIMA$ model, white noise residuals are expected. Most of the non-stochastic information in the response variable y is either filtered out (by differencing the series) or is disguised in the residuals.

Intervention analysis, following the concept of Box and Tiao (1975), modifies (25.3) to

$$y_i = f(I(t_i)) + c + e_i,$$

where f is commonly called an intervention function (i.e., a transfer function with deterministic input) and $I(t_i)$ a non-stochastic design (intervention) variable which satisfies $I(t_i) \in [0,1] \cap Z$ (Z set of integers). More flexibility in modelling and better interpretability can be achieved following suggestions from Schimek (1988). For instance, applying a so-called process input is recommended, i.e., $I(t_i) \in [0,1] \cap Q$ (Q set of rational nubers), in cases with sufficient prior information on the intervention. But all these amendments cannot remedy the fundamental drawback that only the remaining non-stochastic information among the residuals is analysed. Most of the information concerning the trend g in (25.1) is not available any more when prefiltering has been undertaken.

What kind of embedded episodes can be analysed by means of intervention functions? Suppose (X_t, Y_t) are time-dependent pairs of variables, X_t being the input, and Y_t being the output of a linear dynamic system. The general form of a transfer function model of order (r, s) is

$$\delta_r(B)Y_t = \omega_s(B)X_{t-b}, \tag{25.4}$$

with two polynomials in B,

$$\delta_r(B) = (1 - \delta_1 B - \delta_2 B^2 - \ldots - \delta_r B^r)$$

and

$$\omega_s(B) = (\omega_0 - \omega_1 B - \omega_2 B^2 - \ldots - \omega_s B^s),$$

and a parameter b indicating the discrete system lag.

The stability of the linear dynamic system is guaranteed when the roots of $\delta_r(B) = 0$ lie outside the unit circle.

Substitute y_i for Y_t, and $I(t_i)$ for X_t. The intervention function is then defined by

$$y_i = \frac{\omega_s(B)B^b}{\delta_r(B)} I(t_i) + c + e_i.$$

The pairwise similarity between the polynomials $\Phi_p(B)$ and $\Theta_q(B)$ on the one hand, and $\delta_r(B)$ and $\omega_s(B)$ (see 25.2 and 25.4) on the other, is obvious. So the type of embedded episode which can be modelled is very specific. The standard intervention function is able to describe intervention effects of stepwise exponential behaviour with a discrete input variable $I(t_i)$. A single abrupt change from one mean level to another forms the simplest case (zero-order intervention function). A number of examples of zero- and first-order intervention models are displayed in Schimek (1988, p. 56), and second-order examples are found in McCleary and Hay (1980, pp. 185–188), though there are, of course, many other types of embedded episodes in single-case studies as well. Hitherto, non-parametric approaches have been the only methods able to cope with them.

When analysing impacts on an empirical time series, a theory of change is needed in order to select an appropriate intervention model. There is no parametric method of exploring data with respect to intervention functions which might be applied. Furthermore, the properties of the agent of change represented by the time-dependent intervention variable, especially the onset of the influencing event, must be known *a priori*. Again, exploratory data analysis cannot be carried out by means of parametric models.

All the information in the data that can be attributed to an error structure is used to estimate the noise component. Only the remaining information can be modelled by an intervention function. In most cases, this situation is characterised by insufficient knowledge concerning the interventions to be analysed. So, the model selection has to be a trial-and-error procedure.

25.4 The dependent error regression smoothing approach

We now introduce an alternative non-parametric data-driven regression approach for time series which allows one to explore arbitrary trends (embedded episodes other than instantaneous events in the terminology of Schimek, 1988, p. 53), possibly related to interventions in an experimental or at least quasi-experimental design. Understanding of trend comprises probabilistic and deterministic drifts in the time series (see, e.g., Glass, Willson, and Gottman, 1975, p. 114). The approach is called Dependent Error Regression Smoothing (DERS). Under the assumptions made in Section 25.2, we extend the ordinary least squares regression concept by what is called a "roughness penalty" (penalized least squares). This leads to the problem of minimizing a penalized sum of squares with respect to g; in matrix notation,

$$SS(g) = (y - g)^T \mathcal{D}^{-1}(y - g) + \alpha g^T \mathcal{Q} g, \tag{25.5}$$

where y is an n-dimensional column vector of observations, g is an n-dimensional column vector of curve estimates $g(t_i)$, and $\mathcal{D}^{-1}$ the inverse of $\mathcal{D} \neq \mathcal{I}$, the general variance-covariance matrix of size n x n. $\mathcal{Q}$ is a n x n-dimensional operator matrix (second-order difference quotient), explicitly given (for $n \geq 6$) by

$$\mathcal{Q} = \begin{bmatrix} 1 & -2 & 1 & & & & \\ -2 & 5 & -4 & 1 & & & \\ 1 & -4 & 6 & -4 & 1 & & \\ & \ddots & \ddots & \ddots & \ddots & \ddots & \\ & & 1 & -4 & 6 & -4 & 1 \\ & & & 1 & -4 & 5 & -2 \\ & & & & 1 & -2 & 1 \end{bmatrix}.$$

Finally, $\alpha (\alpha > 0)$ denotes the smoothing parameter, controling the exchange between the local variability of the curve as measured by the roughness penalty (right-hand term of (25.5)) and the infidelity to the data as quantified by the ordinary sum of squares expression.

The non-parametric approach defined above is a discrete version of cubic spline smoothing (as described in Silverman, 1985), which can easily be motivated as a standard choice when one does not require specific smoothness properties. It is the optimum choice if the mathematical expectation of g is a linear function.

A generalized least squares estimator for (25.5) related to ridge regression is

$$\hat{g} = (\mathcal{D}^{-1} + \alpha \mathcal{Q})^{-1} \mathcal{D}^{-1} y. \tag{25.6}$$

Its linearity (linear projection matrix) is straightforward.

The minimizer $\hat{g}$ can be evaluated for any $AR(p)$ or $MA(q)$ error process under the assumption of weak stationarity or invertibility of a known error structure (variance-covariance matrix $\mathcal{D}$) and a fixed α value. Under the necessary assumption of weak stationarity, $\hat{g}$ is an asymptotically unbiased estimator of trend g. For technicalities, we refer to Schimek (1994).

Very important are aspects of numerical stability and time complexity. Throughout the process of estimation only symmetric, bandlimited matrices are involved. Explicit matrix inversions can be avoided. The numerical stability (critical factorization algorithms involved) is satisfactory. It has been shown (Schimek, 1992) that the estimation of $\hat{g}$ requires only $O(n)$ operations for $n \gg m$, where n is the number of observations of the time series, and m is the order of the error process (p respectively q). Hence DERS is a linear time algorithm.

Slightly more expensive is the calculation of the projection matrix $\mathcal{H}$, fulfilling $\hat{g} = \mathcal{H}y$. It is required for certain diagnostic purposes and for the calculation of exact variance estimates associated with the model fit.

In smoothing spline regression applied to independent observations, the α values are usually determined by the method of cross-validation (see Silverman, 1985, for details). But for time-dependent observations, the cross-validatory choice is biased. Schimek and Schmaranz (1993) provide simple alternative techniques based on exact variance estimates or on pseudo-variance estimates. The latter approach is computationally much simpler. In both cases, the decision for a specific α value is made graphically. For that purpose, we have to plot the variance estimates against the model complexity represented by a sequence of values $\ln(\alpha)$. The advantage of applying exact variances is that they are unbiased for a linear reference trend under arbitrary AR and MA error processes (weak stationarity assumed). However, our experience is that the pseudo-variance based approach works very well and yields much the same α values at almost no extra computational costs (evaluation of the hat matrix is not required).

What are the main achievements of DERS when applied to single case studies? DERS is specifically designed to cope with noisy observations without stressing the noise component of the regression problem. This has become feasible because of the separation of the (now external) parametric identification and estimation of the error model (performed in a way similar to time series analysis) from the non-parametric trend curve estimation, for a given error model and a predetermined α value. By means of α value adaptations and specific error variance assumptions, various signal-to-noise ratios can be considered.

Another advantage is the possibility of searching for an intervention effect within the class of all smooth trend curves. This makes it possible to assume a data exploration perspective based, of course, on the experimental or quasi-experimental background of the single case study. With a known intervention effect which can be adequately described by an intervention function, not too frequent a situation, the non-parametric approach can be adopted as well, but is not worthwhile. The usual Box and Tiao technique is then the right choice.

In cases where the trend structure changes with the duration of the intervention (as in the example in Section 25.5 below), no parametric approach is viable. In some cases, non-parametric techniques can be successfully applied and DERS is a good device for the exploration of embedded episodes. It provides the shape information necessary for a final parametric fit in, for instance, the framework of Box and Tiao. If a highly complicated (e.g., combined intervention functions) model proves necessary, a parametric fit should be avoided because of the lack of psychological interpretability. Trend curves obtained by non-parametric means are more easily understood with respect to specific intervention designs. Limited information about the duration and intensity of interventions is less critical when applying an exploratory technique such as DERS.

If the interventions are known, if sufficient information to select standard intervention functions is available, and if the features of the time series enable this knowledge to be used, parametric estimation should be the final goal. The advantage is that a few parameters, which can be tested statistically, represent the model. Furthermore, the results are easy to communicate.

There are only two limitations of DERS. First, the errors follow a non-stationary (e.g., explosive) stochastic process, or a process other than linear, the variance-covariance matrix $\mathcal{D}$ cannot be specified. In such a case, when the aim is data exploration, it is acceptable to assume white noise errors (i.e., $\mathcal{D} = \mathcal{I}$). Second, if the error variance is heteroscedastic, biased estimates of the trend curve are obtained. All parametric techniques of intervention analysis fail under such conditions.

25.5 An example

This section analyses a time series from a single case study subject to interventions, first with the time series intervention approach, and then with the dependent error regression smoothing approach.

In this single case study the personal wellbeing of a client was assessed twice a week during a period of changing psychotherapeutic interventions. There were 82 equally spaced observations, the first 9 constituting the baseline period. The time series is plotted in Figure 25.1. An initial analysis showed that weak stationarity could safely be assumed for that series. Within the framework of the Box and Jenkins (1976) methodology, an $AR(1)$ error model was identified and estimated from the total series, with parameter $\phi = 0.39$. The total series was used for two reasons: First, the number of observations belonging to the baseline are too few and, second, for DERS an error model is required that holds true for the whole time series. No additional seasonal component was needed. The residual series did not differ statistically from white noise, which fully supported the model fitted. The residual variance assumed the large value $\hat{\sigma}_r{}^2 = 1336.81$; the reason for this can be seen in Figure 25.1. There is a remarkable discrepancy between the data y_i and the $AR(1)$ fitted $\hat{y}_i$. A considerable amount of variation remains unexplained, and the existence of embedded episodes cannot be ruled out.

Following the ABABAB design (A free of interventions) of the single case study, the simplest intervention hypothesis was a zero-order function in one ω (level) parameter with binary input $I(t_i)$, putting t_i equal to zero within the A segments and to one within the B segments. This is equivalent to hypothesising abrupt mean level changes from one segment to the next. Alternatively, gradually changing mean levels could be considered. They can be modelled by a first-order

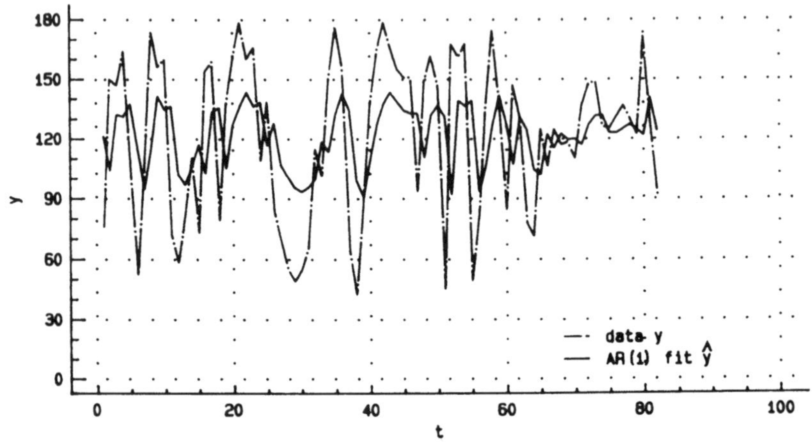

FIGURE 25.1. The time series data and the AR(1) fit with $\hat{\phi} = 0.39$

intervention function in one ω and one δ (rate) parameter based on the same input. Applying the Box and Tiao technique, neither of these two hypotheses led to a fit with significant intervention parameters (i.e., statistically different from zero) and a worthwile reduction in the residual sum of squares relative to the single $AR(1)$ fit. Hence, from a parametric point of view, there is no evidence for any sort of intervention effect.

The alternative non-parametric DERS was tried next. The $AR(1)$ error model with $\hat{\phi} = 0.39$ estimated previously was used to specify the $\mathcal{D}$ matrix. Furthermore, based on initial data analysis, an error variance of $\hat{\sigma}_\epsilon{}^2 = 100$ was taken to be reasonable; this is much smaller than the residual variance previously obtained. A set of α values was required for the purpose of trend estimation. A standard choice is then the sequence of exponentials of the integers $-10 \leq \ln(\alpha) \leq 20$. A very efficient way to choose a proper α value is to calculate the pseudo-variance for each fit, given the sequence of values above, and to plot the variance estimates against the model complexity represented by $\ln(\alpha)$. For an adequate or more than adequate (i.e., non-parsimonious) model, the variance stabilises and reaches a (local) minimum (Schimek and Schmaranz, 1993).

The relationship between the pseudo-variance estimate and the model complexity for this data set is displayed in Figure 25.2. Interpreting the graph from left to right: the function value obtained for $\ln(\alpha) = -10$ is associated with a fit almost interpolating the data. For $\ln(\alpha) = 4$, the function is almost down to its minimum level. For a $\ln(\alpha)$ larger than, say 6, the model structure cannot be improved any further in terms of complexity. Thus, where the function gets very close to a minimum asymptote, adequate smoothing parameter values can be found, representing appropriate model complexities. Values further to the right force the regression approach to oversmooth the data. An arbitrarily large α produces a straight line fit.

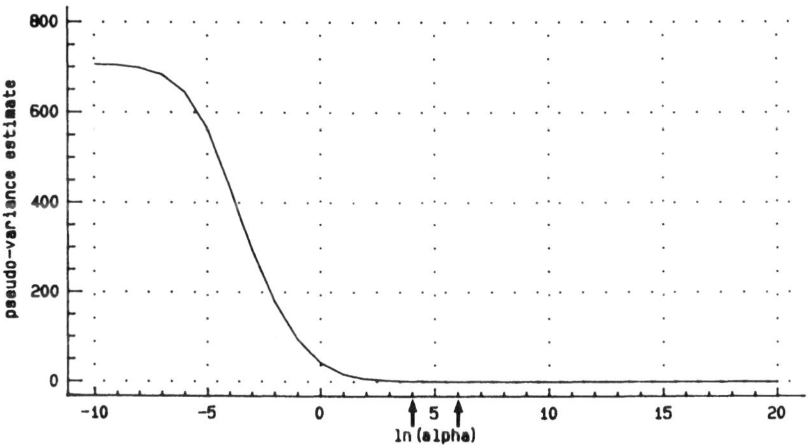

FIGURE 25.2. Plot of pseudo-variance estimates against the model complexity $\ln(\alpha)$

The best choice seems to be $\ln(\alpha) = 6$ (approximately $\alpha = 403.43$). The estimated trend $\hat{g}$, under the error model $AR(1)$ with $\phi = 0.39$, is displayed in Figure 25.3, for $\ln(\alpha) = 4$ and $\ln(\alpha) = 6$ together with the time series data. The α value near the minimum level (i.e., $\ln(\alpha) = 4$) is still not entirely satisfactory. But the chosen α value (i.e., $\ln(\alpha) = 6$), gives a proper estimate of the function g. Notice that the mean level changes over the sample period are in contradiction to the result obtained from the time series approach above. This is due to the emphasis on the noise component in the parametric fit.

In addition, the quasi-experimental ABABAB design of the single case study is displayed in Figure 25.3. It can be immediately recognized that the trend curve changes its direction whenever the intervention switches from A to B, and vice versa, and also depends on the duration of the embedded episodes. This is a good example to show how the DERS approach improves the identifiability of otherwise hidden intervention effects by means of non-parametric trend estimation.

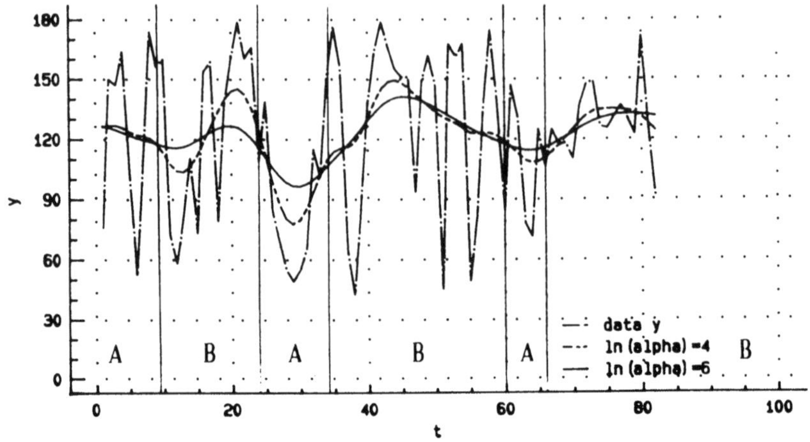

FIGURE 25.3. The time series data, estimated trends for $\ln(\alpha) = 4$ and $\ln(\alpha) = 6$, and the quasi-experimental ABABAB design

25.6 References

Box, G.E.P., & Jenkins, G.M. (1976, 2nd edition). *Time series analysis. Forecasting and control.* Oakland, CA: Holden-Day.

Box, G.E.P., & Tiao, G.C. (1975). Intervention analysis with applications to economic and environmental problems. *Journal of the American Statistical Association, 70,* 70–79.

Chatfield, C., & Schimek, M.G. (1987). An example of model-formulation using IDA. *Statistician, 36,* 357–363.

Glass, G.V., Willson, V.L., & Gottman, J.M. (1975). *Design and analysis of time-series experiments.* Boulder, CO: Colorado Univ. Press.

Gottman, J.M., & Glass, G.V. (1978). Analysis of interrupted time-series experiments. In T.R. Kratochwill (Ed.), *Single subject research. Strategies for evaluating change* (pp. 197–235). New York: Academic Press.

Gregson, R.A.M. (1983). *Time series in psychology.* Hillsdale, NJ: Lawrence Erlbaum.

Kratochwill, T.R. (1978). *Single subject research. Strategies for evaluating change.* New York: Academic Press.

McCleary, R., & Hay, R.A. (1980). *Applied time-series analysis for the social sciences.* Beverly Hills, CA: Sage.

Schimek, M.G. (1988). How to differentiate between impact and effect using Box and Tiao intervention functions: Some zero and first order cases. *EDV in Medizin und Biologie, 19,* 49–57.

Schimek, M.G. (1992). The DERS algorithm in comparison with related spline smoothing algorithms: Numerical aspects and time complexity. In F. Faulbaum (Ed.), *SOFTSTAT '91. Advances in Statistical Software 3* (pp. 267–274). Stuttgart: Fischer.

Schimek, M.G. (1994). Penalized least squares regression smoothing with dependent errors. (Submitted for publication.)

Schimek M.G., & Schmaranz, K. (1993). Dependent error regression smoothing: Technique, PC program and application. (Submitted for publication.)

Silverman, B.W. (1985). Some aspects of the spline smoothing approach to non-parametric regression curve fitting (with discussion). *Journal of the Royal Statistical Society, B, 47,* 1–52.

26

Effects of Collapsing Data from Crossover Designs

John W. Cotton[1]

ABSTRACT Much behavioral research using within-subject designs employs special balancing methods to control for carryover effects and for period effects due to practice and fatigue. Unfortunately, analysis of data from these studies often does not make full use of the information available. Rather, data are analyzed after collapsing scores for a given treatment into a total for each subject. This method of data analysis is shown to produce bias in estimation of treatment effects and of the standard errors of such estimates. Therefore, both the nominal significance levels and the nominal power values for these analyses may be in error. A numerical example illustrates these difficulties. Recommendations are given for changes in experimental design and analysis methods that can obviate these difficulties.

26.1 Orientation

26.1.1 Nature of the Problem

Often experimenters wish to make within-subject comparisons of performance on different treatments, such as kinds of stimuli to be recognized rapidly. One hopes to estimate *treatment effects* without bias even if the performance measure also reflects time-related factors such as practice effects. Such effects will usually be given the generic label of *period effects* in this chapter. Biasing effects may also stem from some feature of a prior treatment. These are *carryover effects*: Behavior may carry over effects of that treatment. Some authors call carryover effects *indirect* or *residual treatment effects.*

Experimenters attempt to control for nuisance factors such as period effects and carryover effects by using a variety of means of balancing the temporal ordering of the different stimuli (treatments) presented to a given subject. Unfortunately, these experimenters are less likely to employ statistical controls parallelling these experimental controls. This point may be clarified by thinking about a variety of ways in which a representative psychological question could have been explored experimentally and obtained data could have been analyzed statistically. For that purpose I make repeated reference below to a temperature perception experiment of Refinetti (1989) involving period and carryover effects in addition to effects of the current temperature.

---

[1]Departments of Education and Psychology, University of California, Santa Barbara, CA 93106
I thank the editors of this volume and their consultants, as well as Richard J. Shavelson, Dean of the Graduate School of Education, University of California, Santa Barbara, for their very helpful suggestions about this chapter.

26.1.2 Models

Let us call $Y_{jk(m'm)}$ the score for Subject k on period j with treatment m presented currently and treatment m' in the immediately previous period. The following statistical model is proposed:

$$Y_{jk(m'm)} = \mu + s_k + \pi_j + \tau_{[j,m]} + \gamma_{[j-1,m']} + e_{jk(mm')}, \tag{26.1}$$

where μ is an additive constant, s_k is the random effect of Subject k, π_j is the effect of period j, $\tau_{[j,m]}$ is the direct effect of treatment m presented on period j, and $\gamma_{[j-1,m']}$ is the classical carryover effect of treatment m' presented on period $j-1$. Because no experimentation precedes period 1, the carryover present on the first period, $\gamma_{[0,0]}$, is assumed to be 0. Also $e_{jk(m'm)}$ is the random error operative for Subject k receiving treatments m and m' on the stated periods. The $e_{jk(m'm)}$ are assumed to be independently and identically distributed (i.i.d.) with $N(0, \sigma^2)$; the s_k are assumed i.i.d. $N(0, \sigma_s^2)$.

In a simpler model one deletes carryover effects and the m' symbol from this equation, obtaining:

$$Y_{jk(m)} = \mu + s_k + \pi_j + \tau_{[j,m]} + e_{jk(m)}. \tag{26.2}$$

Or one may also delete period effects, yielding:

$$Y_{jk(m)} = \mu + s_k + \tau_{[j,m]} + e_{jk(m)}. \tag{26.3}$$

26.1.3 Some Definitions

A *crossover design* is a within-subject design in which at least one subject receives different treatments on two or more periods of observation.

Period (trial) and treatment effects have conventional psychological meanings here. I consider two possible kinds of carryover effects: (a) *Classical carryover effects.* Delayed effects of prior treatments may occur, unaffected by the current treatment. For example, a classical carryover from treatment A may occur on a subsequent occasion either in the presence of treatment A or of some other treatment such as treatment B. (b) *Transitional carryover effects.* Effects of treatment changes from one period to another, such as from treatment A on period 1 to treatment B on period 2. Note that transfer effects during training may be defined as a composite of period effects and carryover effects.

The efficiency with which a design estimates a difference between two parameters such as τ_{m1} and τ_{m2} will be defined in relation to the minimum possible variance of such an estimator, $2\sigma^2/r$, where r is the number of replications per treatment. The efficiency, E_t, for estimating $\tau_{m1} - \tau_{m2}$ with a specific treatment effect difference unadjusted for carryover effects becomes (Jones and Kenward, 1989, p. 194):

$$E_t = \frac{2\sigma^2/r}{V[\hat{\tau}_{m1} - \hat{\tau}_{m2}]} \times 100. \tag{26.4}$$

Similar definitions apply to E_d, the efficiency in estimating treatment effect differences adjusted for carryover effects, E_c, the efficiency of a design for assessing carryover effect differences, and E_p, the efficiency for estimating period effects, with r = number of replications per period. Notation such as $\overline{E}_d$ represents efficiency averaged over all pairs of effects being examined; it is needed if efficiency varies for different pairs of effects. For all types of efficiency used here, $E = 100$ is the maximum possible value; a design with that value for a certain effect is called *optimal* for that effect.

In seeking to improve an experimental procedure, one may either (a) seek a design approximating efficiency of 100 for a given σ^2, (b) try to reduce the value of σ^2 by improving the precision of measurement or by shifting from a between subjects design to a within subjects design, or (c) do both approaches. Since it is often useful to compare designs of a certain class (e.g., all designs with K treatments, N subjects, and p periods), it is standard practice also to call a design *optimal* for assessing an effect if it has maximal efficiency (not necessarily 100) among all designs of that class. Note that some designs may simultaneously be optimal for assessing two or more kinds of effects.

26.1.4 Should We Collapse Crossover Design Data Before Analysis?

A full trial-by-trial or period analysis is rare in crossover design studies by psychologists. Instead, most investigators seem to employ some variant of the following procedure: First, find a treatment average, $\overline{Y}_{.k(.m)}$, for Subject k receiving treatment m, doing so for all k and m values. Then, make an $N \times K$ table of these treatment totals, and assess subject and treatment effects with a t-test (for $K = 2$) or a 2-way ANOVA with one observation per cell, plus related analyses. In some cases, collapsing is not complete: Although individual period means cannot be compared and carryover effects cannot be assessed in that case, one can analyze treatment means for each of two or more sessions, as in a music perception report by Cohen, Trehub, and Thorpe (1989), permitting measurement of session differences at least.

Is a collapsed data approach legitimate? Is it as efficient as a full General Linear Model (GLM) analysis in the tradition described by Searle (1987)? On the one hand, it is legitimate to combine an individual subject's scores into such a summary score as a contrast for assessing a specific effect (Jones and Kenward, pp. 160-163, pp. 247-260). On the other hand, I know of no serious statistical discussion of the effects of collapsing data in the present manner prior to performing an ANOVA with them. This chapter attempts to assess the effects of such collapsing procedures.

26.1.5 ILLUSTRATIVE CONSTRAINTS ON CROSSOVER DESIGNS

I distinguish between four degrees of experimental constraint, illustrating the first three with recent psychological experiments which simultaneously could have assessed treatment, period, and carryover effects if appropriate statistical techniques had been readily available. The fourth kind of constraint leads to an optimal design. One such optimal design will be compared in some detail with the second or block randomization design type. Later I examine for that second design (a) how period or carryover effects can be extracted from error estimates as a means of increasing the efficiency of data analysis, (b) whether collapsing of data before statistical analysis leads to bias in means and variances, and (c) whether an optimal design would be beneficial in the sense that the efficiency gain from its use would be substantial.

Complete Block Designs (Random Order, Constant Frequency of Each Treatment Per Subject)

In this design type, each subject receives the same number of presentations of each treatment, in random order. For example, Shoben, Sailor, and Wang (1989) used their Experiment 1 partly to establish whether humans could judge the comparative magnitude of two animals on a dimension more easily if its value on that dimension were salient for both animals being compared (as in the case of size for two small animals such as "mole" and "worm") than if it were nonsalient. Given five binary within-subject factors, it was possible to present 32 stimulus pairs ($K = 32$) four times each in random order ($p = 32\times4 = 128$), with each trial having a subject see a comparative term such as "smaller", and a stimulus pair such as "mole-worm", and then respond with the appropriate animal name. Reaction time and error rate were the principal response measures.

Block Randomization Designs (Random Arrangement of Treatments within Successive Blocks)

This type is the same as for complete block designs except that periods 1 through K include one instance of each treatment, as do all other successive blocks of K periods.[2]

For example, in Refinetti's (1989) Experiment 1 on temperature judgments, mentioned earlier, something approximating block randomization of 30 temperatures as stimuli ($K = 30$) was employed with seven of the subjects (N) in three blocks ($p = 30\times3 = 90$). The author informally reported that, when presented immediately after a high temperature, a moderately low temperature sometimes received a number lower than the anchor judgment of "20" assigned to the lowest temperature actually presented. This is an indication of a carryover effect whose

[2]Cox (1951, p. 320) discussed the origins of block randomization designs (called blocking by replicates by him) in agricultural research. Psychologists (e.g., McBurney, 1983, p. 99) frequently recommend block randomization as a means of correcting for unwanted sequence effects.

TABLE 26.1. A randomly selected 2-block randomization design with $N=16$, $K=4$, and $p=8$

Subject	Period							
	1	2	3	4	5	6	7	8
1	B	C	A	D	B	C	A	D
2	B	D	A	C	A	C	D	B
3	D	A	C	B	B	C	D	A
4	A	B	C	D	B	D	C	A
5	C	D	B	A	A	C	D	B
6	A	D	B	C	C	B	A	D
7	B	A	D	C	D	B	C	A
8	C	B	A	D	B	A	D	C
9	A	C	D	B	A	D	B	C
10	B	A	C	D	D	C	A	B
11	B	A	D	C	A	C	B	D
12	D	A	C	B	C	A	D	B
13	D	C	A	B	D	A	B	C
14	B	A	C	D	D	C	A	B
15	D	A	B	C	C	B	A	D
16	A	D	B	C	B	C	D	A

magnitude depends on the size of the difference between the prior and present treatments. Such an effect has previously been considered briefly by D'Amato (1970, pp. 52-54).

Table 26.1 above is an example of a block randomization design of a smaller size, with the order of treatments in each of two blocks of four trials for each subject being randomly selected.

Uniform Designs (Blocking by Subjects and by Periods, as in Latin Squares)

Hedayat and Afsarinejad (1978) define a broad class of designs, including Latin square designs as special cases: A *uniform* cross-over design is one in which, for each subject, each treatment appears the same number of times as every other treatment, this number being constant across subjects. Subjects receive any given treatment on any specific period. Similarly, the Williams (1949) balanced Latin squares are examples of *completely balanced* (or *strongly balanced*) designs (Jones and Kenward, 1989, p. 210), ones in which every possible pair of treatments such as AD appears equally often, pairs being on adjacent periods for a given subject.

Uniform designs are similar to block randomized designs, but in addition, the number of subjects, N, is a multiple of K, and every period has exactly

TABLE 26.2. Berenblut's optimal design for $K = 4$ and $p = 8$

Subject	Period							
	1	2	3	4	5	6	7	8
1	A	D	C	B	B	C	D	A
2	B	A	D	C	C	D	A	B
3	C	B	A	D	D	A	B	C
4	D	C	B	A	A	B	C	D
5	D	D	B	B	A	C	C	A
6	A	A	C	C	B	D	D	B
7	B	B	D	D	C	A	A	C
8	C	C	A	A	D	B	B	D
9	C	D	A	B	D	C	B	A
10	D	A	B	C	A	D	C	B
11	A	B	C	D	B	A	D	C
12	B	C	D	A	C	B	A	D
13	B	D	D	B	C	C	A	A
14	C	A	A	C	D	D	B	B
15	D	B	B	D	A	A	C	C
16	A	C	C	A	B	B	D	D

N/K subjects receiving each treatment. For example, this was accomplished in a 60-subject, 6-period, 3-treatment study of the coordination of information in a computer display of two moving objects (Morrin, 1989). This design may be said to have treatment orders as rows, periods 1 through 3 as columns, and treatments as letters inside two different 3×3 Latin squares. Ten subjects received each of the six treatment orders, and periods 4 through 6 for a given original treatment order had the following structure: One male and one female subject out of the 10 for that order each received a new order of the three treatments, with period 4 being assigned a treatment different from that of period 1 and different for each of the five pairs of subjects.

Optimal Designs, for Maximizing Efficiency of Estimating Some Effect

Berenblut (1964, 1967) showed how to construct an optimal design appropriate to K treatments, $2K$ periods, and K^2 subjects. Jones and Kenward (1989, Table 5.10, p. 208) present an example of a 4-treatment Berenblut design, appearing below as Table 26.2.

This Berenblut design has $E_d = 100.00$, $\overline{E}_c = 85.94$, and $\overline{E}_p = 95.52$, making it an excellent design to employ for assessing these three kinds of effects. ($\overline{E}_t$ need not be computed because the easy computational procedure for assessing carryover effects makes it appropriate to capitalize upon the maximal E_d value instead.) Since every possible pair of treatments appears exactly seven times in Table 26.2, this Berenblut design is also completely balanced and, by Cheng and Wu's (1980, 1983) Theorem 3.1, optimal among designs with 4 treatments, 16 subjects, and 8 trials for the estimation both of direct and carryover effects. The calculation method for finding E and $\overline{E}$ values like these is detailed in Sections 26.1.3 and 26.2.2.

26.2 Assessment of Effects of Collapsing Data from a Block Randomization Design

26.2.1 Information Needed

For a specific design constraint type such as a complete block, block randomization, or uniform design, we may ask what is the effect of variations in K, N, and p. For a GLM analysis we would need to show results of estimation and hypothesis testing and to make efficiency calculations in order to assess the relative attractiveness of a given design, compared to an optimal one. To assess a collapsed analysis of treatment effects, we would need comparable calculations plus measures of mean bias; the variance of that bias; the ratio of the variance of the biased estimator to the variance of the unbiased estimator from a between-subjects comparison of treatments with the same number of observations and the same population standard deviation; the mean bias divided by the population standard deviation; the correct significance level conditional upon a given degree of bias; and the power function conditonal upon that bias.

26.2.2 Findings

Suppose that Refinetti (1989) had sought perceived temperature judgments for 4 temperatures encountered 2 times each in a total of 8 periods for 16 subjects. The statistical properties of the relevant block randomization design (Table 26.1) are examined in detail, comparing it to the relevant optimal design (Table 26.2).

Statistical Properties of a Specific Block Randomization Design

Since the frequencies of the 16 pairs, AA AB ...DD, are not constant in Table 26.1, this design is not completely balanced. Nonetheless, the method of analyzing data from the design is straightforward: First, the standardized estimates of variances of treatment effect differences are computed. Those values and (26.4) together imply an average efficiency ($\overline{E}_d$) for estimating treatment effect differences of 86.44. Similar calculations yield an average efficiency ($\overline{E}_c$) of 73.96 for carryover effects estimation and an average efficiency ($\overline{E}_p$) of 91.15 for period effects estimation. Ratios of $\overline{E}_d$, $\overline{E}_c$, and $\overline{E}_p$ for this design compared to those for the Berenblut design are 86.4%, 86.1%, and 95.4%. This means, for example, that 87 subjects with the Berenblut design give as much information about treatment effect differences as 100 subjects with this block randomization design. This suggests that choosing the Berenblut design has a small, practical advantage, except when primary emphasis is placed upon estimation of period effects.

If there are no carryover effects, making (26.2) apply, corresponding calculations lead to an average efficiency, $\overline{E}_t$, value of 95.06 for identifying treatment effects without extracting carryover effects. Also $\overline{E}_p$ is 97.85. Finally, if there are neither carryover nor period effects, making (26.3) apply, the present design has an efficiency of 100 for estimating treatment effects. These calculations presume use of a full GLM analysis of the data, as when appropriately using SAS GLM (SAS Institute, 1988) or the GAUSS mathematical programming language (Aptech Systems, 1988).

The Full Analysis: General Linear Model Analysis

Effects of Collapsed Analysis on Mean and Variance of Estimators of Treatment Effects

First, assume that we want to test the hypothesis that the effects of treatments A and B are equal, taking into account the possibility that period and carryover effects are also present. For Subject 1 for treatment A, involving treatment effects from periods 3 and 7 in Table 26.1, with a carryover effect of C from both periods 2 and 6, the current model's average in the collapsed data summary is

$$\overline{Y}_{.1(.A)} = \mu + s_1 + .5(\pi_3 + \pi_7) + \tau_A + \gamma_C + .5(e_{31(A)} + e_{71(A)}). \tag{26.5}$$

For Subject 1, treatment B, involving treatment effects from periods 1 and 5 and a carryover effect only from period 4, the average is

$$\overline{Y}_{.1(.B)} = \mu + s_1 + .5(\pi_1 + \pi_5) + \tau_B + .5\gamma_D + .5(e_{11(B)} + e_{51(B)}). \tag{26.6}$$

The difference between these two values could be used in estimating the difference between two treatment effects:

$$\begin{aligned}\overline{Y}_{.1(.A)} - \overline{Y}_{.1(.B)} = \; & \tau_A - \tau_B + [.5(\pi_3 + \pi_7 - \pi_1 - \pi_5) + \gamma_C - .5\gamma_D] \\ & + .5(e_{31(A)} + e_{71(A)} - e_{11(B)} - e_{51(B)}).\end{aligned} \tag{26.7}$$

The possible bias introduced by the bracketed part of this equation has a parallel for every other subject in this table. These terms affect our experimental conclusions in two ways: (a) Average bias across all subjects may distort our view of the size and direction of treatment effect differences. (b) Since random error will be independent of these individual biases (bracketed terms as in (26.7)), it follows that a desired variance will be the sum of variance for error and variance of bias across subjects:

$$\text{Var}(\overline{Y}_{..(.A)} - \overline{Y}_{..(.B)}) = \text{Var(individual-biasing-terms)} + \sigma^2/N. \tag{26.8}$$

This value is typically greater than σ^2/N, thus possibly reducing the power of a collapsed data test for treatment effects.

Let us assign arbitrary values to the π_j, $\eta_{[j,m]}$, and $\gamma_{[j-1,m']}$ parameters of (26.1) and study how this controls relative bias and variance of treatment effect estimation for a variety of σ^2 values. It will be useful to employ great generality, exploring relative sizes of bias (from 0 to ∞) due to a fixed pattern of period effects, compared to error variance. To do this, one fixes a pattern of π_j values, weights each by a non-negative multiplier, Q, to vary the size of bias, and computes the resulting variance of bracketed terms just discussed: Let the π_j values be expressed in standard deviation units times an arbitrary multiplier, Q, becoming: 0, $1.46Q\sigma$, $2.62Q\sigma$, $3.55Q\sigma$, $4.28Q\sigma$, $4.86Q\sigma$, $5.32Q\sigma$, and $5.69Q\sigma$, respectively for successive periods. So there is an exponential increase with periods. In the first analysis, let us assume that all carryover values ($\gamma_{m'}$) are zero; there are period effects but no carryover effects.

Period Effects as Nuisance Effects

Now compute the net contribution of π_j values from (26.7), finding a value of $1.83Q\sigma$ for Subject 1. Similar calculation for each other subject of Table 26.1 gives a set of 16 biases (1.83, 0.61, ..., −0.35, −0.61) with a mean of $-0.084375Q\sigma$ and a variance of $Q^2\sigma^2$. The importance of a specific mean bias depends in part on the variance of $\overline{Y}_{..(.A)} - \overline{Y}_{..(.B)}$ and in turn on the value of Q.

From the above and (26.8), the variance of $\overline{Y}_{..(.A)} - \overline{Y}_{..(.B)}$ for this example will be $Q^2\sigma^2 + \sigma^2/16$. The ratio of this variance to $\sigma^2/16$ is $16Q^2 + 1$; it is called VarRatio in the top half of Table 26.3 where it and other statistics are related to the value of Q. Note that 4MeanBias/σ, using the mean bias value from above, proves to be $-0.3375Q$. The tabled quantity $\lambda = 16Q^2$ will be discussed shortly. In the absence of carryover effects, Q values of less than 0.088 yield less than a 12.5% error in the variance of the estimate, $\widehat{\tau_A - \tau_B}$. Larger Q values are of more concern.

Carryover Plus Period Effects as Nuisance Effects

Now keep the period effects just discussed and also include carryover values $\gamma_{[j-1,m']}$ of $0.7Q\sigma$, $1.4Q\sigma$, $2.1Q\sigma$, and $2.8Q\sigma$, for period $j-1$ treatments A through D, respectively, yielding a combined mean bias of $+0.046875Q\sigma$ and a variance of biases of $2.99Q^2\sigma^2$ for an experiment with this design, set of period effects, and set of carryover effects. Note that the mean bias has changed its sign and also has a smaller absolute value than before; carryover effects partly counteract period effects here.

Now the variance of $\overline{Y}_{..(A)} - \overline{Y}_{..(B)}$ for the biased cases will be $2.99Q^2\sigma^2 + \sigma^2/16$. The value of VarRatio will therefore be $16(2.99)Q^2 + 1$. The bottom half of Table 26.3 shows how VarRatio and related statistics vary with Q for the specific combination of period and carryover effects assumed above. So a Q value of less

TABLE 26.3. Relations between scale factor Q and variances of estimators with and without carryover effects

(Period Effects Present)				
Q	$\lambda = 16Q^2$	VarRatio	s.e.Ratio	δ_B = 4MeanBias/σ
$0.5\sqrt{2}$	8.000	9.000	3.00	-0.24
0.500	4.000	5.000	2.24	-0.17
0.354	2.000	3.000	1.73	-0.12
0.250	1.000	2.000	1.41	-0.08
0.177	0.500	1.500	1.22	-0.06
0.088	0.125	1.125	1.06	-0.03
(Period and Carryover Effects Present)				
Q	$\lambda = 16(2.99)Q^2$	VarRatio	s.e.Ratio	δ_B = 4MeanBias/σ
0.433	8.970	9.970	3.16	0.08
0.354	5.980	6.980	2.64	0.07
0.177	1.500	2.500	1.58	0.03
0.125	0.750	1.750	1.32	0.02
0.088	0.370	1.350	1.16	0.02
0.051	0.125	1.125	1.06	0.01

than 0.051 is required to prevent having a 12.5% or greater error in the variance of the parameter difference being estimated.

Effect of Collapsed Analysis on Power and Probability of Type I Error

Computation of E(MStreatments) and E(MSerror) values conditional on size of bias makes it possible to employ doubly noncentral F tables (Tiku, 1974) in assessing power and probability of Type I error in testing an H such as $\tau_A = \tau_B = \tau_C = \tau_D$. Alternatively, a t-statistic comparing $\overline{Y}_{..(.A)}$ and $\overline{Y}_{..(.B)}$ in the presence of bias has a doubly noncentral t-distribution (Bulgren, 1974) because t reflects a numerator noncentrality δ associated with average bias of the mean difference (plus a possible difference between τ_A and τ_B) and denominator noncentrality λ associated with an inflated $\mathrm{Var}(\overline{Y}_{..(.A)} - \overline{Y}_{..(.B)})$. The 4Mean Bias/$\sigma$ values of Table 26.3 are also δ values for the experimental design in question. Also, what is labeled VarRatio in that table is $\lambda + 1$.

Table 26.4 below, computed from Bulgren (1974), shows how the actual significance level for a 2-tailed t-test of H: $\tau_A - \tau_B = 0$ for collapsed data from the Table 26.3 design varies over Bulgren's boundary values for δ and λ, given a criterion $|t| > 2.20$ with a nominal 0.044 significance level. Note that the set of δ values in Table 26.3 was not usable for Table 26.4. Therefore, a later table based on Table 26.4 will require nonlinear interpolation, as described below.

TABLE 26.4. Actual significance levels when $|t| > 2.20$, $df = 15$, as a function of $|\delta_B|$ and λ

$\|\delta_B\|$	λ=0	λ=4	λ=8
0.0	0.044	0.025	0.015
1.0	0.141	0.095	0.064
2.0	0.440	0.345	0.268
3.0	0.782	0.697	0.613
4.0	0.956	0.924	0.883
5.0	0.996	0.990	0.982

In this range of parameter values, Table 26.4 shows that the actual significance level increases with increased $|\delta|$ values or decreased λ values. For any δ, I presume that there is a large enough λ to make the significance level be equal to the nominal value of 0.044 for the 2.20 t-criterion. Comparison to Table 26.3 shows that our example design has $|\delta|$ and λ values covarying positively; this seems not to produce large significance levels. For example, the first row of Table 26.3 data, with $\lambda = 8.00$ and $\delta = -0.24$, has an 0.017 significance value obtained by nonlinear interpolation between the first two entries of the final column of Table 26.4. All Q values from Table 26.3 yield δ and λ combinations between the first and second rows of Table 26.4. Therefore, I would expect Type I error rates based on collapsed data from this design to approximate the nominal values assigned, usually being too small if not correct.

The above is only half the truth about Type I errors for this example. For $\delta = +1$ and $\lambda = 0$, all but 0.002 of the 0.141 comes from positive t values. All other combinations with $\delta > 1$ yield no more than 0.001 significant negative t values. Corresponding results occur for negative values of δ. Accordingly, the overall significance levels for small biases may be reasonable, but the apparent direction of effects will often be misleading. There will be a built-in preference for hypotheses in the direction of the bias when Type I errors occur.

Table 26.4 is also useful as a power table: Let δ_U be the unbiased δ reflecting true treatment effect differences. Depending on the nature of the correct model, there may or may not be a bias δ_B due to collapsing. Then a composite δ may be designed as $\delta = \delta_U + \delta_B$. Table 26.5 presents four power values that come directly from Table 26.4 (with the δ_B there now interpreted as the composite δ). The fourth and sixth power values in Table 26.5 were obtained by nonlinear interpolation using Bulgren's (1974) tables.

For the $\delta_U = 2.00$ case, the correct power of 0.440 with no bias is substantially increased for $\delta_B = 1.00$ and very substantially decreased for $\delta_B =$ -1. For the $\delta_U =$

TABLE 26.5. Representative power values when $|t| > 2.20$, $df = 15$, as a function of λ and a composite δ

δ_U	δ_B	δ	λ	Power
2.00	0	2.00	0	0.440
2.00	1.00	3.00	8.0	0.613
2.00	-1.00	1.00	8.0	0.064
3.24	0	3.24	0	0.841
3.24	-0.24	3.00	8.0	0.613
3.24	0.24	3.48	8.0	0.764

Power obtained in the presence or absence of bias from collapsing data for analysis

3.24 case, however, positive and negative bias examples both produce reduction of power. Furthermore, for a small true δ and a biasing quantity in the opposite direction, the sign of an obtained significant t may suggest the wrong direction for the treatment effect. It appears that the analysis of collapsed data may distort power and the inferred direction of a treatment effect difference. The principal other objection to analysis of collapsed data is failure to test for and find period or carryover effects when they are present.

Analyzing Collapsed Data: No Nuisance Effects Present

When there are no period and carryover effects, analogues of (26.7) have no bracketed term for any subject in the experiment. In this case the probability of a Type I error is exactly the value intended by the investigator; also, E_t for the block randomization design of Table 26.1 can be shown to be 100.

26.3 Other Issues

26.3.1 Examining Factorial Arrangements of Within-Subject Variables

The problem of factorial variation of within-subject treatments is discussed in Jones and Kenward (1989, pp. 225-228), but with little emphasis on period effects. Shoben, Sailor, and Wang's (1989) Experiment 1 is a psychological example of this kind of design.

26.3.2 Re-analysis of Data from a Block Randomized Design

Cotton and Othman (1991) have re-analyzed data for Refinetti's (1989) Experiment 1 on temperature judgments, with 30 different temperatures in a "block

TABLE 26.6. Recommendations about crossover designs in psychology

1. Use efficient designs with easy correct calculation methods, such as Williams's balanced latin squares and the Berenblut designs.
2. Search for aditional experimental designs having optimal properties.
3. Perform the correct statistical analyses even when difficult, as in some general linear model procedures.
4. Expand known analysis methods to be more readily applicable to experiments with many observations per subject.

randomization" design and 3 periods for each treatment, making a total of 90 periods. (The 30 temperatures employed each day were not quite identical, making this label only approximately correct.) Partly because of a desire to assess data from one subject at a time, the method of analysis went beyond conventional GLM approaches, using Hastie and Tibshirani's (1990) theoretical work on nonparametric smoothing of effects within generalized additive models, as implemented in the Almudevar and Tibshirani (1990) GAIM software package and more recently in S (Chambers and Hastie, 1991).

Examining the only three available subjects' data, Cotton and Othman (1991) found significant period effects with each subject. (Declining linear trend in estimates as periods increased for Subject 1 but nonlinear with two changes of direction for Subject 2 and three changes of direction for Subject 3.) All three subjects exhibited significant effects of temperature, with increased estimates of temperature effects as actual temperatures increased (nonlinear but monotonic trend for Subject 1 only.) Subject 1 also showed significant carryover effects generally increasing with the size of the discrepancy between the preceding and current temperatures, as expected by Refinetti (1989). A significant nonlinear carryover trend appeared for Subject 3 (Three reversals).

26.4 Recommendations for Data Analysis When Period or Carryover Effects are Likely

Table 26.6 presents four points relevant to the design and analysis of crossover designs in psychology. In this section, (26.1) or (26.2) is assumed to hold.

26.4.1 Use Efficient Designs

The first recommendation of using efficient designs with easy, correct methods of statistical analysis, can often be done if one is sensitive to the need for better research techniques in psychological experimentation.

26.4.2 Find Optimal Designs

The second recommendation, to seek further experimental designs with optimal properties, may be illustrated by Cheng and Wu's (1980, Theorem 3.2) use of a Berenblut design as a building block for a larger, also optimal, design. Development of optimal designs may require extensive study of the literature cited above, plus other sources such as Dodge, Fedorov, and Wynn (1980).

26.4.3 Perform Correct Statistical Analyses, Even When Difficult

The third recommendation, to perform the correct statistical analysis, even when difficult, seems intellectually obvious. The safest and most powerful data analysis strategy will avoid collapsed analyses unless one is so convinced that no period effects and no carryover effects exist that one is willing to present all treatment A trials first, then all treatment B trials second, etc. When using continuous measures where one believes that either period or carryover effects may be present, then one usually should perform a general linear model analysis of variance. Jones and Kenward (1989), Milliken and Johnson (1984), and Ratkowsky, Evans, and Alldredge (1993) are the principal books describing such analyses in the contexts of crossover designs; the SAS manual section on GLM (SAS Institute, 1988) and part of Searle (1987, especially Chapter 12) can also be of help in this area.

26.4.4 Expand Known Analysis Methods to Be More Readily Applicable to Experiments with Many Observations Per Subject

When behavioral scientists employ designs of the type just described, they often employ very large numbers of periods. In contrast, when example designs and their theory are presented in the statistical literature, they usually involve no more than 12 periods. This suggests a need to examine experimental designs with several treatments and many periods, checking them for optimality and planning appropriate data analyses for them. [3]

26.5 References

Almudevar, T., & Tibshirani, R. (1990). *GAIM. Version 1.0.* Toronto: S. N. Tibshirani Enterprises.

Aptech Systems (1988). *GAUSS. The GAUSS system version 2.0.* Kent, WA: Aptech Systems.

[3]B. Jones of the Institute of Mathematics and Statistics at the University of Kent at Canterbury, England, (personal communication, March 5, 1991) is now administering a 3-year Science and Engineering Research Council studentship award for this purpose. He would welcome additional data sets for study.

Berenblut, I.I. (1964). Change-over designs with complete balance for first residual effects. *Biometrics, 20*, 707–712.

Berenblut, I.I. (1967). The analysis of change-over designs with complete balance for first residual effects. *Biometrics, 23*, 578–580.

Bulgren, W.G. (1974). Probability integral of the doubly non central t-distribution with degrees of freedom n and non-centrality parameters δ and λ. In H.L. Harter & D.B. Owen (Eds.), *Selected tables in mathematical statistics* (Vol. II, pp. 1–138). Providence, RI: American Mathematical Society.

Chambers, J.M., & Hastie, T. (1991). *Statistical methods in S.* Pacific Grove, CA: Wadsworth.

Cheng, C.S., & Wu, C.F. (1980). Balanced repeated measurements designs. *Annals of Statistics, 8*, 1272–1283.

Cheng, C.S., & Wu, C.F. (1983). Corrigendum. *Annals of Statistics, 11*, 349.

Cohen, A.J., Trehub, S.E., & Thorpe, L.A. (1989). Effects of uncertainty on melodic information processing. *Perception & Psychophysics, 46*, 18–28.

Cotton, J.W., & Othman, A.R. (1991, November). *Modeling perception of temperature change using the generalized additive model.* Paper presented at the meeting of the Psychonomic Society, San Francisco.

Cox, D.R. (1951). Some systematic experimental designs. *Biometrika, 38*, 312–323.

D'Amato, M.R. (1970). *Experimental psychology. Methodology, psychophysics, and learning.* New York: McGraw-Hill.

Dodge, Y., Fedorov, V.V., & Wynn, H.P. (1988). *Optimal design and analysis of experiments.* Amsterdam: North Holland.

Hastie, T., & Tibshirani, R.J. (1990). *Generalized additive models.* London: Chapman and Hall.

Hedayat, A., & Afsarinejad, K. (1978). Repeated measurements designs, II. *Annals of Statistics, 6*, 619–628.

Jones, B. & Kenward, M. G. (1989). *Design and analysis of cross-over trials.* London: Chapman and Hall.

McBurney, D.H. (1983). *Experimental psychology.* Belmont, CA: Wadsworth.

Milliken, G.A., & Johnson, D.E. (1984). *Analysis of messy data. Vol. I: Designed experiments.* Belmont, CA: Lifetime Learning.

Morrin, K.A. (1989). *Individual differences in the ability to coordinate information across independent cognitive domains.* Master's thesis. Santa Barbara, CA: University of California, Santa Barbara.

Ratkowsky, D.A., Evans, M.A., & Alldredge, J.R. (1993). *Cross-over experiments. Design, analysis, and application.* New York: Marcel Dekker.

Refinetti, R. (1989). Magnitude estimation of warmth: Intra- and intersubject variability. *Perception & Psychophysics, 46*, 81–84.

SAS Institute (1988). *SAS/STAT user's guide. Release 6.03 edition.* Cary, NC: SAS Institute, Inc.

Searle, S.R. (1987). *Linear methods for unbalanced data.* New York: Wiley.

Shoben, E.J., Sailor, K.M., & Wang, M-Y (1989). The role of expectancy in comparative judgments. *Memory & Cognition*, *17*, 18–26.

Tiku, M.L. (1974). Doubly noncentral F distributions – Tables and applications. In H.L. Harter & D.B. Owen (Eds.). *Selected tables in mathematical statistics* (Vol. II, pp. 139–149). Providence, RI: American Mathematical Society.

Williams, E.J. (1949). Experimental designs balanced for the estimation of residual effects of treatments. *Australian Journal of Scientific Research*, *A2*, 149–168.

27

Natural Sampling: Rationality without Base Rates

Gernot D. Kleiter [1]

ABSTRACT The Bayesian classification of single cases into classes is investigated. The neglect of class base rates is usually considered to be a fundamental violation of the principles of rationality and, in human judgment, is called 'the base rate fallacy'. This paper analyzes an important special sampling condition in which the base rates actually turn out to be irrelevant in estimating classification probabilities. It is shown that the base rates do not affect the precision of the estimates of classification probability. This precision is modeled by a second-order probability distribution. Implications for the design of artificial and for the modeling of natural knowledge systems are discussed.

27.1 Introduction

This paper consists of two parts: (i) the derivation of a formal theorem in probability theory and (ii) a discussion of the relevance of that theorem for the psychology of inductive reasoning. It begins with the second topic, the psychology of inductive reasoning.

The question 'Is human reasoning rational?' has a long history in philosophy and a considerable tradition in psychology. It has been raised in connection with deductive reasoning: 'Does human deductive reasoning obey the laws of logic?' Severe violations of logical inference rules are found, e.g., in Wason's selection task. Reliable deviations of human solutions from the normative standard are called biases or fallacies. For recent reviews, see Macnamara (1986), Evans (1989), and Johnson-Laird and Byrne (1991). Similar questions have been raised in respect to inductive reasoning (Kahneman, Slovic, and Tversky, 1982; Gigerenzer, 1991). Although the consensus about normative references is much less in the field of inductive reasoning, probability theory and Bayes theorem still play a widely acknowledged role. In the sixties, human judgment was supposed to correspond approximately to the normatively prescribed idealizations of decision theory. In psychological models, deviations from optimality were accomodated by the introduction of free parameters which could be adjusted to fit the data. The picture changed decisively in the early seventies after the first papers of Kahneman and Tversky on heuristics in judgment and decision making were published.

---

[1]University of Salzburg, Department of Psychology, Hellbrunnerstr. 34, A-5020 Salzburg, Austria
I wish to thank Maximilian Thaler and Marianne Kardinal for their help. This research was supported by the Fonds zur Förderung der wissenschaftlichen Forschung, Vienna.

For about twenty years now, human inductive reasoning has been described as basically non-rational (Tversky and Kahneman, 1974; Kahneman, Slovic, and Tversky, 1982). It has been claimed that it fundamentally violates the principles of rationality in a qualitative way. An outstanding instance of such a violation is the neglect of base rates in judgmental classification. The judgments of a rational decision maker about diagnostic probabilities should be in accord with Bayes' theorem. That theorem is one of the principles of normative decision theory. It prescribes that base rates should be included in the calculations and combined with likelihoods in a multiplicative way. Kahneman and Tversky have argued that human judgment often ignores base rates totally and thus violates rationality at a very elementary level. According to their view, human judgment is not guided by optimal, but by heuristic principles. Heuristics are often close to optimal solutions, but may fail dramatically under certain special conditions. If we were to explain the neglect of base rates by a heuristic, we would need to show that in most naturally occurring cases the neglect heuristic works well.

It is proved in the Appendix that, under a special sampling scheme, called 'natural sampling' (Aitchison and Dunsmore, 1975), the base rates actually do *not* enter into the estimated Bayesian diagnostic probabilities or their precision. That is to say, there exists a condition in which it is rational to neglect base rates. It will be argued that this condition is typical for many situations in which people acquire their experiences. There is a correspondence between the constraints of the statistical model of natural sampling on the one hand, and the constraints under which human information is acquired on the other. Natural sampling is the way in which experiences about classes and their members is acquired naturally. It represents the 'normal case' for intuitive judgments under uncertainty.

27.2 Example

A hospital performs a diagnostic screening for the presence of a wide spectrum of diseases. Let one of the diseases be D, and let us assume that one of the hospital physicians is an expert on D to whom all suspect patients are automatically sent. The specialist relies only on his or her own experience. During his or her career the physician has seen a total of $n = 36$ such patients, all of whom were sent to him or her after the screening. Of the 36 cases, $g = 12$ were finally diagnosed as being afflicted by disease D (Figure 27.1); $h = 24$ were found to be unafflicted by D. Of the 12 D-cases, $a_1 = 9$ showed a certain symptom S, and $b_1 = 3$ did not show S. Of the 24 healthy cases, $a_2 = 4$ gave false alarms by showing symptom S, and the other $b_2 = 20$ cases did not show the symptom (Figure 27.1, left hand and middle tree). Now, the numbers of 9 hits and 3 missed alarms on the one hand, and 4 false alarms and 20 correct non-alarms on the other hand, add up to 12 and 24, respectively, and finally aggregate to 36. The

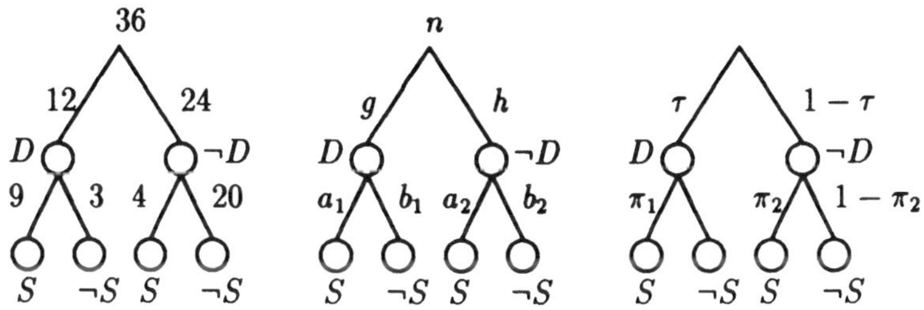

FIGURE 27.1. Structure of an elementary probabilistic classification; D=disease present, $\neg D$=disease absent, S= symptom present, $\neg S$=symptom absent. On the left hand: numerical example; in the middle: frequency notation; on the right hand: parameter notation.

total sample size of the expert's experience cascades down in an additive way from all cases to cases suffering and not suffering from D, and finally to those showing and not showing S. Suppose you are the specialist's case number 37. You show the symptom S. What is the probability that you are suffering from D? The classification probability is obtained from Bayes theorem:

$$p(D|S) = \frac{p(D)p(S|D)}{p(D)p(S|D) + p(\neg D)p(S|\neg D)}.$$

In this formula, $p(D|S)$ denotes the probability of disease D given that the symptom S is present; $p(D)$ is the base rate, i.e., the probability that a patient sent from the screening to the specialist actually suffers from the disease; $p(\neg D)$ denotes the probability that the patient does not suffer from the disease. Finally, $p(S|D)$ is the likelihood of observing symptom S, given that a patient is afflicted, and $p(S|\neg D)$ is the likelihood of the symptom given that the patient is not afflicted. The specialist estimates the probabilities in the Bayes theorem by the respective relative frequencies:

$$p(D|S) = \frac{\frac{ga_1}{(g+h)(a_1+b_1)}}{\frac{ga_1}{(g+h)(a_1+b_1)} + \frac{ha_2}{(g+h)(a_2+b_2)}}.$$

Since $g = a_1 + b_1$ and $h = a_2 + b_2$, this simplifies to

$$p(D|S) = \frac{a_1}{a_1 + a_2}.$$

The numbers in the example give $p(D|S) = 9/(9+4) \approx 0.7$. It can be seen that, under natural sampling conditions, the base rates cancel.

27.3 Natural Sampling

In a statistical model, the base rates and the characteristics of the processes generating the data are described by parameters. The base rate is denoted by $p(D) = \tau$ and the two conditional symptom probabilities by $p(S|D) = \pi_1$ and $p(S|\neg D) = \pi_2$. Knowledge about the base rate and the class conditional symptom probabilities is obtained from sample data (Figure 27.1, right hand). Under natural sampling, *one* sample of N individuals is taken, and sampling is assumed to be identical and independent. Each case is classified as (i) healthy vs. ill, and (ii) symptom present vs. not present. In the terminology of contingency tables, the marginals are random, and no subtotals are fixed. Denote by g and h the number of healthy and ill individuals, respectively. Under natural sampling, g and h contain information about the percentages of healthy and ill individuals in the population of persons to be referred. The relative frequency $g/(g+h)$ is a sufficient statistic for the base rate τ. A different sampling scheme might sample within strata. Here, *two* samples would be taken with g healthy and h ill subjects (often $g = h$). In sampling within strata, the class marginal is fixed, so that g and h do not contain information about the percentage of healthy or ill individuals in the population. Many other sampling schemes (e.g., sequential sampling) may be distinguished. For our purposes, we define natural sampling as follows:

Definition 1 *Consider two classes and two binary features. Let knowledge about the proportion of cases contained in the first class be expressed by the Beta distribution $\tau \sim Be(g,h)$, and knowledge about the class conditional features by $\pi_1 \sim Be(a_1, b_1)$ and $\pi_2 \sim Be(a_2, b_2)$. The natural sampling condition is fulfilled if $g = a_1 + b_1$ and $h = a_2 + b_2$. An important example is sampling with random marginals in a 2×2 table with vague priors.*

Aitchison and Dunsmore (1975) distinguished two types of experiments: ***natural informative*** and ***designed experiments.*** In a natural informative experiment, the sample sizes of the classes are random variables and contain information about the corresponding base rates. In a ***designed experiment,*** the sample sizes are fixed by the experimenter and do not contain information about the base rates. In this case, the specification of the base rates $p(D)$ and $p(\neg D)$ would have to be based on sources other than the experiment.

The observed class and symptom frequencies are additive, under practically all statistical sampling schemes; in the simple example above $g = a_1 + b_1$ and $h = a_2 + b_2$. Additivity is a necessary but not a sufficient condition for the sample sizes to be informative. Conditional independence is the condition that is both necessary and sufficient for g and h to be sufficient statistics for inferences about base rates.

Two further examples of natural sampling came from two less obvious areas: Pattern classifiers constitute an important class of neural networks. A pattern classifier learns its weights by processing a sample of patterns contained in a training set. Each pattern of the set is presented in the training period together with information which class it belongs to. Class and feature frequencies are therefore additive.

The second example is classical conditioning in a probabilistic environment: An animal learns to predict the occurence of an unconditioned stimulus (US) on the basis of a conditioned stimulus (CS). The CS may be intepreted as a predictor of the US (Gallistel, 1990). The numbers of presentations of the US and not-US and CS and not-CS may often be additive.

Natural sampling represents a condition in which a strategy ignoring base rates is perfectly rational. In the approach of Kahneman and Tversky it is the candidate for the 'normal' condition in which the heuristic works well. The situation is intriguing, though, because neglect of base rates under natural sampling is not just close to optimal; it is exactly optimal.

27.4 Member Parameter

In the introductory example we assumed that the specialist estimates the probabilities by relative frequencies. If the number of cases is small, the precision of the estimates is low. If the number of cases is large, the precision is high. To include the precision into the analysis, we consider a parameterized statistical model. The simple Bayes theorem is not adequate to treat the classification problem satisfactorily (Kleiter, 1992). We treat the probabilities in the Bayes theorem as parameters and denote them by $\tau = p(D)$, $\pi_1 = p(S|D)$, $\pi_2 = p(S|\neg D)$, and $\mu = p(D|S)$. The Bayes theorem is then rewritten in parametric form ,

$$\mu = \frac{\tau\pi_1}{\tau\pi_1 + (1-\tau)\pi_2}.$$

The inverse probability μ is expressed as a function of three other parameters, τ, π_1, and π_2. In statistics, inferences about parameters which are a function of other parameters – e.g., the difference between two proportions, $\pi_1 - \pi_2$, or the ratio of two proportions, π_1/π_2 – are quite common (Bishop, Fienberg, and Holland, 1975).

27.5 Prior and Posterior Distributions

We express the knowledge about the parameters by probability distributions. In Bayesian statistics it is common practice to employ Beta distributions to investigate proportions. Let us first adopt for the three distributions improper vague prior Betas $Be(p,q)$ with coefficients $p \to 0$ and $q \to 0$. In this case, the sample frequencies g, h, a_1, b_1, a_2, and b_2 lead to the following posterior Beta distributions:

$$p(\tau|g,h) = Be(g,h), p(\pi_1|a_1,b_1) = Be(a_1,b_1), \text{and} p(\pi_2|a_2,b_2) = Be(a_2,b_2).$$

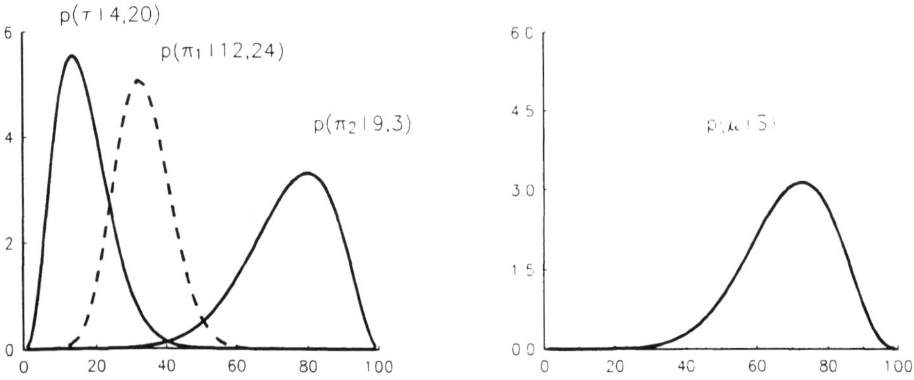

FIGURE 27.2. Left panel: symptom distributions $Be(4,20)$ (solid left) and $Be(9,3)$ (solid right), base rate distribution $Be(12,24)$ (middle); right panel: member distribution $Be(9,4)$.

The means of these distributions are equivalent to the maximum likelihood estimates of τ, π_1 and π_2. In the general case, let the coefficients of the prior Beta distributions be denoted by g^0, h^0, a_1^0, b_1^0, a_2^0, b_2^0, and the sample frequencies by g^f, h^f, a_1^f, b_1^f, a_2^f, b_2^f. The coefficients of the posterior Beta distribution are then given by $g = g^0 + g^f$, $h = h^0 + h^f$, $a_1 = a_1^0 + a_1^f$, $b_1 = b_1^0 + b_1^f$, $a_2 = a_2^0 + a_2^f$, $b_2 = b_2^0 + b_2^f$. We suppose that the coefficients of the two prior symptom distributions add up to the coefficients of the prior base rate distribution: $g^0 = a_1^0 + b_1^0$ and $h^0 = a_2^0 + b_2^0$. For example, suppose that, if the prior symptom distributions are non-informative uniform $p(\pi_1) = Be(1,1)$ and $p(\pi_2) = Be(1,1)$, then the prior base rate distribution is $p(\tau) = Be(2,2)$. The important condition is that the posterior symptom coefficients add up to the base rate coefficients.

The left hand side of Figure 27.2 shows the three Beta distributions for the numerical example given in the introduction, i.e., the base rate distribution $\tau \sim Be(12,24)$ and the two conditional symptom distributions $\pi_1 \sim Be(9,3)$ and $\pi_2 \sim Be(4,20)$.

27.6 Member Distribution

The introduction of probability distributions for the parameters τ, π_1, and π_2 implies a distribution for the member parameter μ. We denote the probability density function of μ, given the symptom, by $p(\mu|S)$. We call this distribution *member distribution*. The distribution is a second order probability distribution (Sahlin and Gärdenfors, 1983; Gärdenfors and Sahlin, 1982). It can be used to express the ambiguity (Einhorn and Hogarth, 1985) about the member probability. The member distribution is given by (cf. Kleiter, 1992)

$$p(\mu|S) = \frac{1}{B(g,h)B(a_1,b_1)B(a_2,b_2)}\mu^{g-1}(1-\mu)^{h-1} \times$$
$$\int_{\pi_1=0}^{1} \pi_1^{h+a_1-1}(1-\pi_1)^{b_1-1}\left(\int_{\pi_2=0}^{1} \frac{\pi_2^{g+a_2-1}(1-\pi_2)^{b_2-1}}{[(1-\mu)\pi_1+\mu\pi_2]^{g+h}}\,d\pi_2\right)d\pi_1 .$$

The equation is obtained from the product of the three posterior Betas whith a Jacobian factor

$$\frac{\partial\tau}{\partial\mu} = \frac{\pi_1\pi_2}{[\pi_1+\mu(\pi_2-\pi_1)]^2},$$

after integrating out π_1 and π_2.

The member distribution is complicated. We proposed an approximation by a Beta which is based on the variances of the three parameters τ, π_1, and π_2 (Kleiter, 1992). The approximation is asymptotically optimal. Under the natural sampling condition, however, the equation for the member distribution greatly simplifies. The cumbersome double integration can be solved analytically, whence we obtain the following result: Under natural sampling, the member distribution is a Beta distribution with shape parameters a_1 and a_2, i.e., $p(\mu|S) = Be(a_1, a_2)$. More formally, we state the following theorem:

Theorem 1 *Consider two binary events, $D/\neg D$ and $S/\neg S$, and the following three probabilities:*

$$p(D) = \tau, \quad p(S|D) = \pi_1 \quad \textit{and} \quad p(S|\neg D) = \pi_2 .$$

Suppose the second order probability density functions of the three probabilities are given by the Betas:

$$\tau \sim Be(g,h), \quad \pi_1 \sim Be(a_1,b_1), \quad \pi_2 \sim Be(a_2,b_2) .$$

If $g = a_1 + b_1$ and $h = a_2 + b_2$, then

$$\mu_1 \sim Be(a_1,a_2) \quad \textit{and} \quad \mu_2 \sim Be(b_1,b_2),$$

where $\mu_1 = p(D|S)$ and $\mu_2 = p(D|\neg S)$.

Corollary 1 *The marginal symptom distribution for $\beta = p(S)$ is given by the Beta distribution $\beta \sim Be(a_1 + a_2, b_1 + b_2)$.*

The proof of the theorem is given in the Appendix. The corollary is obtained by interchanging D and S and applying the theorem in the reversed direction. Figure 27.3 represents the result in a two 2×2 table: If the conditional probabilities in the columns, π_1 and π_2, and the marginal of the colums, τ, are Beta distributed, then the conditional probabilities in the rows, μ_1, μ_2, and the

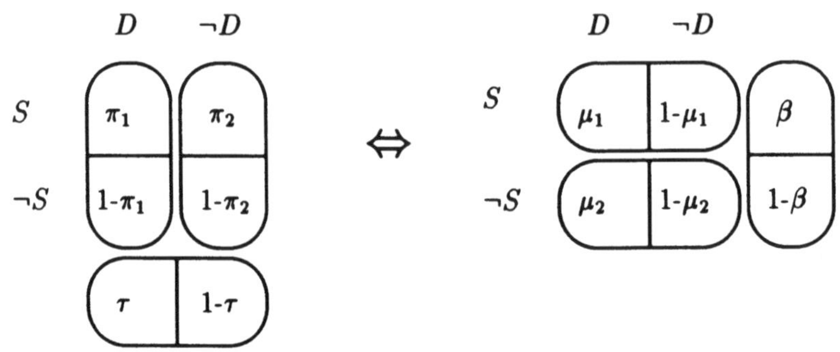

FIGURE 27.3. Representation of the theorem on natural sampling in a 2×2 table. If the conditional probabilities in the columns and the column marginals are Beta distributed, then the conditional probabilities in the rows and the row marginals are also Beta distributed, and *vice versa.*

marginal of the rows, β, are also Beta and *vice versa*. The coefficients of the Beta distributions are obtained from an analogous table with the cell entries a_1, b_1, a_2, b_2 and with the according sums in the margins.

The theorem and the corollary can directly be generalized to systems of Dirichlet distributions in multiway contingency tables.

For the numerical example given in the introduction, the right panel of Figure 27.2 gives the member distribution $Be(9, 4)$ for a single case showing symptom S. The mean of the distribution is approximately 0.7. The precision of the member distribution is less than the precision of the base rate distribution. For a discussion of such relationships, see Kleiter, (1992).

27.7 Properties

The result shows that under natural sampling conditions all the probabilistic information about the single case is contained – and contained only – in the symptom frequencies a_1 and a_2, i.e., the hit and the false alarm frequencies. In particular, the base rate frequencies g and h do not enter the member probability. Sahlin (1990) proposed to call a condition c to be irrelevant for an event e if the first and the second order probabilities of e, given c, are equal to those of e alone. In this sense, the base rate frequencies are irrelevant, for the member probabilities. We note that not only the base rate is irrelevant but also b_1 and b_2. This fact reflects the likelihood principle which states that only the data actually observed are relevant for probabilistic inference, but not hypothetical data which might have been observed. We further note that the mean of the Beta distribution is of course identical to the frequentistic result of the simple Bayes Theorem given in the introduction:

$$E(\mu|S) = \frac{a_1}{a_1 + a_2}.$$

This relationship expresses de Finetti's (1974) conception that probabilities can be conceived as expected values. Note that, in the terminology of signal detection theory, $E(\mu|S)$ is given by *HR/(HR+FAR)*, where *HR* denotes the hit rate, and *FAR* the false alarm rate.

The result makes the choice of a Beta distribution in our previous approximation reasonable even for the general case. The natural sampling constraint represents a special case in which the result exactly matches a Beta. It is a good approximation to many unimodal, symmetric or non-symmetric distributions which are not exactly a Beta. We also note that the result can easily be extended to more than two classes. If the Betas are replaced by Dirichlet distributions, then the classes can be collapsed in such a way that the problem reduces to a two class problem (Kleiter, 1992), and the previous results become directly applicable.

27.8 Empirical Investigations

As it is not the aim of the present article to review the empirical studies on the base rate fallacy; we make a few remarks only. Since the original demonstration of the base rate fallacy (Tversky and Kahneman, 1974), numerous investigations have shown that human judgment under some circumstances is responsive to base rate changes. Perceived causality (Ajzen, 1977), specificity (Bar-Hillel, 1980), degree of representativeness (Fischhoff and Bar-Hillel, 1984), procedural changes and the task format (Christensen-Szalanski and Beach, 1982; Fischhoff, Slovic, and Lichtenstein, 1979; Manis, Dovalina, Avis, and Cardoze, 1980) induce a different base rate utilization. Ofir (1988) has shown that also under the original experimental conditions of Kahneman and Tversky the base rates can have large effects.

If natural sampling is 'natural', we may suppose that human subjects overgeneralize the according judgmental strategy. The subjects do not realize the non-natural sampling conditions in the experiments and continue to neglect base rates. The neglected base rates reflect the fact that the subjects are insensitive to the unusual sampling conditions. All the investigations on the base rate fallacy I am familiar with investigated percentages, not frequencies. Thus, we can only compare point estimates, not the imprecision. In the case of *non-diagnostic evidence,* the normative model predicts responses equal to the base rates, while an overgeneralizing natural sampling model predicts a 50 % judgment. In their Engineer/Lawyer study Tversky and Kahneman observed an average response close to 50 %. Similar results were obtained by Bar-Hillel (1980), Fischhoff and Bar-Hillel (1984). Quite contrary, Ofir (1988) found strong base rate effects in judgments based on nondiagnostic evidence. In an investigation of Bar-Hillel (1982), the *diagnostic* class-conditional symptom rates were equal to 5 % and 15 %, respectively; the mean estimated classification probability was 25 %. This matches exactly the natural sampling prediction! It seems quite difficult, though, to obtain a clear overall picture from the mosaic of results reported in the literature.

27.9 Discussion

The literature emphasizes that, in order to be rational, we should be sensitive to the base rates. We have shown, however, that under natural sampling conditions we can be rational without the base rate frequencies. In medical diagnosis, for example, under natural sampling conditions all the relevant information is contained in the symptom frequencies. This result has consequences for the design of artificial and for the modeling of natural knowledge systems. Let us first turn to the artificial systems. It has been shown that the processing of uncertainties in knowledge based systems can be modelled in an efficient way by Bayesian methods (Pearl, 1988; Spiegelhalter, 1986; Spiegelhalter and Lauritzen, 1990). We proposed to weight the relations in a knowledge base by actual or hypothetical frequencies and to use Bayesian inference to calculate the diagnostic probabilities and their second-order distributions (Kleiter, 1992). For some subdomains of a knowledge base it may be highly plausible to assume a natural sampling structure. For such domains it is necessary to assess the frequencies at the most specific data levels only. This leads to a simplification of the knowledge acquisition procedure and helps to avoid the exponential growth of the number of combinatorial possibilities that must be considered in the problem space (combinatorial explosion).

Under natural sampling, the knowledge about the classes is much larger than the knowledge about each of the conditional symptoms. The natural sampling condition is not satisfied if we have extra knowledge about the classes or if we have extra knowledge about the symptoms. In the first case, we may analyze a well-known disease with a new diagnostic test. The sample drawn to investigate the new diagnostic test is much smaller than our surveys about the epidemiology of the well-known disease. This is the condition typically used in the formulation of cover stories which provoke the base rate fallacy in human judgment.

We now turn to the natural knowledge systems. In the investigations about the base rate fallacy the subjects are presented three percentage numbers – the base rate and two class-conditional symptom probabilities – invoked in a short story that provides a semantic intepretation of the abstract problem structure. This is not the kind of input format the subjects' cognitive inference module is used to obtain from its own personal knowledge base. The actual presentation of a series of events seems to be a more realistic experimental alternative (Shanks, 1992). We know that many mammals encode the absolute frequencies of important events, but they do not encode their proportions (Gallistel, 1990). The representation of numerosity allows living organisms to derive the statistical uncertainty or precision associated with probability estimates. This is important in a non-stable environment. Recently, Gigerenzer (1991) and Cosmides and Tooby (1992) have shown that human subjects are less ignorant of the base rates when they are given frequency information instead of proportions. There is evidence that the human knowledge base encodes numerosity and that the human inference module works closer to Bayesian principles if it is provided with frequencies instead of proportions. Human inferences may still have a tendency to ignore base rate information because they are best tuned to process data from the own personal knowledge base. Such data are acquired under natural sampling conditions, the storage and processing of base rate information would be redundant in this case.

Those who are experts acquire their knowledge from an array of multiple sources: from their own experiences, from written documents, or from interaction with other experts. Before the emergence of cultural tradition, the invention of external memory, and the evolution of the mutual exchange of knowledge through spoken language, the knowledge bases of living organisms only contained data acquired through their own experiences. Such a highly individual knowledge base is constrained by a specific structural feature: the total number of observations made decomposes top-down in a strictly additive way. The same structural constraints are often present in diagnostic classification problems. We suspect that our judgments tend to ignore base rates because we are well adjusted to natural sampling conditions. If we process the information of our episodic memory, which by definition is a highly personal knowledge base of our own experiences, we are rational without base rates.

27.10 Appendix

We show that, if $g = a_1 + b_1$ and $h = a_2 + b_2$, the member distribution $p(\mu|S) = Be(a_1, a_2)$. Let us begin with the member distribution equation from Section 1.6:

$$p(\mu|S) = \frac{1}{B(g,h)B(a_1,b_1)B(a_2,b_2)}\mu^{g-1}(1-\mu)^{h-1}\times$$
$$\int_{\pi_1=0}^{1} \pi_1^{h+a_1-1}(1-\pi_1)^{b_1-1}\left(\int_{\pi_2=0}^{1}\frac{\pi_2^{g+a_2-1}(1-\pi_2)^{b_2-1}}{[(1-\mu)\pi_1+\mu\pi_2]^{g+h}}\,d\pi_2\right)d\pi_1 .$$

We first solve the integral involving π_2. We substitute

$$z = (1-\mu)\pi_1 + \mu\pi_2,$$

so that

$$\pi_2 = \frac{z-(1-\mu)\pi_1}{\mu}, \quad 1-\pi_2 = \frac{\mu+(1-\mu)\pi_1 - z}{\mu}, \quad \frac{dz}{d\pi_2} = \mu, \quad d\pi_2 = \frac{dz}{\mu}.$$

The lower limit of the inner integral, $\pi_2 = 0$, becomes $z_1 = (1-\mu)\pi_1$, and the upper limit, $\pi_2 = 1$, becomes $z_2 = (1-\mu)\pi_1 + \mu$. This leads to the expression

$$\begin{aligned} \int_{\pi_2=0}^{1} \frac{\pi_2^{g+a_2-1}(1-\pi_2)^{b_2-1}}{[(1-\mu)\pi_1+\mu\pi_2]^{g+h}} d\pi_2 &= \int_{z_1}^{z_2} \frac{\left[\frac{z-(1-\mu)\pi_1}{\mu}\right]^{g+a_2-1}\left[\frac{\mu+(1-\mu)\pi_1-z}{\mu}\right]^{b_2-1}}{z^{g+h}} \frac{dz}{\mu} \\ &= \frac{1}{\mu^{g+a_2+b_2-1}} \int_{z_1}^{z_2} \frac{(z-z_1)^{g+a_2-1}(z_2-z)^{b_2-1}}{z^{g+h}} dz\,. \end{aligned}$$

Because of the natural sampling condition $h = a_2 + b_2$, the integral takes on the form of the left hand side of the following integral given by Gröbner and Hofreiter (1966, p. 175, eqn. 4):

$$\int_a^b \frac{(x-a)^{k-1}(b-x)^{\lambda-1}}{(cx+d)^{k+\lambda}} dx = \frac{(b-a)^{k+\lambda-1}B(k,\lambda)}{(bc+d)^k(ac+d)^\lambda},$$

where in our case $c=1$ and $d=0$. The application of the integral leads to

$$\frac{1}{\mu^{g+h-1}} \frac{\mu^{g+h-1}B(g+a_2,b_2)}{[(1-\mu)\pi_1+\mu]^{g+a_2}[(1-\mu)\pi_1]^{b_2}} = \frac{B(g+a_2,b_2)}{[(1-\mu)\pi_1+\mu]^{g+a_2}[(1-\mu)\pi_1]^{b_2}}.$$

Using this result, we now solve the integral involving π_2. Because of the natural sampling condition $h = a_2 + b_2$, we obtain:

$$\begin{aligned} &B(g+a_2,b_2) \int_{\pi_1=0}^{1} \frac{\pi_1^{h+a_1-1}(1-\pi_1)^{b_1-1}}{[(1-\mu)\pi_1+\mu]^{g+a_2}[(1-\mu)\pi_1]^{b_2}} d\pi_1 \\ &= \frac{B(g+a_2,b_2)}{(1-\mu)^{b_2}} \int_{\pi_1=0}^{1} \frac{\pi_1^{a_1+a_2-1}(1-\pi_1)^{b_1-1}}{[(1-\mu)\pi_1+\mu]^{g+a_2}} d\pi_1\,. \end{aligned}$$

We substitute

$$z = (1-\mu)\pi_1 + \mu,$$

so that

$$\pi_1 = \frac{z-\mu}{1-\mu}, \quad 1-\pi_1 = \frac{1-z}{1-\mu}, \quad \frac{dz}{d\pi_1} = 1-\mu, \quad d\pi_1 = \frac{dz}{1-\mu}.$$

The lower limit, $\pi_1 = 0$, becomes $z_1 = \mu$, and the upper limit, $\pi_1 = 1$, becomes $z_2 = 1$. This leads to

$$\frac{B(g+a_2,b_2)}{(1-\mu)^{b_2}} \int_{z_1}^{z_2} \frac{\left(\frac{z-\mu}{1-\mu}\right)^{a_1+a_2-1}\left(\frac{1-z}{1-\mu}\right)^{b_1-1}}{z^{a_1+b_1+a_2}} \frac{dz}{1-\mu}.$$

We apply the integral from Gröbner and Hofreiter once more. This yields

$$\frac{B(g+a_2,b_2)}{(1-\mu)^{b_2}}\frac{(1-\mu)^{a_1+a_2+b_1-1}B(a_1+a_2,b_1)}{(1-\mu)^{a_1+a_2+b_1-1}\mu^{b_1}}=\frac{B(b+a_2,b_2)B(a_1+a_2,b_1)}{(1-\mu)^{b_2}\mu^{b_1}}.$$

So we finally obtain

$$p(\mu|S)=\frac{\mu^{g-1}(1-\mu)^{h-1}B(g+a_2,b_2)B(a_1+a_2,b_1)}{B(g,h)B(a_1,b_1)B(a_2,b_2)\mu^{b_1}(1-\mu)^{b_2}}.$$

In the exponents, we apply the natural sampling condition $a_1 = g - b_1$ and $a_2 = h - b_2$; after simplifying the incomplete beta terms, $B(p,q) = \frac{\Gamma(p)\Gamma(q)}{\Gamma(p+q)}$, we have

$$p(\mu|S)=\frac{1}{B(a_1,a_2)}\mu^{a_1-1}(1-\mu)^{a_2-1}=Be(a_1,a_2).$$

27.11 References

Aitchison, J., & Dunsmore, I. R. (1975). *Statistical prediction analysis.* Cambridge: Cambridge University Press.

Ajzen, I. (1977). Intuitive theories of events and the effects of base rate information on prediction. *Journal of Personality and Social Psychology, 35,* 303–314.

Bar-Hillel, M. (1980). The base rate fallacy in probability judgments. *Acta Psychologica, 44,* 211–213.

Bar-Hillel, M. (1982). The base rate fallacy controversy. In R. W. Scholz (Ed.), *Decision making under uncertainty* (pp. 39–61). Amsterdam: Elsevier.

Bishop, Y. M. M., Fienberg, S. E., & Holland, P. W. (1975). *Discrete multivariate analysis.* Cambridge, MA: MIT Press.

Christensen-Szalanski, J. J. J., & Beach, L. R. (1982). Experience the base rate fallacy. *Organizational Behavior and Human Performance, 29,* 270–278.

Cosmides, L., & Tooby, J. (1992). *Are humans good intuitive statisticians after all?* Santa Barbara: Department of Psychology, University of California, Santa Barbara.

de Finetti, B. (1974). *Theory of probability* (Vol. 1). London: Wiley.

Einhorn, H. J., & Hogarth, R. M. (1985). Ambiguity and uncertainty in probabilistic inference. *Psychological Review, 92,* 433–461.

Evans, J. St. B. T. (1989). *Bias in human reasoning.* Hove: Erlbaum.

Fischhoff, B., & Bar-Hillel, M. (1984). Diagnosticity and base-rate effect. *Memory and Cognition, 12,* 402–410.

Fischhoff, B., Slovic,B., & Lichtenstein, S. (1979). Subjective sensitivity analysis. *Organizational Behavior and Human Performance, 23,* 339–359.

Gallistel, C. R. (1990). *The organization of learning.* Cambridge, MA: MIT Press.

Gigerenzer, G. (1991). How to make cognitive illusions disappear: Beyond "heuristics and biases". *European Review of Social Psychology, 2,* 83–115.

Gärdenfors P., & Sahlin, N.-E. (1982). Unreliable probabilities, risk taking, and decision making. *Synthese, 53,* 361–386.

Gröbner, W., & Hofreiter, N. (Eds.)(1966). *Integraltafel* (zweiter Teil). Bestimmte Integrale. Wien-New York: Springer-Verlag.

Johnson-Laird, P. N., & Byrne, R. M. J. (1991). *Deduction.* Hove: Erlbaum.

Kleiter, G. D. (1992). Bayesian diagnosis by expert systems. *Artificial Intelligence, 54,* 1–32.

Kahneman, D., Slovic, P., & Tversky, A. (Eds.). (1982). *Judgment under uncertainty: Heuristics and biases.* Cambridge: Cambridge University Press.

Macnamara, J. (1986). *A border dispute.* Cambridge, MA: MIT Press.

Manis, M., Dovalina, I., Avis, N. E., & Cardoze, S. (1980). Base-rates can affect individual predictions. *Journal of Personality and Social Psychology, 38,* 231–248.

Ofir, C. (1988). Pseudodiagnosticity in judgment under uncertainty. *Organizational Behavior and Human Decision Processes, 42,* 343–363.

Pearl, J. (1988). *Probabilistic reasoning in intelligent systems: Networks of plausible inference.* San Mateo, CA: Morgan Kaufmann.

Sahlin, N.-E. (1990). *On higher order beliefs.* Lund, Sweden: Department of Philosophy, University of Lund.

Sahlin, N.-E., & Gärdenfors, P. (1983). Decision making with unreliable probabilities. *British Journal of Mathematical and Statistical Psychology, 36,* 240–251.

Shanks, D. R. (1992). Connectionist acounts of the inverse base-rate effect in categorization. *Connection Science, 4,* 3–18.

Spiegelhalter, D. J., & Lauritzen, S. L. (1990). Techniques for bayesian analysis in expert systems. *Annals of Mathematics and Artificial Intelligence, 2,* 353–366.

Spiegelhalter, D. J. (1986). A statistical view of uncertainty in expert systems. In W. Gale (Ed.), *Artificial intelligence and statistics* (pp. 7–56). Reading: Addison Wesley.

Tversky, A., & Kahneman, D. (1974). Judgment under uncertainty: Heuristics and biases. *Science, 185,* 1124–1131.

28

A Dichotomization Method for Boolean Analysis of Quantifiable Co-Occurrence Data

Peter Theuns[1]

ABSTRACT Boolean analysis is a partial order generalization of scalogram analysis. It enables one to build a family of implication schemes that bears a 1-1-relationship with a data set. Unlike scalogram analysis, this method does not impose any constraints on the data structure. As a result, implication schemes may become extremely complex and may contain errors. Therefore, methods are needed for the selection of subsets of data to be modeled (dichotomization methods). In this chapter, a dichotomization method is introduced which uses the relative frequency of data patterns with the same number of correct responses as a given criterion. It is argued that the present approach yields implication schemes which comply better with the underlying data structure, especially when the subjects show similar abilities in solving the items.

28.1 Introduction

Observations in human sciences often concern the joint presence of different *features,* or *items* (qualities, abilities, properties, symptoms, attributes, traits, etc.) in one *subject* (a person, organisation, social system, etc.). Co-occurrence data consist of binary *patterns,* profiles or vectors of observed co-occurrences of features. The term "quantifiable co-occurrence data" refers to a class of data in which the (binary) response categories, according to which features are scored, are the same for all observed features (e.g., correct/false, or yes/no). In this case the count of equal responses in a pattern is meaningful (e.g., the number of items correct, or the number of symptoms observed). Several authors remark that in mainstream statistical analysis there is a tendency not to treat co-occurrence data as such, but to cut patterns of co-occurrences down to separate variables and distributions before analysing them (Ragin, 1987; Feger, 1991). Information on the combined occurrence of features is often lost by such practice.

A number of data-analytic techniques exist which allow the analysis of co-occurrence data (see, e.g., Guttman, 1950; Flament, 1976; Ragin, Mayer, and Drass, 1984; Feger, 1988). One method is the Boolean analysis of questionnaires (Flament, 1976), which is a partial order generalization of Guttman-analysis, based on Boolean algebra. In this contribution Boolean analysis is considered as a method that enables one to represent any set of binary patterns as a family of

[1]Center for Manpower Planning and Studies, Vrije Universiteit Brussel, Pleinlaan 2, B-1050 Brussel, Belgium e-mail:ptheuns@vub.ac.be

(more or less) comprehensive implications between responses, called *implication schemes.* These schemes are in a 1-1-relationship with the original data (Flament, 1976). For the sake of brevity, the actual data-analytic method and the processing of patterns resulting in implication schemes will not be treated in full. Instead, the interested reader is referred to Flament (1976), Van Buggenhaut and Degreef (1987), Theuns (1989), and Theuns (1992).

28.2 Basic theory

In Boolean analysis, responses to a set of items, a, b, c, d, e say, are encoded in *patterns,* i.e., in ordered sets of 0/1 responses. For example, "01100" denotes a negative response (0) to items a, d, and e, and a positive response (1) to items b and c. Alternatively, as in Boolean algebra, this pattern can be denoted $a'bcd'e'$, where positive responses are denoted by characters and negative responses by primed characters. The term *subpattern* refers to a pattern obtained by discarding a number of responses in a pattern; for example, the subpattern $01xx0$ is obtained by discarding the responses to items c and d, these responses being replaced by an "x". Alternatively, this subpattern can be denoted $a'be'$. The number k of characters (or 0's and 1's) in a subpattern is called the *length* of that subpattern.

Theoretically, a set of n items generates a set R consisting of 2^n patterns. Out of this set, only a subset R^* will be observed, leaving a (possibly empty) set R° of unobserved patterns. Formally, we have

$$R^* = \{\text{patterns with frequency} > 0\}, \text{ and}$$

$$R^\circ = \{\text{patterns with frequency} = 0\}.$$

Data in Boolean analysis consist of a set R^* of observed patterns. Obviously, as $R = R^\circ \cup R^*$, R° can easily be recovered from R^*, and *vice versa.* A central notion in Boolean analysis, introduced by Flament (1976), is a PCU (short for the French *"Projection Canonique Ultime"*), which is a *prime implicant* (see, e.g., Birkhoff and Mac Lane, 1977; Lipschutz, 1976; Ragin, 1987) of the Boolean expression obtained as the Boolean sum of patterns in R°.

All PCUs are, in fact, subpatterns with minimal lengths which figure in some patterns in R° but not in any of the patterns in R^*. It turns out that a PCU is a (minimal) non-observed subpattern. Several algorithms using Boolean algebra allow the enumeration of all PCUs in any Boolean expression (i.e., for any R°). Moreover, from the set of PCUs one can easily reconstruct R°, and thus R^*, in a unique way.

As any PCU is a non-observed subpattern, it enables one to construct *implications,* which is the subject of the following definition.

Definition 1 *Given a PCU, say p, and a non-empty subpattern s, such that $s \subset p$ and $p \backslash s$ not empty. Then p defines the implication*

$$s \text{ implies } not \ (p \backslash s),$$

which is formulated as

$$s \rightarrow \neg (p \backslash s).$$

In plain language, given the *PCU* p in the definition, if s is observed in a subject, then for at least one of the responses in $(p \backslash s)$ (the subpattern of p obtained by discarding all responses in s) the inverse response is observed in that subject. Note that, because $(p \backslash s)$ is not empty, there is a complementary implication $(p \backslash s)$ *implies* $\neg s$. According to this observation, any *PCU* with length k generates a family of $2^k - 2$ implications, the number of real, non-empty subsets in a set of k elements.

For example, the *PCU* ab' generates 2 implications : $a \rightarrow b$, and $b' \rightarrow a'$. The *PCU* $ab'c$ generates 6 implications[2]: $a \rightarrow b \oplus c'$, $b'c \rightarrow a'$, $b' \rightarrow a' \oplus c'$, $ac \rightarrow b$, $c \rightarrow a' \oplus b$, and $ab' \rightarrow c'$. More examples can be found in Van Buggenhaut and Degreef (1987), and in Theuns (1989). Clearly, a *PCU* can be reconstructed from each one of the implications that it generates. It can be inferred from the former that not only the number, but also the complexity of implications increases with the length of the *PCU*. Apparently, a *PCU* with length 2 yields the most straightforward implications; *PCU*s with length 1 are trivial, and *PCU*s with lengths > 3 yield implications that may be too complex for interpretation.

Finally, the outcome of a Boolean analysis is an *implication scheme* which is constructed by selecting exactly one implication from each *PCU*; the implications are chained if possible, that is, the right side of one implication is used as the left side of the next implication whenever this is possible. Clearly, according to these heuristics, an implication scheme is not generally unique. Some rules for selecting implications and implication schemes are provided by Theuns (1992). It is remarked that, according to the heuristics, every *PCU* is represented once in the implication scheme, and therefore R^o, and consequently R^*, can always be recovered from every scheme. Selection of one implication per *PCU* does not therefore affect the reproducibility of the data.

An "optimal" or approximated implication scheme is often preferred in practice to a perfect scheme. An optimal implication scheme is defined as an implication scheme that models a subset of the data rather than the complete data set and which complies with criteria of complexity (e.g., all *PCU*s have lengths ≤ 2) and comprehensiveness (e.g., a maximal number of patterns with maximum frequency assigned to R^*).

There are three important arguments for this. First, an implication scheme perfectly matching the data necessarily includes the error that is typical for psy-

[2]The Boolean sum operator ("OR") is denoted $\oplus$, while the Boolean product operator ("AND") is not written explicitly and juxtaposition is used instead.

chological data. For example, people can make a lucky guess or a careless error such that their observed pattern does not match their actual state. Although efforts can be made to minimize such errors in the data, it is hard to imagine empirical data where no errors whatsoever are present. Second, perfect implication schemes can be built, but may be rejected because of complexity. In some cases, a researcher may prefer a simple model that is able to predict most behavior rather well, over a very complex but perfect model. Third, Boolean analysis dichotomizes patterns into real (R^*) and error (R°), but the frequency of the real patterns does not affect the implication scheme; that is, all real patterns are assigned equal importance in the analysis. Especially for data obtained from a large sample, this may imply that patterns with a very high frequency get the same weight in results as patterns with a relatively low frequency. In this case, the researcher may wish to make a number of successive analyses whereby the minimal frequency of real patterns is altered systematically to allow a differentiated interpretation of the implication scheme(s).

Other data-analytic methods like Guttman-analysis (or scalogram analysis) (see Guttman, 1950) and Feature Pattern Analysis (Feger, 1984; Feger, 1991) impose some restrictions to the data to be modeled, such that the dichotomization of R into R^* and R° is inherent in the method. In Boolean analysis, however, such restrictions are not present. Thus, to comply with the previous considerations, methods for determining a dichotomy in the data, such as dividing them into real patterns versus error patterns, are needed in addition to the actual data analytic method. Methods which aim at the dichotomization of observed patterns for obtaining approximated implication schemes are called *dichotomization methods.*

Previous methods (Flament, 1976; Van Buggenhaut, 1987; Van Buggenhaut and Degreef, 1987) dichotomize the set R of all possible patterns into a set R^* (real patterns) of observed patterns with a minimal frequency exceeding some number β (called the *dichotomization threshold*) on the one hand, and a set R° (error patterns and non-observed patterns) on the other. Formally, we have

$$R^* = \{\text{patterns with frequency} > \beta\}, \text{ and}$$

$$R^\circ = \{\text{patterns with frequency} \leq \beta\}.$$

Van Buggenhaut (1987) introduced within this context a method to determine an optimal value for β, which he called the *cohesion threshold* (β_0). He defined the cohesion threshold as the minimal β for which the implication schemes would be intelligible, that is, all PCU have lengths ≤ 2 (or 3). A method for predicting this β-value without calculation of all the PCUs was introduced by Theuns (1989). This latter method affords an important gain in processing speed as it allows to omit time-consuming PCU-calculation for some low β-values for which it is predicted that some PCU do not meet the criterion.

28.3 Shortcomings of "classical" approaches to dichotomization

As previously stated, this present paper is concerned with patterns in which response categories have the same meaning for all items (like correct/false). Patterns are considered for which the number of 1-responses and that of 0-responses is quantifiable. For example, students taking a test in mathematics give either correct or false responses. The number of correct responses is meaningful and may be used as a "score". In clinical psychology, the diagnosis of a depressive disorder is obtained from the presence of some specific combination of symptoms. Patterns of symptoms can be ascertained by means of a symptom list like the DSM III-R (American Psychiatric Association [A.P.A.], 1987), whereby each symptom observed contributes to the final diagnosis.

So, the idea is that response-values have the same meaning for all items and that all contribute in predicting some external dependent variable (score, diagnosis). Data of this type are quite common in the human sciences. Generally, response-values are assigned such that "1" corresponds to a positive contribution to prediction of the dependent variable (1 for "correct", or "symptom observed", and 0 otherwise). According to this specific encoding, the number of ones in each pattern can be used to rank patterns. The number of ones in a pattern is called the *level* of that pattern.

In scalogram analysis, R^* contains exactly one pattern for each possible level $(0, \ldots, n)$. As a result, in the case of a perfect scale, an observed pattern can be recovered perfectly from its level. This idea is basic for what is called "scalability" in scalogram analysis (Guttman, 1950).

I do not, however, adopt the notion of scalability here. In practice, perfect scales are very rare (Guttman, 1950). Consequently, the requirement that a pattern can be recovered from its level is abandoned.

In the next section, an example is presented which clarifies the motivation for developing an alternative method of dichotomization.

Example 1

Suppose an artificial data set to be constructed according to some simple assumptions on item difficulty and on the probability of error. Let a, b, c, and d be four items in X that can be ordered according to item difficulty as $a \gg b \gg c \gg d$ (Guttman scale). Let each item have a probability of 0.05 of eliciting a lucky guess and 0.1 for a careless error. If, in a sample of 500 people, ability at answering these items shows a perfect spread (there are as many subjects capable of answering none of the items as there are capable of answering either 1, 2, 3, or all 4 items), then the *actual* states (1111, 0111, 0011, 0001, and 0000) have the same frequencies, namely 100. However, this is not quite what will be observed: There will be sampling errors. Again, for reasons of simplicity, suppose that a subject makes no more than one error. Every real pattern will yield, according

TABLE 28.1. Expected frequencies of observed patterns under "perfect" ability spread

actual level	4		3		2		1		0	
		f_i		f_i		f_i		f_i		f_i
real pattern	1111	100	0111	100	0001	100	0011	100	0000	100
observed	1111	60	0111	65	0011	70	0001	75	0000	80
patterns	0111	10	1111	5	1011	5	1001	5	1000	5
	1011	10	0011	10	0111	5	1010	5	0100	5
	1101	10	0101	10	0001	10	0011	5	0010	5
	1110	10	0110	10	0010	10	0000	10	0001	5

Expected distribution of real pattern frequencies over observed patterns due to error

TABLE 28.2. Artificial data in case of "perfect" sample spread pattern

pattern	frequency	pattern	frequency
1111	**65**	**0111**	**80**
1110	10	0110	10
1101	10	0101	15
1100	0	0100	5
1011	15	**0011**	**85**
1010	0	0010	15
1001	5	**0001**	**90**
1000	5	**0000**	**90**

Real patterns are printed in boldface

to the specified error probabilities, a number of observed patterns as indicated in Table 28.1. So, under the conditions set out above, the observed data should approximate Table 28.2.

Apparently, the five patterns that determine a Guttman scale are observed the most frequently. Moreover, there is a large gap between these frequencies of the scale patterns and the other error patterns. A Boolean analysis of these data yields, for $\beta = 15$,

$$R^* = \{1111, 0111, 0011, 0001, 0000\}, \quad \text{the set of } PCU\text{s}^{3} \text{ being:}$$

$$P = \{10xx, 1x0x, 1xx0, x10x, x1x0, xx10\}$$

$$= \{ab', ac', ad', bc', bd', cd'\},$$

which yields the following implication scheme[4]:

$$a \longrightarrow b \longrightarrow c \longrightarrow d.$$

This scheme is valid for 82.0% of the subjects; that is, 82.0% of the subjects show patterns contained in R^*. We say: The ***sample fit*** = 82.0% . Clearly, this implication scheme indicates that items ***a*** to ***d*** can be ranked according to item difficulty and that subjects with a positive response on a more difficult item show positive responses on all less difficult items. Thus, for $\beta = 15$, the underlying Guttman scale is recovered by Boolean analysis. However, the cohesion threshold is even lower than $\beta = 15$; the cohesion threshold ($\beta_0 = 10$) yields the following set of PCUs,

$$P = \{1x0x, 1xx0, x1x0\} = \{ac', ad', bd'\},$$

and the corresponding implication scheme (with sample fit = 91.0%)[5]:

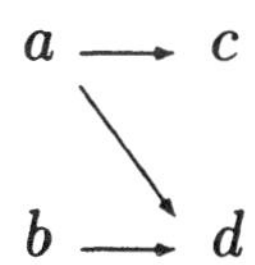

Clearly, the subjects in this sample are equally spread over the 5 levels (actual states) in the questionnaire. This equal spread of actual states is somehow still present in the observed states: The sizes of the groups of people who answered positively to exactly 4, 3, 2, 1, or none of the questions are very similar (65, 115, 115, 115, 90). If, however, the population is not spread – more or less – equally over the performance levels, then the whole underlying response structure can become lost in the observed pattern frequencies. This case is illustrated in the next example.

Example 2

Suppose now that the same 4-item questionnaire is administered again to a sample of 500 subjects, 30% of whom actually are on level 4, 50% at 3, 10% at 2, 5% at 1 and 5% at 0, i.e., the sample is concentrated around levels 3 and 4. In this case, the observed patterns resulting from the distribution are those given in Tables 28.3 and 28.4.

Clearly, these data do not deliver the expected scalogram for any β-value because two error patterns (0110 and 0101) have a frequency that is as great as the frequencies of two scale patterns (0001 and 0000). The following set of PCUs is obtained for the cohesion threshold $\beta_0 = 24$, where, according to its definition,

---

[3]It can be seen that the set of PCU, P, contains all subpatterns with minimal lengths that are contained in patterns with frequency $\leq \beta$, and are not contained in any of the patterns in R^*.

[4]Implications which can be inferred using transitivity of implications are not represented explicitly in the scheme.

[5]Apparently, the cohesion threshold assigns some error patterns to R^*.

TABLE 28.3. Expected frequencies of observed patterns for a sample "concentrated" around levels 3 and 4 actual level

actual level	4		3		2		1		0	
		f_i		f_i		f_i		f_i		f_i
real pattern	1111	150	0111	250	0011	50	0001	25	0000	25
observed patterns	1111	90	0111	167	0011	34	0001	19	0000	21
	0111	15	1111	8	1011	3	1001	1	1000	1
	1011	15	0011	25	0111	3	1010	1	0100	1
	1101	15	0101	25	0001	5	0011	1	0010	1
	1110	15	0110	25	0010	5	0000	3	0001	1

Expected distribution of real pattern frequencies over observed patterns due to error

TABLE 28.4. Artificial data in case of sample "concentration"

pattern	frequency	pattern	frequency
1111	**98**	**0111**	**185**
1110	15	*0110*	*25*
1101	15	*0101*	*26*
1100	0	0100	5
1011	18	**0011**	**60**
1010	0	0010	6
1001	1	**0001**	**25**
1000	1	**0000**	**24**

Real patterns are printed in boldface, errors assigned to R^* in italics.

the set contains all subpatterns with minimal lengths that are contained in patterns with frequency $\leq \beta_0$ but not in the patterns with higher frequencies:

$$\begin{aligned} P &= \{10xx, 1x0x, 1xx0, x1x0, xx00\} \\ &= \{ab', ac', ad', b'd', c'd'\}, \end{aligned}$$

which yields the following implication scheme (sample fit = 83.8%):

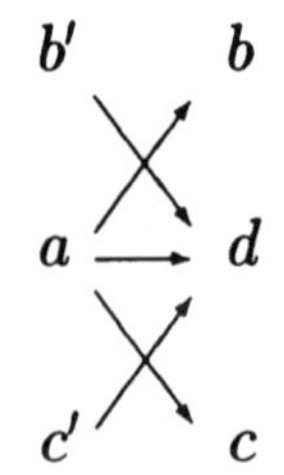

Note that a' and b' appear in this implication scheme which is derived from the set P.

28.4 Preliminary discussion

According to the above examples, the spread of subjects over the actual states is of major importance when a dichotomization method based on absolute pattern frequencies is utilized. Obviously, such methods show little robustness with respect to variations in this spread. The clustering of subjects around certain patterns or levels can originate from several causes. For example, there may be certain items (e.g., corresponding to some notions of knowledge) which subjects generally need a long time to master. However, once such a "difficult" item is mastered, it may be that a significant step in the learning process has been taken, allowing the subject to move on towards mastering the next notion quickly. So this results in a small number of subjects mastering the difficult item while failing to master the next one. In fact, this situation will happen whenever there is a significant difference in "step-length" in moving from one item on a scale to another. Also, in practice, subjects often show a similar ability in some domain, especially when the population under study is rather homogeneous (e.g., students in one class). This frequently occurs in educational research when children in the same class or grade are tested. As a consequence, absolute-frequency based dichotomization methods should be used only in certain specific situations. There is no doubt that such methods are especially useful in purely descriptive research and for the analysis of qualitative data. If the objective of a Boolean analysis is to map the interrelations of responses in a sample, then clearly the absolute frequencies are the most relevant. If, however, a test consisting of quantifiable items is analyzed in order to reveal certain prerequisite relations between responses to different items, and if the data show homogeneity in the ability of the subjects (say, an important subgroup in the sample is concentrated around one level), then these methods may not be appropriate. In the next sections, a dichotomization method is developed which is more robust with respect to sample homogeneity.

28.5 Dichotomization method based on relative level frequencies

An alternative approach might consider the frequency of every observed pattern relative to the frequencies of other patterns at the same level. In this way it might be possible to ignore some patterns that occur with a high frequency and belong to a level with a high density, while assigning some low-frequency patterns to R^* which belong to levels with a low density in the given population. The following method seems to be appropriate:

- Arrange patterns according to level (equal numbers of positive answers).

- Select within every level the maximum frequency as referent.

TABLE 28.5. Artificial data in case of "perfect" sample spread

pattern	frequency	relative level frequency	pattern	frequency	relative level frequency
1111	65	100.0	0000	90	100.0
1110	10	12.5	0001	90	100.0
1101	10	12.5	0010	15	16.7
1011	15	18.8	0100	5	5.6
0111	80	100.0	1000	5	5.6
1100	0	0.0	0011	85	100.0
1010	0	0.0	0101	15	17.6
1001	5	5.9	0110	10	11.8

- Express all frequencies at the same level as a percentage (the *relative level frequency*) of the maximal frequency in the class.

Let $W = \{w\}$ be the ordered set of relative level frequencies corresponding to R. Now any $w \in W$ can serve as a dichotomization threshold β and will yield a different result.

According to this method, which is called *dichotomization on relative level frequencies,* the definitions of R^* and R° are altered to

$$R^\circ \;= \{\text{patterns with relative level frequency} \leq \beta\}, \text{ and}$$

$$R^* \;= \{\text{patterns with relative level frequency} > \beta\}.$$

Once this method is specified, the cohesion threshold can be defined, as before, as the lowest value $\beta = \beta_0$ for which all PCUs have lengths ≤ 2. According to Table 28.5, the first example gives the ordered set of relative level frequencies

$$W = \{100.0, 18.8, 17.6, 16.7, 12.5, 11.8, 5.9, 5.6, 0.0\}.$$

Boolean analysis with dichotomization on relative level frequency yields for $\beta = 18.8$ the following set of PCUs:

$$\begin{aligned} P &= \{x10x, xx10, x1x0, 10xx, 1x0x, 1xx0\} \\ &= \{bc', cd', bd', ab', ac', ad'\}, \end{aligned}$$

and the implication scheme (sample fit $= 82\%$) is:

$$a \longrightarrow b \longrightarrow c \longrightarrow d.$$

TABLE 28.6. Artificial data in case of "concentration"

pattern	frequency	relative level frequency	pattern	frequency	relative level frequency
1111	98	100.0	0000	24	100.0
1110	15	8.1	0001	25	100.0
1101	15	8.1	0010	6	24.0
1011	18	9.7	0100	1	4.0
0111	185	100.0	1000	1	4.0
1100	0	0.0	0011	60	100.0
1010	0	0.0	0101	26	43.3
1001	1	1.7	0110	25	41.7

As before, the presupposed Guttman scale can be inferred from one of the implication schemes with an appropriate β. The cohesion threshold $\beta_0 = 12.5$, however, gives the following set of PCUs,

$$P = \{x1x0, 1x0x, 1xx0\} = \{bd', ac', ad'\},$$

and the implication scheme (sample fit = 91%) is:

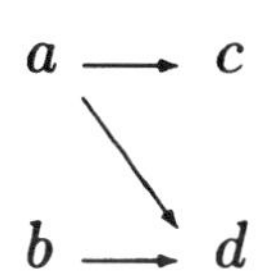

The second example gives the relative level frequencies shown in Table 28.6. Boolean analysis with dichotomization on relative level frequency yields, for $\beta = 43.3$, the following set of PCUs:

$$\begin{aligned} P &= \{x10x, xx10, x1x0, 10xx, 1x0x, 1xx0\} \\ &= \{bc', cd', bd', ab', ac', ad'\}, \end{aligned}$$

and the implication scheme (sample fit = 78.4%) is

$$a \longrightarrow b \longrightarrow c \longrightarrow d.$$

Contrary to previous findings with dichotomization according to absolute frequencies, the presupposed Guttman scale can now be recovered, even in the case of sample homogeneity, if the appropriate value for β is utilized. For the cohesion threshold $\beta_0 = 41.7$, the following set of PCUs is obtained:

$$P = \{10xx, 1x0x, 1xx0, xx10, x1x0\} = \{ab', ac', ad', cd', bd'\},$$

and the implication scheme (sample fit = 83.6%) is:

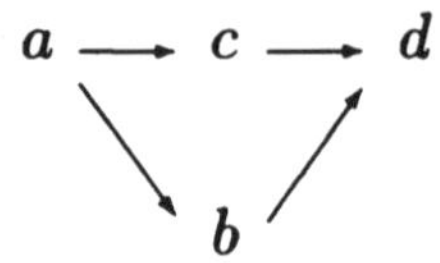

From this example it can easily be seen that the actual scalogram, around which the data have been constructed, emerges under both population conditions as the result of one dichotomy. However, the scalogram does not correspond with the cohesion threshold in either of the conditions.

28.6 Some remarks

The examples show that sample fit may be (and often is) lower for implication schemes obtained with dichotomization on relative level frequencies than with absolute-frequency based approaches. This indicates that, if Boolean analysis is applied for sample description purposes, dichotomization on absolute frequencies yields better results (higher sample fit). When the researcher wants to uncover the item structure and the hierarchical ranking of the respective items, however, dichotomization on relative level frequencies is the more appropriate method. In this case, sample fit is not a good measure for the quality of the implication scheme[6]. Some other measure is needed to permit comparison of the implication scheme directly with the ongoing process, but not with observations in some sample. Clearly, the problem here is that *a priori* the actual process is unknown and Boolean analysis is applied in an attempt to uncover it. Finally, there is a situation where possibly no measures exist which would permit the evaluation of the quality of the obtained implication scheme. Such evaluation may only become possible at a later stage when research on the process is further developed.

28.7 Effects of data recoding

An important consequence of arranging patterns according to levels prior to dichotomization is that the results of Boolean analysis are, contrary to the more classical approaches (see Flament, 1976), not robust for data recoding (Theuns, 1992). Indeed, if data are first encoded yes = 1/no = 0, for example, and afterwards recoded to correct = 1/wrong = 0, this may affect different items differently. As a result, patterns can be assigned to different levels after recoding. While such a result would be unacceptable in qualitative data, it

[6] The idea is that subjects are going through some (learning) process in which mastery of items is acquired, and the goal of the researcher is to describe this process. Therefore it is not the sample but rather the underlying process which is at interest. It follows that sample fit may not be a relevant criterion in this case.

can yield interesting results in quantitative data. Indeed, recoding quantifiable co-occurrence data is meaningful; that is, recoding yes/no, for example, to be correct/wrong indicates a different research stand-point and different research hypotheses. Therefore, it can be argued that it is a valuable feature that data analysis may, in this case, yield different outcomes depending on the research hypotheses under study. A recoding of the data, which affects some items and leaves other items unchanged, necessarily alters the interpretation of the "number of positive responses". If this interpretation is, indeed, different, then there is no theoretical reason why analyses of the same data, but with different encoding, should yield the same outcomes; that is, the same implication schemes.

28.8 Discussion

More 'classical' dichotomization methods for Boolean analysis apply to both qualitative and quantitative dichotomous co-occurrence data in general. For the latter type of data, a specific dichotomization method, called dichotomization based on relative level frequencies, was introduced. This method follows the idea that for quantifiable dichotomous co-occurrence data the count of positive/negative responses is meaningful. Dichotomization on relative level frequencies seems to be a promising approach with quantifiable co-occurrence data, especially when the aim of the analysis is to reveal the underlying hierarchical structure of the responses. If a purely descriptive picture of the sample is preferred, more 'classical', absolute-frequency based dichotomization methods may be more appropriate.

Dichotomization on relative level frequencies proves more robust to sample homogeneity than other dichotomization methods introduced by Flament (1976), Van Buggenhaut (1987), Van Buggenhaut and Degreef (1987), and Theuns (1992). That this method is not robust for data-recoding has the consequence that data must be encoded according to the research hypotheses. An unsolved problem is that none of the 'classical' dichotomization methods for Boolean analysis, nor the method introduced here, incorporate any modeling of errors; that is, patterns are assigned to R° and R^* purely according to their frequencies. Often, however, no information on error-rates is (as yet) available, and so frequency has to be the only criterion. In cases where such information is available or a theory concerning the underlying process exists, dichotomization approaches can be developed which incorporate the theory. The interested reader is referred to Theuns (1992) for a dichotomization method that is in concordance with the theory of knowledge spaces (Doignon and Falmagne, 1985).

Research with empirical data is needed, where Boolean analysis with dichotomization on relative level frequencies is applied together with other data-analytic methods, for testing and evaluating the utility of the dichotomization method introduced.

28.9 References

American Psychiatric Association (1987). *Diagnostic and statistical manual of mental disorders. Third edition. Revised [DSM III-R].* Washington, DC: American Psychiatric Association.

Birkhoff, G., & Mac Lane, S. (1977). *A survey of modern algebra* (4th ed.). New York: MacMillan.

Doignon, J.-P., & Falmagne, J.-Cl. (1985). Spaces for the assessment of knowledge. *International Journal of Man-Machine Studies, 23,* 175–196.

Flament, C. (1976). *L'analyse booléenne de questionnaire.* Paris: Mouton.

Feger, H. (1988). Spatial representation of feature patterns. In H.H. Bock (Ed.), *Classification and related methods of data analysis* (pp. 431–437). Amsterdam: North-Holland.

Feger, H. (1991). *Structure analysis of co-occurrence data.* Unpublished manuscript. Berlin: Freie Universität.

Guttman, L. (1950). The basis of scalogram analysis. In S.A. Stouffer, L. Guttman, E.A. Suchman, P.F. Lazarsfeld, S.A. Star, & J.A. Clausen (Eds.), *Measurement and prediction, studies in social psychology in world war II* (Vol. 4, pp. 60–91). New York: Wiley.

Lipschutz, S. (1976). *Theory and problems of discrete mathematics.* New York: McGraw-Hill.

Ragin, C.C., Mayer, S.E., & Drass, K.A. (1984). Assessing discrimination: A Boolean approach. *American Sociological Review, 49,* 221–234.

Ragin, C.C. (1987). *The comparative method. Moving beyond qualitative and quantitative strategies.* Berkeley: University of California Press.

Theuns, P. (1989). Predicting an optimal threshold in Boolean analysis of questionnaires. In E.E. Roskam (Ed.), *Mathematical psychology in progress* (pp. 329–343). Berlin: Springer-Verlag.

Theuns, P. (1992). *Dichotomization methods in Boolean analysis of co-occurrence data: Applications in psychopathology, organizational psychology, socio-economic research and knowledge theory.* Unpublished doctoral dissertation. Brussels: Vrije Universiteit Brussel.

Van Buggenhaut, J. (1987). Questionnaires booléens: Schémas d'implications et degrés de cohesion. *Mathématiques et Sciences Humaines, 98,* 9–20.

Van Buggenhaut, J., & Degreef, E. (1987). On dichotomisation methods in boolean analysis of questionnaires. In E.E. Roskam & R. Suck (Eds.), *Progress in mathematical psychology* (pp. 447–453). Amsterdam: North-Holland / Elsevier.

Author Index

Subject Index

Made in the USA
Monee, IL
13 September 2026

62978985R10238